Dangerous Motherhood
Insanity and Childbirth in Victorian Britain

Hilary Marland
Centre for the History of Medicine
Department of History
University of Warwick

First published 2004 by
PALGRAVE MACMILLAN
Houndmills, Basingstoke, Hampshire RG21 6XS and
175 Fifth Avenue, New York, N.Y. 10010
Companies and representatives throughout the world

PALGRAVE MACMILLAN is the global academic imprint of the Palgrave Macmillan division of St. Martin's Press, LLC and of Palgrave Macmillan Ltd. Macmillan® is a registered trademark in the United States, United Kingdom and other countries. Palgrave is a registered trademark in the European Union and other countries.

ISBN 1–4039–2038–9 hardback

This book is printed on paper suitable for recycling and made from fully managed and sustained forest sources.

A catalogue record for this book is available from the British Library.

Library of Congress Cataloging-in-Publication Data
Marland, Hilary.
Dangerous motherhood : insanity and childbirth in Victorian Britain / Hilary Marland.
    p. cm
Includes bibliographical references and index.
ISBN 1–4039–2038–9 (cloth)
    1. Puerperal psychoses–Great Britain–History–19th century.
2. Childbirth–Great Britain–Psychological aspects–History–19th century.
3. Motherhood–Great Britain–Psychological aspects–History–19th century.
I. Title.

RG851.M37 2004
618.7′6′09410934–dc22
                                                                2003070728

10   9   8   7   6   5   4   3   2   1
13 12 11 10 09 08 07 06 05 04

Printed and bound in Great Britain by
Antony Rowe Ltd, Chippenham and Eastbourne

To Daniel and Sam
and Mrs Eliza Gipps

Contents

List of Figures

Acknowledgements

Writing this book has been a journey towards a different understanding of nineteenth-century psychiatry, one that was deeply troubling in some ways, but in others revealing in that it opened up a broad emotional, moral and social framework for explaining mental illness in the past that I could never have imagined at the outset of this project. Many sad narratives are recounted in this book, but also many stories of sympathy and recovery. It has led me to think in very different ways about the landscape of mental illness and the complexities of motherhood in the past as well as in the present.

Numerous institutions and individuals, colleagues and friends, have contributed to the writing of this book and the ideas explored in it. The Wellcome Trust provided the funding for the research through a University Award, held initially at the Centre for the Study of Social History at Warwick University and subsequently in the History Department there. Dr David Allen, who set up the Awards, cherished my generation of historians of medicine and added a personal touch to the network he created. This project would not have seen completion without the support of the AHRB, who awarded me a Research Leave Award in 2002–3.

Very kind assistance and support during the reviewing and production of the book was provided by Luciana O'Flaherty and Daniel Bunyard at Palgrave Macmillan. My thanks go to Ruth Willats who carefully read the text and made helpful suggestions at the copy editing stage and to Michele Clarke who compiled the index.

Parts of chapter 6 first appeared in Mark Jackson (ed.), *Infanticide: Historical Perspectives on Child Murder and Concealment, 1550–2000* (Ashgate, 2002) and are reproduced by the kind permission of the publishers. My particular thanks go to the Harry Ransom Humanities Research Center, The University of Texas at Austin and to Priscilla Cassam, the copyright holder, for permission to quote extracts from the diary of Sara Coleridge. I would also like to acknowledge the Wellcome Trust Library, the Bethlem Royal Hospital Archives and Museum, and Warwick County Record Office for permission to reproduce illustrative material.

Many libraries and archives have provided expertise and support. The Inter-Library Loan Department at the University of Warwick chased material, extended loans and dealt with numerous requests. The Modern Records Centre, also at Warwick, provided access to the Sara Coleridge papers and a welcoming work environment. The Wellcome Trust Library, as always, offered excellent facilities and advice on archives, printed material and illustrations. My thanks also go to the staff of the Warwick County Record Office, the Mitchell Library in Glasgow, London Metropolitan Archives, the Department of Special Collections, Glasgow University Library, the British Library and the Women's Library. The Lothian Health Board Archivists at Edinburgh University Library offered assistance on my many visits and chased up details that I had missed; Michael Barfoot generously shared his knowledge of the Royal Edinburgh Asylum and provided detailed commentary on chapter 4.

Seminar papers were delivered at numerous locations in Britain, Europe and North America during the many years I have worked on this project, and I am grateful to the audiences for asking questions, giving me ideas and feedback, and answering many of my own questions. My many debts to historians in the social history of medicine and psychiatry and women's history are acknowledged, adequately I hope, in my references. I consider myself very lucky to be part of such a generous community of scholars, many of whom have read portions of the book, and many of whom have supported me in other ways: Rima Apple, Peter Bartlett, Hal Cook, Anne Crowther, Roger Davidson, Anne Digby, Faye Getz, Cath Quinn, Mark Jackson, Helen King, Irvine and Jean Loudon, Lara Marks, Jo Melling, Margaret Pelling, Marijke Gijswijt-Hofstra, Anne-Marie Rafferty, Jonathan Reinarz, Len Smith, Steve Sturdy and David Wright. The last years have been marred by the sad deaths of two generous colleagues, who are much missed: Joan Lane and Roy Porter. Rob Ellis, Pam Michael and Frank Crompton shared their findings and material with me on their own asylums. At Warwick, I have enjoyed an invigorating work environment and the support and stimulus of colleagues and students past and present, especially Gillian Bendelow, Sarah Hodges, Ros Lucas, Liese Perrin, Molly Rogers, Carolyn Steedman, Claudia Stein and Mathew Thomson. The graduate community of the Centre for the History of Medicine have been a source of ideas and wonderful company. I would particularly like to thank Colin Jones and Margot Finn for their close reading of penultimate chapter drafts and for encouraging me to think about the material in different ways, and to Jonathan Andrews for refereeing the manuscript.

My mother and, until his death several years ago, my father offered a great deal of encouragement during the long saga of completing the research and writing of this book. My dad might have found this a strange topic, but, as a real supporter of women's issues, would have approved of this book. Though it is my hope that Sebastian, Daniel and Sam did not suffer too much during the years of research and writing, it has been a particularly long and sometimes painful delivery. I would like to thank them for their endless support in ways too many to list here. I would like to dedicate this book to my sons, Daniel and Sam, for their patience, affection, friendship and fun. They are also kind enough to share this dedication with another mother, Eliza Gipps, denied her sanity and her child.

Introduction

As the royal family celebrated Christmas at Windsor Castle in December 1841 concern was expressed about Queen Victoria's poor state of health. Following her recent confinement, she was reported to be 'troubled with lowness ... I should say that Her Majesty interests herself less and less about politics'.[1] A year earlier, in September 1840, the author William Makepeace Thackeray had set off with his young wife, Isabella, already dejected and melancholic following the birth of their third child, to visit her mother in Ireland. During the voyage Isabella flung herself into the sea, where she remained for twenty minutes before she was discovered 'floating on her back, paddling with her hands'.[2] In November of that year, Thackeray, at his wits' end, surrendered Isabella, by then in a desperate and violent state, to the asylum of the celebrated psychological physician Dr Jean-Etienne-Dominique Esquirol at Ivry in France.[3] Several years later in the Welsh countryside a young, unmarried farmer's daughter gave birth to a 'fine' child. All went well at first, but three days after her delivery she alarmed the household waking from sleep exclaiming that she was dying. She was convinced that she was carrying another child, declared hatred for the infant that she had just borne and was 'scarcely a moment without raving'.[4] Into the next decade, in April 1853, Mrs Janet Smith was admitted to the Royal Edinburgh Asylum ten days after a difficult labour, which had required the use of forceps:

a few days after this her mind became gloomy & she wept much; this was followed by sleeplessness, restlessness and a disposition to be noisy & obstinate & destructive of clothing: she rhymes over the letters of the alphabet for hours at once, she often sings, she inclines to run about the room ...[5]

1

These four women, worlds apart from each other in terms of position, lifestyle and experience, shared a common threat to their well-being and ability to take up their role as mothers, having fallen prey to a mental disorder associated with that most female of functions, childbearing.

Queen Victoria displayed a mild though, for her advisers, troubling attack of depression after the birth of her second child, when she became languid and miserable. Her subdued episodes of distress and despondency, which recurred after each of her confinements, would barely warrant attention among less elevated patients, with low spirits and anxiety being recognised as a more or less normal consequence of childbirth. Victoria's significance, particularly as she was projected as a model of bourgeois domesticity and family values, lay rather in her boldness in declaring childbirth a horrible experience in a society where it was held up as an almost sacrosanct function. She intensely disliked pregnancy and childbirth, and would struggle to come to terms with nine confinements in all, turning with enthusiasm to chloroform anaesthesia for the birth of her eighth child in 1853. Her negative feelings about childbearing, which she continued to voice for years after her last child was born, would be echoed to some extent by Victorian doctors, concerned about the strains and rigours imposed on women in childbed.[6] Victoria's attacks of depression, however, were not severe and she was also buffered by her wealth and rank. This would be otherwise for many women, although the majority would eventually recover, restored to health and to their role as mothers, even if suffering from more serious manifestations of mental disturbance.

The violent outburst of the Welsh farmer's daughter, on top of the usual trials and tribulations of giving birth, was attributed to remorse at her unmarried state and the consequent coolness of her friends towards her. However, her affliction was short-lived. She was treated with purgatives and opiates 'until sleep could be induced' and within days 'her demeanour towards the child was that of an affectionate mother'.[7] Janet Smith was one of the numerous impoverished patients to pass through the doors of the Edinburgh Asylum suffering from a mental disorder attributed to childbearing. Though remaining noisy and disruptive, confused and incoherent after admission, she responded well to treatment with laxatives, bed rest, warm baths, stimulants and blisters applied to her neck. She gained weight, slept better, became tranquil and started to work, and was discharged in October after seven months in the asylum.[8] In October 1855 Janet

Smith was brought again to the Edinburgh Asylum, fourteen days after her second confinement, sleepless, taking no notice of her child, violent and abusive, and believing her dead mother to be alive, but once again she made a good recovery.[9] Isabella Thackeray was less fortunate. Her lapse into insanity followed three traumatic deliveries within four years of marriage, aggravated by unpleasant domestic circumstances; the Thackerays struggled financially and Isabella's mother-in-law rebuked her constantly for her ineptness as a housekeeper. Isabella showed occasional flickers of improvement in the first years of her illness, but she was never to recover. She spent the rest of her life either in an asylum or under the care of nurses, separated from her home and family.[10]

* * *

In the early nineteenth century a new term – 'puerperal insanity' – would find its way into medical texts and language, which would encompass diverse forms of mental illness associated with childbirth. Prior to the nineteenth century, observers had remarked on the propensity of pregnant women and new mothers to fall prey to mental disorders, describing a miscellany of disturbing, odd, sometimes aggressive behaviour. However, puerperal insanity was very much a disorder of the nineteenth century, when it was named, defined and avidly debated by the medical profession. An atmosphere of raised anxiety about the dangers of childbirth and threats to the sanctity of the bourgeois home offered an ideal medium for it to take hold and flourish. Yet the definition of puerperal insanity would be wide-ranging, partly because of the difficulties encountered by the medical profession in deciding on its causes and who exactly would be susceptible. Though the rigour of childbearing and women's intrinsic biological weakness were paramount, as the examples cited above have demonstrated, domestic troubles, poverty, fear and anxiety, and an inability to adapt to the demands of maternity were also implicated in causing mental distress in new mothers. Puerperal insanity could strike the highest born and the poorest, the most esteemed to the least respectable. Women were believed to be particularly at risk shortly after childbirth when they were physically weak and mentally susceptible, but they could also become mad during pregnancy or several months after delivery. It encompassed relatively brief attacks, nervous upsets, violence or delusions, as well as long-term manifestations of mania or deep and protracted melancholia, which could put at risk the life of the mother and child. While milder forms of postnatal depression engage much medical interest today, nineteenth-century physicians dwelt almost exclusively on violent mania and severe melancholia. There was little discernible interest in what is now termed 'baby blues' or postpartum dysphoria, and mild depression and gloomy behaviour seem to have been accepted as a common and not particularly worrying aspect of childbirth and attracted little comment.[11] Puerperal insanity in many ways is a precursor of puerperal psychosis, and nineteenth-century descriptions of the condition are recognisable today, even though the language of psychiatry has changed dramatically. But, as I will argue in the following chapters, puerperal insanity was very much a disorder that 'belonged' to the nineteenth century in terms of its medical and social setting.[12]

It was the ferocity of mania that stunned observers, attracted most attention and was the focus of much writing, though melancholia was recognised by doctors as being more intractable and less likely to be cured and in some ways potentially more threatening. The massive contravention of social norms and feminine behaviour demonstrated by women suffering from puerperal mania was particularly shocking, striking so soon after the woman had become a mother when her young infant most needed her care and protection. Robert Gooch, obstetric physician and arguably the foremost authority on puerperal insanity in the early nineteenth century, described how

> When puerperal mania does take place, the patient swears, bellows, recites poetry, talks bawdy, and kicks up such a row that there is the devil to play in the house ... every precaution must be taken to prevent her doing injury to herself, to the infant, or her friends.[13]

The eminent specialist in mental disorders, George Man Burrows, echoed this grim assessment. When delirium occurred during nursing,

> There is a change in temper, an irascibility inducing snappish remarks, and a peculiar hurried manner; sleep is much disturbed, the countenance betrays distrust, the pulse is rapid, and the patients are generally voluble. They often become suddenly negligent of their infant. At length both behaviour and language are incoherent, and delirium is fully developed. Acts of violence, sometimes suicide, are in this stage committed before the nature of the malady is suspected.[14]

Melancholia was depicted as a treacherous disorder, with many of its victims lapsing into permanent insanity, and insidious as it crept rather than stormed into the bosom of the family. In a treatise published in 1810, the celebrated man-midwife Thomas Denman can be credited with producing what remains perhaps the most touching and evocative description of melancholia following childbirth, which he described as

> great and positive prostration of strength with the most inflexible obstinacy, or total insensibility of what is said or done, and to every thing which passes around them. No object, however beautiful or interesting, gives pleasure to their eye, no music charms their ear, no taste gratifies their appetite, no sleep refreshes their wearied limbs or wretched imaginations; nor can they be comforted by the conversation or kindest attention of their friends. With the loss of every sentiment which might at present make life tolerable, they are destitute of hope which might render the future desirable.[15]

Conceptualised as a menacing and prevalent condition, women suffering from puerperal insanity challenged notions of domesticity and femininity and flouted ideals of maternal conduct and feeling. The 'hand that rocked the cradle' was also the hand that slapped, smothered or strangled the infant, as women suffering from puerperal mania and melancholia put at risk the newborn and other household members, as well as their own lives. They contravened their vows of matrimony; far from loving, honouring and obeying, women turned against their husbands, neglected themselves and the household, bullied their servants, broke the china, tore their clothes, roamed the streets and displayed an overt sexuality, making vulgar and suggestive comments to complete strangers. Yet so common was this disorder claimed to be by medical men that it came to be seen as an almost anticipated accompaniment of the process of giving birth. It rapidly became part of the language of various groups of medical practitioners staking claims of expertise to treat it. The disorder was also recognised and discussed by a broad spectrum of lay people, from aristocrats describing the turmoil and neglect of the household following their wives' confinement to the court witnesses who would starkly and accurately describe the infanticidal acts of their distraught and crazed neighbours.

Around the time that Queen Victoria was crowned in 1837 a new woman was well on the way to being constructed in the medical literature, which gave meaning to puerperal insanity and built into the language used to describe and understand it.[16] Far from the 'New

Woman' so feared at the century's end for her ambition, drive, independence of mind and efforts to construct a new morality,[17] early nineteenth-century woman was depicted increasingly as sickly, unhealthy, a victim of her fragile nervous system and unpredictable reproductive organs, and likely to have difficulty in performing her most important function of giving birth. Childbirth was defined as woman's paramount duty and most rewarding purpose in life, but also as a challenge to her body and mind, as a 'new female atlas was started in which the provisional frontiers of new countries of frailty, disease, and nervous instability were charted'.[18] It was into this fertile ground that puerperal insanity was seeded and grew, developing at the meeting point of the two emerging medical fields of nervous disorders and obstetrics. It became one of the numerous dangerous and debilitating conditions to which women could fall prey, in increasing numbers, in connection with pregnancy, childbirth and lying-in. The disorder would also be of great interest to psychiatrists, forging their specialisation, developing their theories and moving into their vast, lowering asylums, who tied women's mental disturbances to their intrinsic weakness and the rigours of reproduction, but also to a wider range of social, environmental and moral factors.

Puerperal insanity has received little scholarly attention.[19] Yet its exclusivity as a female disorder and strong links with reproduction, as well as its apparently meteoric rise and prevalence in the nineteenth century, offer the historian considerable potential in giving new insights into ideologies of domesticity, femininity and maternity, the 'medicalisation' of women, the doctor–patient relationship and ideas of female madness among the wealthy and the poor in the Victorian period. Elaine Showalter, in her groundbreaking book *The Female Malady*, devotes only a short section to this most female of mental disorders,[20] while it has made the briefest of appearances in other studies of women and mental illness.[21] Based on medical literature, asylum case books, court records, the archives of maternity hospitals, accounts of private practice and women's own writings, this book traces different interpretations of puerperal insanity, different agendas and different ways of explaining its existence. It is a book built on sad accounts of broken lives. But it also has unexpected twists in the story, which offer insight into the relationship between women and their doctors and a surprisingly understanding and informed take on a condition set in broad social, moral and emotional landscapes, with stories of recovery and restoration of sorts, bounded, however, by the conventions of Victorian society.

Puerperal insanity was many things to different people. It caused intense suffering to the women who fell prey to it, robbing some of their senses and liberty for many years, and was especially dreaded by women who became disturbed with each successive pregnancy or birth. There were many ways of explaining its appearance in the homes of the upper and lower classes, for it could strike poor women, debilitated by want, arduous work and hardship, as well as wealthy ladies, cushioned, but also seduced, by luxury and refinement. Puerperal insanity was also linked, as I will explore in chapter 1, to a much broader and older set of anxieties surrounding pregnancy and birth. These anxieties were removed from the female sphere and the office of the midwife to be taken over by male practitioners, who offered protection in return for recognition of their expertise to monitor and take action against the dangers of childbearing. Puerperal insanity was subject to varied interpretations by those with a stake in treating and preventing it, practitioners of midwifery, the psychiatric profession, general practitioners and midwives, with disagreement on causality, susceptibility, diagnosis, outcome, and place and form of treatment. These discussions and the impressive rise of the condition to take an important position in medical debate and practice will be discussed in chapter 2. Throughout the nineteenth century many cases of puerperal insanity were treated in private homes, which gives the historian the rare opportunity to explore mental disorder in domestic spaces as well as institutions and among rich and poor women. Chapter 3 focuses on the devastating impact of puerperal insanity on domestic life. Even if the new mother was not violent and destructive, she was often incapable and unwilling to take care of her infant; in some cases she neglected the household and became dirty, abject and neglectful. The wider cultural meanings of puerperal insanity are also explored in the second section of chapter 3, giving voice to detailed personal accounts of this condition, through the case studies of Sara Coleridge and Isabella Thackeray.

The asylum, however, would come to be the locus of care and confinement for numerous women during the Victorian era, and in chapter 4 the case notes of the Royal Edinburgh Asylum open up the world of these expanding institutions, exploring attitudes towards patients, and ways of understanding and treating puerperal insanity. This theme is continued into chapter 5, which utilises a broader selection of case histories of private and institutional practice to examine the rich variety of explanations put forward to explain the onset of puerperal insanity, encompassing moral, social and environment factors. In the worst-case scenario, puerperal insanity came to be closely associated, as chapter 6 will demonstrate, with the potential to harm, and even destroy, the newborn infant. The majority of women, however, even those who murdered their infants, recovered under the mild therapies adopted to treat the disorder, and for many puerperal insanity elicited sympathy and resulted in respite, a break from work, household cares and mothering. In the final chapter it will be shown how these regimes and attitudes were challenged towards the end of the century, under the aegis of hereditary explanations of mental disorder, and with increased scepticism about the existence of puerperal insanity as a distinct condition.

During much of the nineteenth century, however, puerperal insanity reflected the strains, the challenges and the conflicts Victorian women faced, their troubles in settling to a life of domesticity and adapting to the arrival of children, to living with men who disappointed them, and of being deprived of meaningful activity. Lower-class women also faced the challenges of poverty and hardship, financial struggle to provide homes for their families while keeping their own bodies and souls together. Exploring puerperal insanity, particularly through case histories and personal accounts, reveals much about the relationship between doctors, patients and their families, of medical practitioners' ambitions to establish a reputation and their practice, but also their concern to cure their patients and relieve families of this dreadful affliction. It gives us the opportunity too to explore the implications of the maternal ideal in Victorian society. It was largely because notions of femininity and maternity began to be so clearly expressed, that puerperal insanity became so visible and appeared to be so threatening. Yet, what at first appears to be a disorder that confronted and contradicted notions of domesticity and maternity, reveals on closer reading the extent to which madness and menace were construed not as offending or opposing female nature, but among its very components.

1
The Birth of Puerperal Insanity

A slow gestation

The erratic, crazy and disruptive behaviour of women around the time of childbed had been described long before the nineteenth century. In a unique account of madness by a medieval woman, Margery Kempe expressed in detail the mental torment and spiritual crisis that followed the birth of her first child.[1] She was troubled with severe sickness during her entire pregnancy, tortured by dreadful labour pains and following the birth,

> she saw, as she thought, devils opening their mouths all alight with burning flames of fire, as if they would have swallowed her in, sometimes pawing at her, sometimes threatening her ... and bade her that she should forsake her Christian faith and belief, and deny her God ... And so she did. She slandered her husband, her friends, and her own self ... She would have killed herself many a time as they stirred her to, and would have been damned with them in hell ...[2]

Kempe's words are steeped in reflective Christianity, but they would be echoed in eerily similar descriptions of puerperal insanity centuries later, with sufferers relating how they were provoked and tormented by devils.[3]

Seventeenth-century women also wrote in spiritual meditations about their suicidal feelings, melancholy and fears for their own life and that of their children in connection with childbearing. Elizabeth Walker reflected back on her eighth delivery in 1690, 'In this Lying-in I fell into a Melancholy, which much disturbed me with Vapours, and was very ill'.[4] For many of the patients of the astrological physician Richard

9

Napier, the suffering of childbirth was said to have contributed to their mental disorder, including, very unusually, one man, whose fear at his wife's awful pain drove him mad.[5] The physician Charles Lepois noted in the mid-seventeenth century that some women developed mental disorders during childbirth or delirium immediately after,[6] and in the 1690s John Pechey described 'a very Beautiful Lady, that presently after delivery fell melancholy and was mad for a Month, but by the use of a few Medicines recovered her Senses'.[7] By the eighteenth century 'both the general public and medical practitioners acknowledged that pregnancy and labour were associated with physical and emotional stress', with local newspapers reporting on women performing bizarre acts during pregnancy.[8] One woman, whose dreadful and astonishing display of madness was reported in the *York Courant* in 1742, ripped her womb open, 'which done, she took out the Child, and cut it in Pieces'.[9] Comments on mental disturbances following childbirth were most likely to be made in connection with cases of infanticide, with women being said to have destroyed their infants while in a 'temporary phrenzy' or 'out of her senses'.[10]

In 1716 an extraordinary case attended by the physician John Woodward linked mental disorders to maternal imaginings, the idea that women could influence the development and appearance of the infant in their womb through their thoughts and actions.[11] Woodward's patient, Mrs Holmes, had suffered several miscarriages and stillbirths, and, when pregnant with her sixth child, became 'persecuted almost incessantly' with the notion that her child had been harmed by her seeing a porpoise swimming in the River Thames, a sight that had at first much delighted her. Mrs Holmes was ill with 'Pain of the Stomach', 'Disturbance of the Thought' and 'Fits of Weeping' during pregnancy, and after delivery, following a few days of improvement, she relapsed and became suicidal.[12] Woodward's description of her terrors echoed that of Margery Kempe several centuries earlier:

> Amongst others, she had Thoughts of the Devil, as tempting and vehemently urging of her to ill; particularly to fling her Child into the Fire, beat its Brains out, and the like; to which she had the utmost Horror and Aversion ... She had frequent Temptations to lay violent Hands on herself. These Molestations and Suggestions of the Devil, as she apprehended, became continually more frequent and troublesome, so as almost to distract her; and she was neither capable of Business, nor any regular Thought ...[13]

Woodward purged Mrs Holmes heavily, and '[i]n the operation of the third Purge, which was very plentiful, her Stomach became wholly easy, her Thoughts free, and what she called the Suggestions of the Devil, wholly ceased'.[14] John Leake, Physician to the Westminster (later General) Lying-in Hospital in Westminster, writing in 1773, made little reference to mental disorders affecting women in childbed, but was still pointing to the dangers of maternal imagination, which tended 'to disturb their repose and fill their minds with horror and dreadful apprehensions'.[15]

Early modern midwives wrote too of the psychological afflictions of childbearing women, but framed these as fears and anxieties rather than as a form of mental illness. In 1671, Jane Sharp, midwife and author of *The Midwives Book*, described a range of female mental disturbances, including 'fits of the Mother, strangling of the Womb, Rage of the Matrix, extreme Melancholly'.[16] While such conditions could, Sharp concluded, affect women at different times, most often maids and widows, in childbed 'they have swoonding and epileptick fits, watching and dotings', but Sharp has little to say on these conditions.[17] One midwife, Martha Mears, writing in 1805, was eager to remove the dread and 'mother's terrors' felt by many pregnant women, and wrote too of the increased sensibility of mothers-to-be. In a woman of a 'very irritable frame and temper … she grows more impatient and fretful: her fears as well as her angry passions are more easily excited'. In those of weak and nervous habits quickening could produce 'hysterical affections', due to the increased sensibility of the womb, while fainting fits 'to which women of weak nervous habits are very subject even in the unimpregnated state' grew in frequency 'by the greater sensibility of their new condition'. These women, Mears insisted, required great care and tenderness, especially on the part of their husbands. Knowledge of nature's ways and of how to care for themselves during pregnancy, she explained, would 'soon expel the dark phantoms of the brain'. They were to be encouraged 'to acquire a habit of serenity, cheerfulness, and good humour, which we have shewn to be so essential to their own health and that of their offspring'.[18] There was no attempt by midwives to slot these nervous disturbances, ranging from mild irritability, crossness, upsets and fretfulness through to melancholia or a state of frenzy and hysteria, into any kind of framework, and they remain very different from the textbook definitions of puerperal insanity which developed during the nineteenth century. These conditions were described as being a natural response to the excitement, discomfort and trepidation associated with childbirth, rather than an indication of the onset of severe mental illness.[19]

Before 1800 mental disturbances associated with childbirth received only scant mention in medical or midwifery texts written by male practitioners.[20] William Smellie's three-volume *Treatise on the Theory and Practice of Midwifery* referred to the 'passions of the mind' during the puerperal period and to the difficulties these could cause during labour. Smellie remarked on the need for rest and quiet after delivery, but the mental state of the woman was related more than anything to the efforts of parturition and preparation for this rather than to the actual impact of pregnancy and birth.

> The patient's imagination must not be disturbed by the news of any extraordinary accident which may have happened to her family or friends; such information hath been known to carry off the labour-pains entirely, after they were begun, and the woman has sunk under her dejection of spirits: and even after delivery, these unseasonable communications have produced such anxiety as obstructed all the necessary exertions, and brought on a violent fever and convulsions, that ended in death.[21]

Smellie cited several cases where the labour had been disturbed by anxiety, fright or bad management. In 1751 he attended a case in Fenchurch Street, London where an 'old blundering pretender' and his midwife associate told the patient that she was in great danger and the child already dead: 'The woman's pains had been vigorous, but these dismal operations frightened her so much, that when I arrived, they were quite gone off.' A quarrel broke out between the various parties, and Smellie threatened to bring the old pretender before 'the college', before taking over the management of the case himself. The woman was given medicine to help her sleep and she was safely delivered the next day.[22]

William Hunter, the leading obstetrician and surgeon-anatomist of the late eighteenth century and an associate of Smellie, was intrigued by the temporary insanity that could overwhelm women in the month following childbirth and its connection to infanticide, but in his 1784 treatise was sanguine about the chances of their making good recoveries.[23] In his *Practical Essays on the Management of Pregnancy and Labour*, first published in 1793, John Clarke rejected the idea that pregnancy was a state of disease and commented on the generally good health of pregnant women, but also referred to 'some complaints which occur during their situation at that time, which are at least troublesome, and in a few instances dangerous'.[24] Like Smellie, he believed that labour could be disturbed and retarded by fear and lack of confidence: 'the state of the patient's spirits will depend very much on the conduct of those who are about her; therefore cheerfulness in the demeanour of her medical or other attendants is of much importance.'[25]

Most of the literature on childbirth and mental disorder would remain in the domain of midwifery into the early nineteenth century, but there were exceptions, and by the late eighteenth century cases linked to reproduction were finding their way into publications on mental disease and into the small number of existing asylums. John Ferriar, physician to the Manchester Asylum, became interested in puerperal disorders in the 1790s, considering 'the puerperal mania as a case of conversion':

> During gestation, and after delivery, when the milk begins to flow, the balance of the circulation is so greatly disturbed as to be liable to much disorder from the application of any exciting cause. If, therefore, cold affecting the head, violent noises, want of sleep, or uneasy thoughts, distress a puerperal patient before the determination of the blood to the breasts is regularly made, the impetus may be readily converted to the head, and produce either hysteria or insanity, according to its force and the nature of the occasional cause.[26]

The psychiatric doctor William Pargeter related among his patients' histories published in 1792 a case of 'Mania furibunda', which he blamed on a recent 'unfavourable parturition',[27] and in his 1808 study of nervous diseases, Dr Thomas Trotter described how miscarriages, premature labour and long and severe labours could result in nervous disorders, while many women, in his view, were too weak for the office of breastfeeding.[28]

In his *Observations on Insanity*, published in 1798, John Haslam, apothecary to Bethlem Hospital, pondered why women appeared to be more frequently afflicted with insanity than men.[29] In Bethlem, 4,832 women had been admitted between 1784 and 1794 compared to 4,042 men. Haslam concluded,

> The natural processes which women undergo, of menstruation, parturition, and of preparing nutriment for the infant, together with the diseases to which they are subject at these periods, and which are frequently remote causes of insanity, may, perhaps, serve to explain their greater disposition to this malady.[30]

Haslam was particularly impressed with the number of women admitted whose 'insanity supervenes on parturition'. Between 1784 and 1794, 80 patients had been admitted to Bethlem who had become ill during the puerperal period. He suggested that '[t]he first symptom of the approach of this disease after delivery is want of sleep; the milk is afterwards secreted in less quantity, and, when the mind becomes more violently disordered, it is totally suppressed'.[31] The prognosis, however, was good: 'Women affected from this cause recover in a larger proportion than patients of any other description of the same age. Of these 80, 50 have perfectly recovered.'[32] Haslam described the descent into insanity as a potentially lengthy process, which could commence in early pregnancy, but which would result in actual mania only following birth or during breastfeeding.

Thomas Denman was the first British midwifery practitioner to dedicate serious and detailed attention to the subject of mental disorders associated with childbearing and how these could be managed. Denman, one of the outstanding men-midwives of his period, a pupil of Smellie and accoucheur to the Middlesex Hospital, was an advocate of a conservative, non-instrumental approach in childbirth.[33] In his updated *An Introduction to the Practice of Midwifery* published in 1801 he included a section on 'mania' in pregnancy and among new mothers,[34] and went on to devote many pages to this topic in a treatise published in 1810.[35] He claimed that of the complaints to which women in childbed were liable none was 'more distressing than that aberration of the mental faculties which sometimes occurs in that State'.[36] He gave the disorder the generic name 'mania lactea', arguing that it could appear in the earliest stages of pregnancy, as labour approached, soon after delivery, or during suckling or weaning seven or eight months after delivery. He argued that 'in every case', the disorder was occasioned by 'an uncommon irritation' of the uterus or breasts, which extended its influence to the brain, and was unrelated to 'former disposition or habits'.[37] Denman warned that mania should not be confused with the excited state produced by childbed fever, and argued against adopting the same kinds of treatment, especially copious bleeding.[38] Denman also argued that mania lactea bore little relation to the ease or difficulty of labour, and, while many later authorities would recommend weaning as a remedy, he concluded that the disorder could occur not only in women who breastfed, but also in those who did not suckle their babies as well as after weaning.

The attack, according to Denman, could be instantaneous and unexpected, or it crept up gradually and was barely perceptible in its early stages. Mania, Denman concluded, was a rather imperfect definition of a disorder resting on symptoms such as altered general conduct, unusual actions and incoherent or irrational conversation interspersed with periods of lucidity. Denman spoke of the need to keep lying-in women quiet and free from disturbance as they were in an especially irritable state after delivery and prone to disease. On becoming disturbed, the women became flushed, made violent exertions, had a rapid pulse and their eyes were red and glistening: Denman likened their state to 'delirium without fever'. As to treatment, restraint was sometimes necessary in Denman's view, but attendants must be 'mild and kind' as well as watchful, steady and knowledgeable, acquiring an 'almost irresistible control over patients without exerting any needless authority'.[39] He recommended that visits from the family should be curtailed and strict regularity imposed over the patient's meals and rest. He considered medicines of little use, though he recommended the judicious use of mild doses of opiates, very careful and limited bleeding, purgatives to remove obstructions and to improve the general health, warm bathing to make the patient comfortable, shaving the head and sponging with cold water, and moderate activity. Patients could be 'artful' and 'violent' and had a propensity towards suicide, so required careful watching. Like those writing subsequently on the disorder, Denman was optimistic that most women would recover over a period of several months.

The setting for delivery

Denman in many ways was on the cusp, the boundary of change in midwifery. Lauded by midwives and male practitioners for his common-sense and sympathetic approach and advocacy of limited medical intervention, his comments were based on experience and observation, his advice centred on practical approaches that appeared to work. Denman did not see mental disorder as something that could be closely defined, nor was it an anticipated sequel to childbirth; rather he argued that pregnancy, labour and childbed represented 'an altered, but not a morbid state',[40] just as he also suggested that '*some* women *might* suffer from "various hysteric and nervous affections"' and played down the association between menstruation and disease.[41] Yet, as Anne Digby has argued, descriptions such as Denman's would take on new meanings in the Victorian era.[42] A few decades later, the social context and the link forged between obstetrics and psychiatry, together with redefined notions of women's proper role and functions, would

bring a new perspective to Denman's observations. The suggested likelihood of mental collapse among childbearing women would fit in
with deeper fears about their general vulnerability at this time. The end
of Denman's time of influence would also mark the arrival of a new
phase in thinking about childbirth practices and intervention in the
birth process.

In 1817 a tragic event 'struck Britain like a thunderbolt' and would
impress itself on medical thinking as well as public consciousness, the
death of a princess in childbirth.[43] Richard Croft, who attended at the
confinement of Princess Charlotte, acted according to current best
practice and established ideas on non-intervention, allowing nature to
take its course, an approach shared and propagated by Thomas
Denman, Croft's father-in-law. Though suffering acute and distressing
pain, Princess Charlotte was allowed to continue in labour for fifty
hours without interference; she delivered a large stillborn child and
died five hours later after suffering a massive postpartum haemorrhage.
Though forceps were kept in readiness no attempt was made to use
them.[44] The nation went into deep and collective mourning, and the
medical profession, urged on by the apparent fears of expectant
mothers precipitated by this calamity, entered a new mind-set. Princess
Charlotte's death legitimated and added urgency to trends already in
place, to intervene in childbirth to speed delivery and to predict problems and fret increasingly about the fate of women passing through
what was envisaged more and more as a dangerous and threatening
event.

Women as patients

During the first half of the nineteenth century women's bodies and
minds came under closer scrutiny from doctors, who delivered more of
them in childbirth, who spent more time in middle-class homes as
their general practitioners, and who cared for increasing numbers of
women in asylums. Considerable energy would be devoted to describing and discussing female disorders and responses to them in their
practices and their publications, and ample opportunity would be
offered of contact with childbearing women. Motherhood would be
the experience of most nineteenth-century women and dominate
much of their adult life. Ellen Ross has pointed out that while, from as
early as the 1850s, the middle classes began to practise family limita-
tion, of those women marrying in England and Wales in 1860, 63 per
cent would subsequently bear five or more children.[45] For many
women this was an experience marked by dread of being harmed by

the birth, or even of the death in childbirth of themselves or their off-spring. Many mothers-to-be knew of family or acquaintances who had died during the birth or shortly afterwards, and Princess Charlotte's experience was one which resonated amongst many women.[46]

Though the majority of poor women continued to be delivered by midwives, increasingly in the nineteenth century better-off women were attended by male practitioners.[47] The reasons for this were complex.[48] From the eighteenth century onwards midwifery reinvented itself as a new, male-dominated speciality and 'science', a science that made major advances in terms of elaborating on the anatomy and mechanics of labour, but which would also focus increasingly on diseases and disorders associated with reproduction and the female life cycle. By early in the nineteenth century, building on an impressive corpus of writing, new ways of teaching obstetrics in private midwifery schools, and the reputations of esteemed men-midwives such as William Hunter, William Smellie, William Osborn, John Leake and Thomas Denman, the man-midwife, accoucheur or, by the nineteenth century, obstetrician had marked out an area of practice, which came to include claims to deliver normal cases of childbirth as well as emergency cases and the ever-expanding field of women's diseases.[49] The 'bait' of forceps and the impact of lying-in hospitals as explaining the shift to male practitioners have been played down by some historians.[50] Yet instruments and, after the mid-nineteenth century, the introduction of obstetric analgesia, could be very attractive to women who dreaded a protracted and painful labour.

Lying-in hospitals, small-scale institutions set up particularly in the metropolis and other major cities after the mid-eighteenth century, catered specifically for poor women unable to pay for their deliveries, and were also important in exposing medical men to the medical and social problems of giving birth, enabling them to build up experience of normal and difficult deliveries.[51] Attracting sponsorship from the rich and influential, lying-in hospitals bolstered reputations, and many early authors on the diseases of women, for the most part men prominent in the field of obstetrics, held posts at these institutions, as well as having prestigious private obstetric practices. Similarly, positions in the new women's hospitals and wards set up after the mid-nineteenth century, advertised expertise as well as concern for the poor, as in the case of the fashionable obstetric physician Samuel Ashwell who was instrumental in establishing a ward for gynaecological patients at Guy's Hospital in 1831.[52] Those who held posts at lying-in institutions or the burgeoning women's hospitals also tended to be successful in obtaining the first appointments as lecturers or professors in midwifery and the diseases of women. A striking example was Robert Ferguson who worked with London's poorest women at the General Lying-in Hospital and was appointed Professor of Obstetrics at the newly founded King's College London in 1831 and as Physician-Accoucheur to Queen Victoria in 1840. Increasingly, a group of doctors, particularly in London, found that they were doing enough business as accoucheurs or obstetric physicians specialising in female disorders to devote themselves to this field of practice, delivering the poor in hospital, the rich at home.

Victorian doctors were eager to offer their services to a new client group of well-off women. In order to make a living, general practitioners, whose numbers expanded dramatically in the nineteenth century, focused their attention increasingly on middle-class homes where they delivered babies and attended sickness in the family.[53] The middle class, particularly women and children, comprised an expanding market,[54] although Patricia Branca has shown that health care was also carefully budgeted for even among this social group.[55] This may have led to delay in obtaining treatment for numerous ailments, including perhaps puerperal insanity, a delay much lamented by doctors. It was not just specialist obstetric physicians or accoucheurs who were attending women in childbed, and Anne Digby has argued that 'the key to Victorian general practice' was midwifery, which, though often underpaid and undervalued, opened the door to employment opportunities as the family doctor.[56] General practitioners became more involved in women's procreative activities, advising them to employ a medical man for their confinements and treating a widening range of female disorders.

During the early nineteenth century psychiatry began to emerge from an era where it was practised by a miscellaneous collection of clergymen, mad-house proprietors and doctors in a diverse collection of often ramshackle institutions to become a distinct branch of medicine. More 'alienists' or 'psychological doctors', as they came to style themselves, found employment in this field, which was given a major boost in England and Wales by legislation passed in 1808 and more particularly the Lunatics Act of 1845 which made the establishment of county asylums, catering predominantly for pauper patients, compulsory. Similar legislation was passed in Scotland in 1857.[57] From a handful of asylums in England and Wales at the beginning of the century, by 1850 there were 24, and by 1870 51, each catering for several hundreds of patients.[58] While the county asylums robbed many private mad-houses of their pauper patients, forcing them to close, others continued to compensate for the shortfall in public provision, and some private institutions for the rich flourished and provided an important source of income and esteem for psychiatrists.[59] It was the poor, however, who became particularly visible to the psychiatric profession, as a substantial body of asylum doctors were confronted with ever-growing numbers of patients facing economic hardship and ill health as well as mental breakdown. This exposure to the poor through institutions would have important implications for explaining mental disturbance and superimposing other ideas about causality on purely physical or biological ones.

Elaine Showalter has argued that asylums were increasingly populated by women in the nineteenth century, a factor largely determined by constructions of proper feminine behaviour.[60] There was, however, considerable variation between institutions and more complex demographic and social reasons than suggested by Showalter explain the rise in female incarcerations. Much of the difference between male and female admissions was due to the higher mortality of male inmates and women's economic dependence, which made them more liable to end up in public institutions – whether the workhouse or an asylum. David Wright has recently argued that in public asylums gender did not play a dominant role in diagnosis and that women were institutionalised in numbers commensurate with their representations in the adult population.[61] However, a great many women were admitted to asylums during the course of the nineteenth century and they became more visible through increased contact with asylum superintendents.

Of these female admissions a considerable proportion were for 'puerperal' disorders or other conditions related to the female life cycle, from the onset of menstruation to its cessation. Of the 114 female admissions to the Royal Edinburgh Asylum in 1855, six were said to result from childbearing, six from amenorrhoea or disordered menstruation, three from climacteric change and one from the 'secret vice'. This compared with four men whose disease was claimed to result from the 'secret vice' and one from the 'arrest of spermatorrhoea' out of 109 male admissions.[62] Private asylums for the rich were more likely to admit women who had been treated by specialist gynaecologists or had received more assiduous attention from their family doctor, and many of these had diagnoses linked to female disorders. Ticehurst Hospital in Sussex admitted 194 women between 1845 and 1885 and of these 27 (14 per cent) had symptoms related to childbirth and menstruation or its cessation.[63] Puerperal insanity, itself responsible for around 10 per cent of admissions in many asylums, found a warm welcome in psychiatry as well as obstetrics, particularly as it came to be associated with a good rate of recovery. Therein lay a conundrum, for while a bridge between obstetrics and psychiatry was laid down in the form of puerperal insanity, as we will see in chapter 2, it would also be claimed by specialists in both fields as their own area of expertise.

Female disorders

Reproduction was increasingly defined during the early nineteenth century as taxing and full of risk for mothers. Childbirth and the care of children were women's natural functions, but women at the same time were described as weak and unstable, unfit to withstand the physical strain and mental stress of maternity. Feeble women, subject to the vagaries of their reproductive systems, were compared unfavourably with rational, robust, muscular and stable men, and women's inability to regulate their bodies linked resolutely to an inability to regulate their minds.[64] Bolstered by an outpouring of writing on midwifery and female diseases, as scrutiny of women's life cycle intensified, nervous disorders would be attributed more and more to reproductive processes and an environment created in which puerperal insanity would find a natural home. Puerperal insanity came to be associated by some doctors with morbid disease or a predisposition to mental disorders. However, many others envisaged it as one of a whole spectrum of anticipated dangers linked to childbirth; it was quite simply a straightforward risk of becoming a mother, a result of the very act of giving birth and that was an alarming prospect.

By the early nineteenth century an entire battery of diseases and disorders associated with reproduction was elaborated on, and puerperal insanity would fit neatly into this context as a link was 'forged' between obstetrics, gynaecology and psychiatry.[65] Puerperal insanity would find its place in the growing literature on the diseases of women, often the last chapter among an impressive collection of obstetric complications, disturbances and disasters.[66] John Burns, Glasgow Lecturer on Midwifery and Professor of Surgery, in his 700 plus-page, best-selling *Principles of Midwifery* outlined the danger of puerperal mania in Book III, chapter XIX, following chapters on premature, preternatural, tedious, instrumental, impracticable and complicated labours, situating it among a wide range of postpartum disorders, including puerperal fever and phrenitis, and disorders of the breast, womb and lochia.[67] Women's hospitals were established from the 1840s onwards, and general hospitals also opened wards specialising in women's complaints. Through increased contact with women patients, these institutions gave doctors more experience of the medical as well as social problems faced by women, particularly poorer women.

Gynaecology served to legitimate the notion that women were naturally fitted for activities in the private sphere of the family and childbearing, as well as the view that they would face many challenges in carrying out these functions.[68] But its practitioners were divided and its literature racked by debates between conservatives and interventionists for the rest of the century. This was demonstrated famously by Dr Robert Lee, Professor of Midwifery at St George's Hospital, who fiercely criticised excessive instrumental interference, and specifically the use of the speculum, at a meeting of the Royal Medical and Chirurgical Society in 1850, as well as the outrage sparked by ovariotomies, with proponents of this risky procedure being described as 'sow-gelders and butchers in frock-coats'.[69] There was also a considerable gap between rhetoric and practice, and what was emphasised in the gynaecological literature in terms of 'fashionable diseases', including hysteria and vague uterine disorders, was almost certainly not representative of the practice of doctors seeing cases on a daily basis in large city hospitals, where they were more likely to be confronted with childbirth injuries and complications, prolapsed womb, varicose veins, vaginal tearing, menstrual disorders or discomfort, fibroids or cancers, aggravated in many cases by the poor health of their patients and long-term neglect of their conditions. In the 1850s the Hospital for Women in London admitted patients suffering from chronic ill health or occupational diseases as well as gynaecological complaints.[70] Repeated childbearing, 'living under the shadow of maternity', took its toll on women who were malnourished and weak, as testified to in the case notes of maternity hospitals, as well as asylums admitting patients suffering from puerperal insanity.[71]

Puerperal insanity was of course just one of a number of mental disorders associated with the female life cycle and reproduction that would capture the attention of the Victorian medical profession, though its reputed prevalence would make it particularly significant. The link between menstruation and hysteria had been developed by the end of the eighteenth century by Alexander Hamilton, Professor of Midwifery in Edinburgh, with hysteria said to occur 'most frequently about the time of the periodical evacuation'.[72] John Burns would echo this: 'all women, at the menstrual period, are more subject than at other times to spasmodic and hysterical complaints.'[73] Several medical authors subsequently offered advice on how to retard and temper menstruation, with its associated pain, strain on the spirits and temper, nervousness and hysteria, through nursery diets and regimes, and a rejection of 'sofas to lounge on – the absence of novels fraught with harrowing interest' and 'laborious gaiety, of theatres, and of operas'.[74] Ovarian mania, climacteric insanity, nymphomania and hysteria were, it was concluded, triggered by dramatic changes in the reproductive cycle, with women liable to outbreaks of insanity from adolescence until the cessation of menstruation.

Hysteria is probably the female condition about which most was written in the nineteenth century and, subsequently, by historians.[75] Hysteria had a durable link with the reproductive organs, particularly the womb; it was deemed likely to affect those with a 'susceptible Nervous System' while 'Hysteric Diseases appear only during that Period of Life in which the reproductive Organs perform their Functions'.[76] However, in stark contrast to puerperal insanity, hysteria was responsible for few asylum admissions, although 'hysterical' was used frequently to describe symptoms and behaviour, alongside such expressions as 'erotic' or 'hypochondriacal'.[77] Analogies can be made between hysteria and puerperal insanity, but what distinguished the latter was its 'realness', the fact that it was classified as a distinctive condition, albeit with varied ideas on causality, symptoms and prognosis, and was very prevalent in terms of asylum admissions and cases occurring in private practice. The gap between rhetoric and reality appears to have been narrower for puerperal insanity than was the case for hysteria, 'a name without a disease'.[78] And perhaps to an even greater extent than hysteria, with its long and complex history and language and symptoms adapting over time – 'the womb travels, vapours rise, sympathy transmits symptoms up the body' – puerperal insanity was a disorder linked inexorably to the nineteenth century, though both shared the message offered by Helen King in her analysis of hysteria: 'women are sick, and men write their bodies'.[79]

Fear and responsibility

By the mid-nineteenth century puerperal insanity had become established as one of the many possible disasters associated with childbirth. As childbirth was depicted increasingly as dangerous, and women more subject to bouts of illness, debility and insanity, there is a powerful sense in writings on the subject that this condition was only to be expected among women too feeble to give birth naturally and without major disruption to their reproductive organs and nervous systems. Advice and midwifery books intended for a lay audience, while skating over the process of delivery itself, offered counsel on deportment, diet and exercise during pregnancy, but more than anything warned women to be pre-emptive, to anticipate the risks they were taking and to employ an expert physician to safely deliver them. This rhetoric was directed at a particular group of women, the wealthy, who were described as being weakened by their lives of luxury, who had access to advice literature and who were in a position to employ a male attendant in childbirth. This reconfiguration of childbirth as the responsibility of doctors, combined with increased exposure to written sources of knowledge, may have enabled women to learn more about the potential hazards and created a responsiveness to them, as well as encouraging their midwifery practitioners to anticipate difficulties.[80]

Until the eighteenth century pregnancy and childbirth remained very much part of female culture;[81] thereafter the man-midwife, or accoucheur, began to be the childbirth attendant of choice for many wealthy women.[82] Male practitioners were only too happy to shift responsibility in childbirth from female midwives and helpers, and midwives were eased out of particularly well-to-do practice. Although doctors could potentially become women's allies in childbirth, responding to requests for pain relief and intervention to shorten labour, they also stripped away a source of emotional and practical support, banishing female helpers as outmoded and potentially dangerous. The seventeenth-century French accoucheur François Mauriceau was not alone in warning of the dangers of boisterous and numerous visitors, 'the Gossips and all Comers', during the mother's lying-in, who could endanger her well-being.[83] Smellie had directly related this in the eighteenth century to mental disturbance and the disruption of labour, but this view would be voiced more frequently and vociferously by the nineteenth century. Although by the late nineteenth century in the homes of the well-to-do a monthly nurse would often be employed to take care of the new mother, for other women the isolation of the lying-in room, which resulted from efforts to exclude female helpers, may have made them more anxious and vulnerable to depression. Advice books and health manuals emphasised women's inability to obtain good advice or to ask for help as well as the increased isolation of mothers, and suggested reading their own publications as a way out of this impasse. Many of the authors of obstetric textbooks also wrote a 'popular' version of their directives for managing pregnancy, delivery and child-care. John Burns' *Popular*

Directions for the Treatment of the Diseases of Women and Children was intended to provide 'a little medical knowledge' for 'the comfort, or even the life of the patient, before regular assistance could be procured' and gave plain instructions 'yet not so minute, as to bewilder those readers, for whom they are intended'.[84] The advice books also stressed that motherhood was women's noblest function and that they would be failing in their duties if they did not prepare themselves as best they could for pregnancy and delivery.

Thomas Bull, author of one of the most popular advice books of the nineteenth century,[85] wrote much in his book that was plain common sense and useful advice, but prefaced it with the injunction:

> A woman may consider herself a mother, not only from the birth of her child, but even from the moment of conception. From that important epoch her duties commence – duties amongst the most sacred and dignified which humanity is called upon to perform ... Should she, however, be careless and negligent upon this head, and fail in attention to the measures which her new condition demands ... her child will inevitably be variously and injuriously affected, these causes operating through her system upon that of the child.[86]

He went on to instruct:

> Engage your medical attendant early. You will then be able to seek his direction and guidance in every doubt that may arise, and, confiding your fears and anxieties to him, will derive from his experience and knowledge that rational and kindly explanation of your difficulties which may instantly dispel them.[87]

Authority that had been female-centred and based on midwives, mothers and female friends passed slowly but surely to doctors as an integral part of their move into midwifery work. Medical men built on existing fear linked to birth – dread of the pain, injury or the death of the mother or infant as the time of delivery approached had long been expressed by women in anxious letters to family members and friends[88] – and in various ways doctors professed ownership of this fear and the ability to deal with the dangers they presented to women. Increasingly, pregnancy and birth were conceptualised as routes full of danger and obstacles. This was not particularly new, but what was new was the emphasis on careful monitoring, 'expert' guidance and the possibilities of medical intervention. Martha Mears wrote in 1805 about the naturally occurring anxieties of pregnant women, inspiring them 'with a just reliance on the powers of nature ... See the benignity of nature in all her other works!'[89] Her successors, talking increasingly of 'patients' rather than of 'mothers' and 'women', listed in grim detail the various disorders associated with childbirth or following hard on the heels of delivery in textbooks for colleagues and in some of the advice literature. In addition to preternatural labours, with breech, footling, knee or transverse presentations or tedious deliveries, women could experience difficult labours as a result of distortion of the pelvis, swelling of the soft parts, convulsions, retention of the placenta, uterine haemorrhage, lochial discharge, numerous disorders of the breast, prolapse or inversion of the uterus, rupture, puerperal convulsions, puerperal fever or puerperal mania. Robert Lee offered little sense of optimism in his *Lectures on the Theory and Practice of Midwifery*, published in 1844, and we can only hope that few mothers-to-be ever read his introduction:

> There are many diseases peculiar to women; they are all exposed to great suffering and danger during pregnancy and child-bearing, and many die from acute disorders following delivery. Not only are the functions of the uterine system before conception often disturbed, but all the different parts of this system undergo morbid alterations of structure, which are frequently of a malignant nature, and excite, in their progress towards a fatal termination, the most acute and protracted pain.[90]

Nature apparently had deserted at least some women in childbirth; other authors accused women themselves of deserting nature. Michael Ryan, author on midwifery and medical jurisprudence, summed up the dangers of childbirth in 1841:

> Happily for humanity, the process of labour, in a vast majority of cases, is safe and free from danger, especially when women live according to nature's laws; but among the higher and middle, indeed all classes in civilised society in which these laws are frequently violated or forgotten, or when the constitution is impaired by the luxury or dissipation of modern times, the process of child-bearing is attended with more or less danger; both before and after it is completed ... It is, however, fortunate for suffering humanity,

that the process of parturition may now be greatly accelerated, and the greatest of mortal suffering relieved by the advice and skilful exertions of the obstetrician or medical attendant, and with the most perfect safety to the parent and offspring ... there are few intelligent women who do not prefer medical attendance during labour, to that of midwives.[91]

There was still hope for poorer women living closer to nature's laws, who, Ryan concluded, could be attended by midwives. For the rest preventive action was required through the employment of a doctor trained in obstetrics. Ryan warned too of the dangers of puerperal mania, which could occur soon after delivery or during lactation when the system was 'exhausted', but believed it to be 'rarely incurable'. He stressed, like other authors, mild treatment, careful handling and removal of the patient from her home and its associated distress.[92] Ryan predicted problems and in almost the same breath assured his readers that he could solve them.

* * *

It could be argued that there are traces of puerperal insanity to be found in earlier centuries, but these are traces only.[93] Until the nineteenth century there was little in the way of rigorous interest on the part of any branch of the medical profession, and mental disorders associated with childbirth were conceptualised in a variety of ways, linked to the risk of retarding labour, maternal imaginings, anxiety about childbirth and the threat of harm to the infant. Though they could be inconvenient and troubling for the women afflicted, and potentially dangerous for those around them, they were not seen as serious in the sense that women were expected to get over these disturbances and it was anticipated that few would be affected.

By the turn of the nineteenth century, authors like Thomas Denman, John Haslam and John Burns were taking more interest in mental disorders linked to pregnancy, childbed and lactation and began to classify these as distinct conditions, yet links with older notions of anxiety, fretfulness or despondency persisted. John Burns included a section on puerperal mania in his textbook first published in 1809, yet also had a separate passage on 'despondency', describing women as commonly 'very desponding during pregnancy, and much alarmed respecting the issue of their confinement ... others suffer chiefly during lactation'. He too concluded that little could be done by medicine; 'the mind is to be cheered and supported by those who have most influence with the patient'.[94] It was the welcoming context of the nineteenth century that would give puerperal insanity its rationale and substance, with several branches of the medical profession engaging themselves with female disorders as a source of professional credit and patients, ready to link them to women's frailty and reproductive cycles, and for others their poverty and neglected health, and, as we will see in chapter 2, to develop the idea that they were best fitted and able to cure them. Though it could be argued that there was a period of slow gestation until the nineteenth century, it is the rapid delivery of puerperal insanity in the early decades of the nineteenth century, linked to an intensified concern about women's minds and bodies, that is most striking.

2
Boundaries of Expertise and the Location of Puerperal Insanity

Fondly anticipating the joy, perhaps of a first-born, a beloved wife patiently submits of all the inconveniences and restraints of pregnancy, however irksome, and the pains and dangers of labour, however great. The affectionate husband and relatives await with deep and anxious expectation the event; and at length, when the joyful period arrives, and the happiness of all is completed by a safe delivery, – how dreadfully is the scene reversed, when the happy mother suddenly displays symptoms of delirium![1]

In the first decades of the nineteenth century puerperal insanity was propelled into the medical arena, as attacks of insanity preceding, during or following childbirth became the focus of considerable interest and inquiry, its victims the subject of anxious interjections. As we saw in chapter 1, it was part of a much broader change within medicine, as alienists and experts on obstetrics and the diseases of women began to forge their specialist fields, as new theories enveloped medicine linking female vulnerability, particularly at times of reproductive activity, to nervous disorders, and as women of all social classes became more visible as patients. Childbirth was redefined as abnormal and risky, requiring active intervention and an expectant approach, with doctors increasingly declaring their readiness to deal with complications and prevent disasters after the delivery. The occasional, brief, often matter-of-fact observations of men-midwives, midwives and court witnesses on the depression, disorientation and occasional craziness and violent tendencies of childbearing women described in previous centuries were replaced by more careful definitions and intensive discussion. The disorder was given its name 'puerperal insanity', publications on the topic proliferated, it was quite literally given spin and classified into existence.[2]

Treatises, articles and correspondence to medical journals on the identification and treatment of puerperal insanity and interesting cases encountered in practice proliferated.[3] Obstetric practitioners led the way as the most frequently cited authorities on the disorder, but by the mid-nineteenth century puerperal insanity also had a firmly secured place in the psychiatric literature. It would enter too into the dialogue of general practitioners, who were taking on increasing numbers of midwifery cases. This chapter explores ideas on causality and therapy, as well as the tensions between the different groups who had a stake in treating puerperal insanity, particularly concerning the location of treatment. As the term 'expertise' increasingly entered the English language around the mid-nineteenth century, alongside 'a unifying professional impulse',[4] so too did alienists and obstetric physicians stake their claims to exercise it, bringing into play their training, experience in observing cases and skill in treating them. Obstetric doctors and alienists referred to puerperal insanity as their own territory, not necessarily in a blunt or confrontational way, but implicitly as they wrote up their case studies and outlined their treatment regimes. A divide would open between alienists claiming authority based in the asylum – although there was much ambivalence even amongst this group about the suitability of the asylum as a place of treatment – and midwifery practitioners who pressed for the maintenance of their authority to supervise the lying-in period and recovery of the patient in a domestic setting. This chapter focuses on the early and mid-nineteenth century, a period strongly influenced by ideas of moral management in psychiatry, and it could indeed be argued that two forms of 'moral management' emerged in connection with the treatment of puerperal insanity.[5] In addition to moral management in the asylum, which focused on encouraging the patient to re-establish self-control and proper modes of conduct, to eat, exercise and work, a form of domestic management developed, with similar goals and dominated by the authority of the physician, for the treatment of patients who could afford to pay for private medical care.[6]

Robert Gooch and the naming of puerperal insanity

In 1820 Robert Gooch became the first British physician to write on 'puerperal insanity' in his short but influential treatise, *Observations on Puerperal Insanity*.[7] Though sharing common ground with earlier commentators on mental disturbance and childbirth, including Thomas Denman, William Hunter and the French alienist Jean-

Etienne-Dominique Esquirol, Gooch presented the disorder very much as a new discovery, new territory within the field of midwifery and the diseases of women. Gooch was hardly a self-publicist; he was reserved, given to bouts of self-doubt and constantly anxious about his own failing health, but his work commanded immediate attention and great authority with its engaging and clear writing style and vivid use of case notes. The timing of his treatise was perfect, given the increased interest in disorders linked to reproduction, and the emerging association of obstetrics, women's diseases and nervous disorders. By the time Gooch published his treatise he had built up a flourishing London practice, based largely on obstetric work, had been appointed Lecturer in Midwifery at St Bartholomew's Hospital and Physician to the Westminster Lying-in and the City of London Lying-in Hospitals.[8] Gooch died aged 46 in 1830, but his publications on midwifery, the diseases of women, puerperal fever and puerperal insanity continued to be influential.[9] Gooch was remembered in the decades following his death not merely as being the first to outline puerperal insanity and for the power of his description, but also for the sympathetic and philosophical approach that he brought to his practice, his patients and writings. Though an obstetrician, he had 'a special bent for psychiatry' and 'was an acute observer of the finer points of mental aberration'.[10] His publications would also place puerperal insanity firmly into the context of the middle-class home, with his case histories based on private practice; he discussed the disruption it caused and how the woman of the house could eventually be restored to health.[11] Few writing on the subject after Gooch would fail to acknowledge his findings, even if they were to disagree with them. His book *On Some of the Most Important Diseases Peculiar to Women*, published in 1829, and reissued two years later, shortly after his death, was claimed by the physician Henry Southey to be the most valuable work on the subject in any language, with the chapters on puerperal fever and puerperal madness 'the most important additions to practical medicine of the present age'.[12]

Gooch's published case notes would focus predominantly on patients treated in his private practice, but he claimed to have learnt a good deal about his patients and their disorders through his work in lying-in hospitals. He described his learning experience following his appointment to the Westminster Lying-in Hospital in 1812, where 'the whole task devolved upon me for several years, to attend both the in- and the out-patients in their difficult labours, and their illnesses ... My situation gave me ample opportunities of observing the diseases of lying-in women among the poor of London and its neighbourhood.'[13]

> He who has the care of a Lying-in Hospital, is a Lecturer on Midwifery, and is resorted to by the public as an obstetrical Physician, has opportunities of acquiring knowledge in, and extending the bounds of, obstetric medicine, which no other physician, surgeon, or general practitioner can possess, whatever may be his talents. Your task will go on prosperously, the sooner you have ceased to read, and begun to observe and think ...[14]

In his *Observations*, Gooch described how women who were normally 'perfectly sane' became deranged after delivery, a condition that he was encountering fairly commonly in his own practice. The disorder usually began a few days or weeks after delivery, during nursing or soon after weaning, but Gooch had also seen it at the commencement of pregnancy and during labour itself.[15] The approach of the disease excited little apprehension and the symptoms were very diverse; the women's pulse could be quick, their nights restless and temper short – nothing remarkable. Gooch referred to what followed as 'indescribable hurry' and 'peculiarity of manner'. The women's language became wild and incoherent as they reached a state of mania. When the disease took the melancholic form, which Gooch suggested manifested itself just as often as mania, it usually began some months after delivery and its appearance was much more gradual. These patients typically declined in health while nursing and suffered from failing memory, confusion and a depression of spirits. They found it difficult to concentrate, and were bewildered, anxious and dissatisfied about themselves. After a few weeks the condition became more marked, the women mournful, downcast, silent and thoughtful, imagining that they had a serious disease, accusing themselves of moral depravity and believing that they were objects of punishment and scorn.[16] While his overall description of puerperal insanity was succinct, Gooch elaborated his account with a number of detailed and evocative case notes. Case histories were Gooch's means of building his knowledge, with his patients' stories of their physical discomfort, weariness and mental distress enabling him to place diverse symptoms into a distinct disease category, a technique that would be taken up by many of his successors.[17]

Territories of midwifery and psychiatry

Gooch would remain an authority on puerperal insanity, and the status of his work remained undiminished beyond his death; one enthusiast referred to him as '"a master mind, which we return to again and again, not merely for the knowledge which it contains, but to observe how that mind worked"'.[18] While Gooch was one of the most successful obstetric physicians in London, his experience and background were fairly typical of that of other midwifery practitioners and authors on women's diseases who met puerperal insanity in the course of their work and wrote on the subject, many of whom were also based in the metropolis. The most successful obstetric practitioners combined their experiences of working in lying-in hospitals with expanding private practices among the well-to-do; many had senior hospital appointments and lectureships in midwifery and the diseases of women. A good deal of work on puerperal insanity was written by men specialising in midwifery at a time when opportunities in this field were expanding,[19] and as they were seeking to establish their professional credentials and associations.[20] James Reid, who published a lengthy and influential essay in 1848, based his findings extensively on his long practice experience, including his work as one of Gooch's successors at the General Lying-in Hospital,[21] and by the mid-nineteenth century no textbook on the diseases of women would be complete without a section on puerperal insanity. It was recognised as a common complication of child birth, a condition that students and practitioners were supposed to be familiar with and ready to treat.[22]

In the first half of the nineteenth century obstetric practitioners fairly credited themselves with building up knowledge on the recognition, prognosis and treatment of puerperal insanity. However, as we have seen in chapter 1, a handful of psychological doctors had referred to the relationship between mental disorder and childbirth in the eighteenth century and puerperal insanity would soon find its place in the publications of nineteenth-century alienists, including their textbooks on mental diseases.

One of the earliest to devote serious attention to the subject was George Man Burrows in his 1828 *Commentaries*; he was particularly well placed to do so, having moved from a practice focusing on midwifery to the treatment of mental alienation.[23] Bucknill and Tuke's influential textbook, though its section on puerperal insanity was derivative, acted as a guide to many alienists; it emphasised the prevalence of the disorder, and confirmed its division into the categories of insanity of pregnancy, puerperal insanity and insanity of lactation.[24] Armed with such information, the attention of asylum superintendents was drawn to these disorders, which, with their close link with childbirth, were believed to be easy to recognise. Alienists, in a similar way to their colleagues in obstetrics, were in the process of building up a specialist field, gaining practical experience in expanding asylums catering for the insane poor, which some combined with running private institutions. Lacking formal training or specialised qualifications, and with only fledgling professional bodies, it was the asylum that provided experience and instruction in classifying and treating mental diseases.[25] Claims of expertise with respect to certain disorders became very important. With regard to puerperal insanity, British alienists were no doubt given a boost by influential publications emanating from France, particularly the work of J.-E.-D. Esquirol, star pupil of the famed alienist Philippe Pinel and superintendent of the Salpêtrière Asylum in Paris. In 1819 Esquirol published a paper on the mental alienation of recently confined women and nursing mothers, based on 92 case histories of his asylum patients, 12 per cent of all his female admissions between 1811 and 1814.[26] This was followed several decades later in 1858 by L.-V. Marcé's 400-page monograph on puerperal insanity.[27]

As the two groups of medical practitioners began to build up experience and publish on puerperal insanity, the stakes were high. Puerperal insanity lay at the border between the two disciplines of obstetrics and psychiatry for much of the nineteenth century, with alienists and obstetric doctors keen to bolster their prestige and practices. Although actual skirmishes appear to have been rare, puerperal insanity became a bridge to be taken by the two disciplines. While Gooch saw the lying-in hospital as the best place to learn about female disorders, John Conolly, Medical Superintendent of Hanwell Asylum, argued that most practitioners would meet with only a few cases of puerperal insanity in the course of a long practice, and 'it is only in lunatic asylums that any extensive opportunities can be afforded for studying this affection'.[28] He cautioned the medical students attending his lectures at Hanwell that 'it is an affection which is almost certain to occur, at some time or another, in your practice, and one for which you ought to be especially prepared'.[29] Although puerperal insanity accounted for only 4.6 per cent of female admissions to the Somerset County Lunatic Asylum between 1848 and 1868, its Superintendent Robert Boyd took care of all 63 cases and regarded himself as highly proficient in treating the condition.[30] Alexander Morison meanwhile demonstrated the possibility of seeing large numbers of cases in a private practice specialising in mental alienation, claiming by 1853 to have treated 172 cases of insanity connected to pregnancy, childbearing, lactation and abortion.[31] The stakes were high for another reason: women being treated for puerperal insanity were perceived as having very good chances of recovery. In 1846 John Conolly claimed that 'cases of puerperal insanity appear to afford a better prospect of recovery than any other'.[32] Though the obstetric physician William Tyler Smith opined that parturition stood at 'the boundary between physiology and pathology, being attended by more pain, and being liable to a greater number of accidents, than any other act of the economy',[33] he also declared in 1856 that 'puerperal patients should always be treated as though they were destined to a perfect recovery'.[34] Unlike the dreary prognosis for many insane patients in the nineteenth century, puerperal insanity was recognised by doctors – despite many disagreements about its aetiology and treatment – as a temporary, albeit possibly very serious, aberration and likely to be curable.

Much of the debate – more than for many other forms of mental disorder – would centre on the location of treatment. Many doctors considered puerperal insanity to be a special category of mental disorder, one that lent itself well, in less severe cases and if caught quickly, to domestic management. With most cases following on from delivery at home, it was argued by midwifery practitioners that they should continue to provide treatment. Alienists, meanwhile, embraced the disorder, which accounted for a growing number of female asylum admissions, as a potential boost to rates of recovery and discharge. Yet they were also ambivalent about the necessity of confinement and were more than willing to treat cases in private asylums or private houses with nurses in attendance, advocating this as an alternative to public asylum treatment for those who could afford to pay. Despite the potential difficulties of managing cases at home, with prospects of cure so good, and under a not particularly demanding regime of treatment which relied on patience and intense watching as much as medical intervention, midwifery and general practitioners insisted on retaining control over patients. They also emphasised the fact that the disorder was likely to recur and encouraged women to book them again in future confinements as a preventive measure. As obstetricians argued that they were the authorities on the condition, they were also offering one more piece of evidence to show precisely why they, and not midwives, should deliver babies, and thus extend their remit to include the management of pregnancy, delivery and the lying-in period. Puerperal mania was a challenging disorder, a threat to domestic contentment, maternal well-being and the happy undertaking of women's natural duties. Yet, because it was seen very much as a temporary condition, its darker aspects were also recognised as passing phenomena; women sufferers needed to be closely guarded while ill, but were likely to recover their faculties fully. Medical men who cured women of this vile disorder were due a good deal of credit. If they did this privately away from the public view and public institutions, they also protected the patient and her family from the stigma associated with mental disorder and incarceration.

Ambiguities of puerperal insanity

Though a body of knowledge on puerperal insanity was being built up by the mid-nineteenth century, in many respects – not least because two distinct groups of specialists were generating the literature – it remained poorly coordinated. Puerperal insanity was and has remained an ambiguous category despite its powerful and obvious link with childbirth, in part because it is still not clear whom the condition belongs to in terms of knowledge and treatment; it 'lies uncomfortably somewhere between obstetrics and psychiatry'.[35] However, by the middle of the nineteenth century few obstetricians or alienists doubted the existence or prevalence of puerperal insanity or its sister disorders of insanity of pregnancy and lactation. Mental disorder, it was agreed, could set in at any time from conception to weaning, but it was insanity following hard on the heels of delivery that attracted most commentary and was considered most common. Most physicians offered an optimistic prognosis, with a cure or significant improvement being expected within a few months. The division of the condition into two categories was also adhered to even if the categories were far from rigid, mania being 'attended with great excitement and furious delirium, the other characterized by the features of low melancholy'.[36] There was broad, though not unanimous, agreement that mania occurred much more frequently and showed itself earlier than more insidious melancholia. Although mania was more flamboyant and disturbing, alarming for both sufferers and observers, the melancholic form was considered far more difficult to treat successfully.

Prevalence

Estimates of the prevalence of puerperal insanity varied significantly. Rates of occurrence for general or obstetric practice are hard to come by, though in the mid-nineteenth century James Reid claimed that out of 8,338 labours, 21 were followed by insanity, that is, one in 397.[37] Asylum records provide the best evidence, but the proportion of female admissions recorded as 'puerperal' varied from institution to institution – according to Bucknill and Tuke from around one-eighth of female admissions to as few as one in twenty.[38] These figures, they added, would always be underestimates, as in many instances the patient recovered without being sent to an asylum. While asylum cases increased markedly, it cannot be assumed at any point in the nineteenth century that they represented anything like the true extent of the disorder. Until the establishment of large numbers of county asylums in the mid-nineteenth century many working-class women must have been kept at home or cared for in workhouses, while the well-to-do, with access to private medical attendance and nursing care, continued to resort to treatment outside the asylum. Continuing pressure on families to place women suffering from puerperal insanity in asylums as well as other, somewhat contradictory, comments that some cases were much better off being cared for at home, strongly suggest that asylum figures tell only a partial story about prevalence.[39]

Only seventeen cases of puerperal insanity, accounting for just 2 per cent of female admissions, were recorded at the Buckinghamshire County Pauper Lunatic Asylum between 1853 and 1873.[40] Yet at Leicester Asylum, in the mid-nineteenth century, with a similar patient pool of predominantly pauper women, one-fifth of women for whom a cause was found in the case books were noted to be suffering from insanity caused by childbirth.[41] During the mid-nineteenth century, 6 per cent of female admissions at Abington Abbey, Northampton were claimed to be cases of puerperal insanity, at the Warwick and Hanwell Asylums 11 per cent.[42] Between 1843 and 1848, 899 females were admitted to Bethlem and 111 of these, or 12 per cent, were suffering from puerperal disorders. In the private Bethnal Green Asylum only three patients out of 386 female admissions were linked to puerperal causes, while in a similar institution, Grove House Asylum, 19 out of 467 cases were described as puerperal in origin (less than 1 per cent compared with 4 per cent). Of 703 female cases admitted to the Hanwell Asylum between 1840 and 1848, 79 were broadly related to the reproductive process: 26 were cases of puerperal insanity, four were listed as insanity of pregnancy, seven were connected to nursing, two followed a miscarriage, six suppression of milk, ten were cases of suppressed menstruation, eight related to the 'critical period' (or menopause), nine to uterine excitement and seven were linked to hysteria.[43] A number of asylums claimed around 10 per cent of their female patients as cases of puerperal insanity, but these figures could fluctuate dramatically from year to year. James Simpson was still citing this figure as an average in 1872.[44] Though cases were recorded and treated in lying-in hospitals, the numbers tended to be small: in 1848 James Reid reported only nine cases out of the 3,500 most recent deliveries at his own institution, the General Lying-in Hospital, and eleven out of 2,000 cases at Queen Charlotte's Lying-in Hospital.[45]

Some of the discrepancies in admissions rates were related to problems in distinguishing puerperal mania from other forms of mania or melancholia, the most common categories of admission. Puerperal insanity has been described as one of the most clearly defined psychiatric conditions of the century and also as a 'fashionable label', plumped for by medical officers who needed to enter something in their registers,[46] but it is likely that, while over-recording took place in some institutions, in others the diagnosis was missed. Further problems of classification are discussed in chapter 4 in connection with the Royal Edinburgh Asylum, but suffice it to say here that it seems likely that large numbers of women were admitted to asylums because they were 'believed' to be suffering from puerperal insanity – the likelihood of its occurrence seemed considerable – and because it was a convenient label, even for patients with a rather distant connection to childbearing. Within individual institutions there could be sharp fluctuations in admissions. While no cases of puerperal insanity were recorded in the admissions books of the West Riding Lunatic Asylum in Wakefield in the early 1860s, and only a handful of admissions under the category 'insanity of pregnancy', in 1866, quite abruptly, nine cases of insanity were linked to childbirth, together with one case of insanity of pregnancy and one of cold after confinement. A further increase occurred in the following year to 19 cases of puerperal insanity, making up over 10 per cent of female admissions.[47] It is doubtful that the women of Wakefield and its vicinity developed a sudden propensity to fall victim to puerperal insanity. Much more likely is that the new Medical Superintendent, Dr James Crichton-Browne, who was appointed in 1866, was alert to the 'existence' of the condition and its potential as a category of admission. The absence of the disorder from admission records prior to the late 1860s suggests that women were recorded as suffering from mania or melancholia, and the fact of their having recently given birth was not deemed significant enough to warrant a separate classification.[48] One of the things we can be sure of is that, while no hard evidence can be produced on the prevalence of puerperal insanity, or whether it was most often treated at home or in the asylum, it was perceived as being on the increase and was attracting much attention as a category of mental disorder.

Onset and causality

The onset of the disorder could be insidious and gradual, as it crept unnoticed into the bosom of the family, or remarkable for its suddenness and violence. Many of the initial symptoms were typical of the kinds of complaints that could be expected after childbirth – tiredness, short temper, fretfulness, headaches, and minor digestive and bowel disturbances – and did not arouse much concern, while their sheer variety, as Gooch had suggested, made the disorder difficult to detect. The 'premonitory symptoms' of the disease, as the women became agitated, restless and peculiar, were presented by John Conolly 'as a quick pulse, always an indicator of some mischief when continuing after delivery, together with want of rest and sleep, a remarkable quickness and irritability of manner'.[49] Fleetwood Churchill described how, though 'the premonitory symptoms vary a good deal', puerperal mania developed from 'a degree of exhaustion, conjoined with great excitability, headache, and want of sleep; or the attack may accompany or follow convulsions'.[50] Pallor or flushed skin, vivid eyes, furred tongue and constipation were marked as features of the condition, together with great excitability, expressed through constant chattering, delusions, singing, swearing, picking at bedclothes and tearing clothing, and lewd sexual displays. In the worst-case scenario the mother became 'forgetful of her child', or expressed murderous intent toward the infant, her husband or herself. James Reid described how

> talking becomes almost incessant, and generally on one particular subject, such as imaginary wrongs done to her by her dearest friends; a total negligence of, and often strong aversion to her child and husband are evinced; explosions of anger occur, with vociferations and violent gesticulations; and although the patient may have been remarkable previously for her correct, modest demeanour, and attention to her religious duties, most awful oaths and imprecations are uttered, and language used which astonishes her friends.[51]

A threatening event, often minor or irritating, could act as the final stimulus, which catapulted the women into insanity: the shock of a death in the family, financial problems, a fire in the neighbourhood or an apparently trivial disturbance, ringing doorbells or unexpected visitors. John Burns described how the disorder could appear quite suddenly, 'the patient awakening, perhaps, terrified from a slumber; or it seems to be excited by some casual alarm'.[52]

Writers on puerperal insanity were unable to agree on the causes of the disorder or at least on the relative importance of different influences, whether moral, organic or hereditary. Medical practitioners worked up their own theories from the starting points of the specialities of obstetrics, psychiatry or general medicine, and this led to different emphases. Midwifery practitioners associated the condition closely with the reproductive process and the strain this put on women, while alienists, dealing for the most part with a poorer group of patients, were more likely to emphasise environmental and circumstantial factors, the impact of poverty, need and general ill health. If links with physiology were made by asylum superintendents, this was likely to be in association with weariness and physical decline, a product of simply having too many children and suckling them too long when the mother herself was malnourished and exhausted. Midwifery practitioners tended to deal with more 'nervous' and delicate well-to-do women, alienists with women worn down by hardship. Both groups, however, referred to influences other than biological, and a great deal of emphasis in practice, as reflected in case histories, was placed on stress and environmental and social factors.[53]

The risk of giving birth, however, was, hardly surprisingly, given the lead role, with women seen to be highly susceptible to mental disturbance at this time. Childbearing was described as a dangerous route to be traversed, starting with conception and ending only when the child was weaned. In Gooch's words,

> During that long process, or rather succession of processes, in which the sexual organs of the human female are employed in forming; lodging; expelling, and lastly feeding the offspring, there is no time at which the mind may not become disordered; but there are two periods at which this is chiefly liable to occur, the one soon after delivery, when the body is sustaining the effects of labour, the other several months afterwards, when the body is sustaining the effects of nursing.[54]

George Man Burrows cautioned in 1828 that

> Gestation itself is a source of excitation in most women, and sometimes provokes mental derangement, and more especially in those

with an hereditary predisposition. The accession of mental disturbance may be coincident with conception, and cease on quickening; or it may come on at any time during pregnancy, continue through it, and terminate with delivery, or persevere through all the circumstances consequent on parturition. Some are insane in every pregnancy or lying-in, others only occasionally.[55]

James Reid also remarked, somewhat tantalisingly, how to 'vast changes in the uterine organs during pregnancy, and more rapidly immediately after parturition, there is superadded an acknowledged state of great nervous excitability ... and above all, the influence of *moral* causes to so great an amount'.[56] Moral insanity, as outlined by Dr James Cowles Prichard in the 1830s, redefined madness, not as loss of reason, but as deviance from socially accepted behaviour, a failure to cope with poverty, the temptations of alcohol, domestic crises, disappointments in love, cruelty, misapplied religiosity or, in the case of puerperal insanity, the stresses and strains of childbirth.[57] Agreeing with the influential French alienists Pinel and Esquirol, that it was moral causes that were more likely to produce insanity than physical, Reid expressed surprise, particularly given the potential for the two factors to be combined, that puerperal insanity was not more frequently observed.[58] Disappointment and domestic crises, sleeplessness, hunger, weariness and fright could all, in Esquirol's opinion, result in puerperal insanity, and he claimed that moral causes caused four times more cases than physical.[59] A number of British authors would smugly refute this, claiming moral causes to be far less significant in the British context. Conolly suggested that puerperal insanity was more common in other countries, especially France, because women there continued with their usual employment around childbirth.[60] Others referred to the higher incidence of illegitimate births in France, it being posited that puerperal mania was more likely to occur among unmarried women, and George Man Burrows argued that the disorder was more prevalent in women who neglected to suckle their infants.[61] Hereditary factors, insanity in the family or previous instances of mental disorder were also said to predispose patients to puerperal insanity.[62]

The idea of 'irritability' was applied in diverse ways to explain puerperal insanity, in the sense of irritation of mind and manner as well as physical function, with great emphasis also being placed on the milk, lochia and menstrual flow.[63] It became something of a catch-all, with practitioners also disagreeing about whether some women were more sensitive and susceptible than others. Gooch argued that the 'peculiar state of the sexual system which occurs after delivery' diffused an 'unusual excitement throughout the nervous system' producing the disease in predisposed women.[64] The female constitution was so weakened by labour, and the nervous system so overwrought, that disorders of the mind easily supervened; Gooch likened this state to 'the hysteric affections of puberty' and 'the nervous susceptibility which occurs during every menstrual period'.[65] Burns concluded that 'All women, in the puerperal state, are more irritable, and more easily affected, both in body and mind, than at other times, and some even become delirious',[66] while John Tricker Conquest spoke of the higher incidence of puerperal insanity in 'females of extreme sensibility, whose mental or physical powers dispose them to be inordinately influenced by causes which would scarcely affect other women, or even themselves, but for the susceptibility to disease, and the peculiarity of condition consequent to delivery'.[67]

In 1827 the physician Marshall Hall described puerperal mania as 'a mixed case', caused by 'all circumstances following parturition combined', but chiefly intestinal irritation, loss of blood and exhaustion.[68] If bowel disorders were present, melancholia was more likely to follow.[69] In describing puerperal cases as 'more complicated than any', Hall argued for both somatic and psychological explanations.[70] In 1843 Alexander Morison claimed more straightforwardly that '[t]he changes that take place in the vascular system, and the increased sensibility of women during pregnancy, childbearing and suckling, render them more liable to insanity',[71] while in 1850 Thomas Lightfoot asserted that alterations in the postpartum fluids could disturb the nervous system and mind.[72] Conolly broadly agreed with Gooch's assertion that puerperal insanity was 'a disease of excitement without power' and claimed 'there is no reasonable or manifest physical cause for the occurrence of the malady beyond the mere circumstance of the previous delivery ... it proceeds from a peculiar state of the brain following delivery'. It was sometimes associated with irritations resulting from a retained placenta or with uterine discharge. Puerperal mania could also follow on, Conolly concluded, from extreme uterine haemorrhage or following an imprudent venesection.[73]

Although Gooch had agreed with William Hunter that there were two forms of puerperal mania, one attended by fever and a rapid pulse, which often proved fatal, the other accompanied by only a moderate disturbance in circulation, which usually ended in recovery,[74] by early in the nineteenth century it was broadly recognised – by Denman, Hall and Conolly, for example – that puerperal mania was unrelated to fever.[75] In

1848 Reid argued that puerperal mania was the result 'not of inflammatory action, nor even of congestion of the brain, but to depend more
upon intense *irritability*'.[76] In practice, however, confusion was likely
given that the delirium of puerperal fever, like puerperal mania, struck
women shortly after giving birth, usually in the first or second week after
delivery, and the high death rate from puerperal fever may have distorted mortality figures for puerperal insanity.[77] In 1828 Dr Robert
Ferguson, Gooch's successor at the General Lying-in Hospital, believed
that he recognised 'the countenance of Puerperal Mania' on Harriet Aster
a few days after her delivery. She was irritable and depressed, claimed her
husband had accused her of 'incontinence towards him', was convinced
of her imminent death and attempted to jump out of the window. As
her mental state fluctuated over the next few days, her physical condition gave increasing cause for concern. She vomited constantly, was
feverish, suffered from tenderness around the uterus and became increasingly emaciated and prostrated. She died on 22 June and a post-mortem
examination showed evidence of peritonitis.[78] Asylum records reveal
that on some occasions the medical officers spotted cases where
patients had been admitted by mistake. A woman admitted to the
Royal Edinburgh Asylum in 1848 was recognised to be suffering from
puerperal cerebritis (inflammation of the brain), rather than the puerperal mania recorded in the admission papers. Interestingly enough,
though beyond the asylum's remit, she was not discharged until she
was cured, following treatment with purgatives, cold water ablutions to
the head, blisters and enemas.[79]

While some deaths from 'puerperal mania' were possibly misdiag-
nosed fever cases, other women were recorded as expiring from
exhaustion. Many women were admitted to asylums with puerperal
insanity, but in addition were in an appalling state of health and suffering from other diseases, particularly tuberculosis and bronchitis,
which would sometimes result in their deaths. They died not of puerperal mania, but while suffering from it.

By the 1850s there was still little in the way of agreement about the
causes of puerperal insanity. In 1851 F.W. Mackenzie, Physician to the
Paddington Free Dispensary for the Diseases of Women, concluded
that 'much diversity of opinion prevails' with regard to puerperal
insanity. Many physicians were interested in pinning down an organic
cause, and puerperal insanity was variously attributed to the suppression of the lochia, metastasis of the milk, the 'peculiar condition of the
sexual system which occurs after delivery', local irritation of the
mammae, uterus and 'other parts', disturbances of the vascular system

occasioned by delivery, the combined effects of irritation and loss of blood, nervous irritability and excitement, etc., all, Mackenzie surmised, conditions occurring fairly frequently after delivery set against the 'comparative rarity' of puerperal insanity.[80]

Mackenzie's own contention was that puerperal insanity was related to anaemia. Anaemia in women, particularly poor women such as those treated by Mackenzie at the Paddington Free Dispensary, could go largely unnoticed, even if it resulted in fatigue and an inability to perform 'their allotted and ordinary duties'. 'Should impregnation take place under these circumstances, the blood becomes still further impoverished, and the constitutional powers heavily taxed.' Various functional disorders were likely to occur, with the brain and nervous system becoming 'unduly excitable, and, in some cases, incapable of withstanding the shock and consequences of labour'.[81] Women suffering from anaemia and further depleted by breastfeeding were particularly vulnerable.

There was broad agreement that the six-week puerperal period marked the time within which puerperal mania could develop, with most cases occurring within two weeks of delivery; melancholia took much longer to make itself evident. Burns hedged his bets claiming that '[t]he period at which this mental disease appears is various, but is seldom if ever sooner than the third day, often not for a fortnight, and in some cases not for several weeks after delivery'.[82] Conolly argued that though onset was variable, '[i]t commonly, I think, comes on as early as a day or two after delivery, or within seven days, or within three weeks'.[83] Some doctors cited cases where the women became insane months or even years after giving birth, but Conolly was sceptical about the possibility of this being puerperal insanity proper.

Physicians remarked on feeling particularly distressed when women became manic following a 'normal' delivery; the disorder seemed less perverse when it was related, as it was by many, to difficult births, excessively painful, protracted, attended in an inept way, or following the delivery of twins or a stillborn child. While some authorities linked the condition to frequent childbearing and exhaustion, others claimed that puerperal insanity was most likely to occur among primiparae. And while many wrote of the risk of the disorder recurring in subsequent confinements, some, like Francis Ramsbotham, thought the 'chances are certainly much against such an occurrence', but in the same sentence 'would strongly impress our minds with the *possibility* of a recurrence ... and to use the utmost degree of care for its prevention, not only in the next, but all the following labours'.[84]

Treatment

With respect to treatment there was more, though far from complete, consensus. Both obstetric practitioners and alienists held to the ideal, sometimes flouted in practice, particularly by hard-pressed asylum superintendents, that treatment should be mild and patients not hurried into a rapid cure. Nature would assist in bringing the women to their senses, in a process facilitated rather than enforced by the treatment regime. The eclectic regimes adopted were broad enough to accommodate alienists and obstetric practitioners, who shared an emphasis on physical therapy and moral management of the patient. This was a natural outcome when both moral and physical causes were imputed as causes of puerperal insanity.[85] Treatment was based for the most part on regular purging, tonics, calming medicines, nutritious diet, careful observation, nursing and rest, combined with moral therapy, representing overall a shift away from heroic remedies and depletion towards a more supportive system of therapy. The place of treatment, however, would be contested; asylum versus home consti-tuted a natural line of division between obstetricians and alienists. This debate will be explored in detail in the final section of this chapter.

Unlike the later rest cure treatment, developed in the 1870s and 1880s for the treatment of neurasthenia, the treatment regime was not broken down into detailed daily plans for feeding and exercise, but elements of the rest cure were certainly present in the treatment of puerperal insanity, even in the asylum context where attendants could devote little time to individual patients.[86] The women were urged to eat, carry out tasks, sew and read, re-establish daily patterns of exercise and rest, and in the public asylum to employ themselves in heavier work in the laundry or helping out on the wards. Thomas Denman had stressed the importance of regularity as early as 1810, particularly in treating cases of melancholia, in the taking of meals, rising and sleeping, attention to dress and reminders to attend to the call of nature.[87] The need for patience was emphasised, with the patient reduced in some ways to the status of an infant, watched care-fully in case she harmed herself, and fed, sometimes with a spoon, to make sure she ate enough. As evidenced in many case histories, patients suffering from puerperal insanity were frequently incapable of caring for themselves, needed help dressing and were dishevelled and even incontinent, able neither to control nor care about passing their motions involuntarily.

In his 1820 *Observations*, Gooch made five recommendations con-cerning the management of cases of puerperal insanity: to protect the patient from injuring herself; to evacuate impurities using purgatives; to monitor circulation, and, if congestion or inflammation of the brain should supervene, to remove it by antiphlogistic remedies to reduce inflammation; to procure sleep at night; and to 'manage the mind of the patient, soothing it during irritation, encouraging it during depression, never to attempt the removal of her delusions by argument'.[88] He expanded on these in his later publications, stressing the importance of proper attendance and the avoidance of excitement of any kind. Diet should never be 'very low' and patients should be encouraged to take regular meals. Meat, gruel, milk and wine were recommended, while those suffering from melancholia were encouraged to eat a more nutritious, cordial diet, 'meat every day, with about four ounces of wine'.[89] As to medicines, purges and to a lesser extent vomits were regarded as having inestimable value: Gooch would not be alone in attributing a wonder cure to the production of a good stool. Gooch also recommended the use of narcotics, the 'most valuable' medicines in the treatment of puerperal insanity:

> [i]f given at proper times and in proper doses, they often procure nights of better sleep, and days of greater tranquillity. This calmness is most likely to be followed by some clearing up of the disorder of the mind.[90]

Gooch discouraged bloodletting as unnecessary and, in some cases, pernicious. Stimulants were occasionally valuable if the woman was very pale, exhausted and had a weak pulse.

Gooch's recommendations concerning treatment were followed, if not always to the letter then certainly in spirit, for much of the nineteenth century. A whole range of authorities on insanity agreed that there was no point in trying to reason the insane out of their delusions, and the importance of soothing and supporting the patient, without exciting her, was stressed by many. Gooch's use of narcotics, however, would be disputed. The discussion of a case marked by severe paroxysms and reported to the London Medical Society by Samuel Ashwell in 1829 criticised the heavy use of opiates and venesection. Isolation of the patient, immediate cessation of suckling and dosing with the stimulant camphor were recommended instead.[91] Hefty drug regimens and particularly bleeding were generally disapproved of, and by the mid-nineteenth century there was almost unanimous condemnation of bloodletting as a treatment for mania at the same time that its use declined in general medicine.[92] Purging, however, used for more or less every physical as well as mental disorder in this period, was standard. Francis Ramsbotham declared:

> Purging is of essential service with the view of fully clearing the bowels ... Many cases have occurred where the disease has almost instantaneously given way to the thorough evacuation of the canal. In such, the matters expelled have been in excessive quantity, most unhealthy and offensive, and were evidently the exciting causes of the derangement.[93]

Gooch's advice on the use of narcotics continued to elicit a mixed response; opiates were widely used in asylums and at home if the patient was unable to sleep or rest, as they were for numerous physical disorders, as painkillers, pick-me-ups and nerve-calmers, to treat female complaints and to dull the pain of childbirth.[94] However, though many alienists continued to recommend them as calmatives and to induce sleep, others claimed that opiates could make the patient more excited, and they had the considerable disadvantage of causing constipation.

Mild but eclectic would sum up the approaches of many physicians to treatment. Thomas Graham, in his volume on the diseases of women, recommended that the treatment of mental affection following childbirth should mainly consist of 'soothing the patient by a union of firmness and kindness' and 'acting gently on the bowels'. He also recommended mustard poultices to the legs and thighs and warm baths.[95] Alexander Morison advocated a more varied, and more drastic, range of treatment:

> Shaving the head is often useful, by lessening the heat; the application of cold in various forms; the application of blisters, and the insertion of issues,[96] are all indicated; more or less in different cases, as well as other evacuations tending to diminish determination of blood to the head.

Purgatives were of 'extensive utility', while bloodletting, Morison suggested, could be used with 'great care' if there were symptoms of inflammation or congestion.[97] Bemoaning the delay in admitting such cases that many asylum superintendents complained of, Robert Boyd of the Somerset Asylum stated that most patients suffering from puerperal insanity between 1848 and 1868, the majority of whom fell ill while breastfeeding, were already 'chronic' and exhausted. He argued that opium aggravated their restlessness and advocated opening the bowels, then treating with sedatives, hyoscyamine or tincture of cannabis, combined with a stimulant, ammonia or camphor, in large doses. Boyd used chloroform vapour, sulphuric ether or morphia injections to induce sleep. He also recommended the sedative chloral hydrate, which, unlike opium, did not disturb digestion or 'influence the secretions'. Tonics, cod liver oil, iron, bark and quinine were often found necessary. Packing the patient in a wet sheet was promoted as having a very tranquillising effect. Like many asylum doctors, by the late 1860s, Boyd was still advocating a supportive regime, but also one that drew on a wider range of sedatives.[98]

Bathing and various forms of shower baths were utilised, especially in asylum practice, to soothe or to subdue. Careful monitoring of the breasts, lochia and menstrual flow was also advocated, and the re-establishment of menstruation was seen as a sign of improvement and the restoration of physical health. Keeping patients comfortable and warm was also seen as important. Physical restraint was used rarely and would, in many institutions, be replaced by the use of sedatives. By and large it was recommended only for very disturbed patients, using straitjackets or, as suggested by Morison, long stockings with a bandage, fixing the legs together or 'when disposed to tear clothes or to strike others, leather mitts on the hands, attached to a leather belt around the waist', a device shown in several of his illustrations (see Figure 2.1 below).[99]

Despite an overall emphasis on mild and supportive therapies, John B. Tuke, Assistant Physician to the Royal Edinburgh Asylum, felt prompted to write critically as late as 1867 about the 'routine treatment' of shaving, applying cold to the head, purging, vomiting and blistering, and the use of opiates, cannabis and stimulants, which 'feed the excitement without increasing the bodily strength'.[100] What Tuke was protesting about above all was the application of standard and outmoded or carelessly adopted new remedies to what he suggested was a unique and eminently curable disorder. Sedatives, however, would be resorted to increasingly in asylum practice from the mid-nineteenth century onwards in Edinburgh and numerous other asylums,[101] and, although heroic approaches to the treatment of puerperal insanity were roundly condemned in much of the literature, reports of cases attest to their continuing use. Yet it was widely held that, even in difficult cases, patience, good care, rest and food would the best remedies for puerperal insanity. Tuke believed food was of the greatest importance, not just for its nutritional value but also as a calmative: 'the patient has a better chance of sleep after a dose of beef tea than after a dose of morphia'.[102]

A wholesome diet was suggested by many as the key to the recovery of exhausted women, many of whom were poorly nourished, even semi-starving, either because their families did not have the means to feed them or because they refused to eat and in the latter case force-feeding could be resorted to. Particularly when dealing with poor patients, asylum treatment offered the opportunity literally to fatten patients up, also encapsulating the idea that they were shoring the women up as a preventive to further mental disturbance.

Claims of expertise

Midwifery practitioners and alienists brought into service the tools of their trade to substantiate claims of expertise. A number of obstetric practitioners argued, for example, that the use of forceps to speed delivery could prevent women from becoming insane, describing a transient form of puerperal mania which occurred at the actual moment of birth caused by exhaustion or intense pain. However, here too there was disagreement. William Tyler Smith argued that the transient loss of reason, which occurred most often 'at the moment when the head passes through the os uteri, or the os externum' could be provoked by the use of forceps. It was also more likely to occur if the child had a large head and in first labours. 'This form of mania is so transient,' Smith argued, 'that it generally only requires the necessary amount of restraint, and the removal of means of mischief, for a very short period.' While in general Smith opposed the use of anaesthesia in childbirth, not least because of the promiscuity he claimed it unleashed in women, he recommended that chloroform be used 'before the suffering has reached such a height as to make the patient frantic'.[103] Conversely, Smith's contemporary and lecturer in obstetrics at the London Hospital, Francis Ramsbotham, refused to acknowledge the existence of this transient form of puerperal mania; rather it was merely 'the delirium, – the phrenzy, – of high excitement, produced by intense pain'.[104]

The mid-nineteenth century was an era of intense debate concerning the efficacy and ethics of administering chloroform in childbirth, with some midwifery practitioners citing it as a cause of mania. Aside from the accusation that it led to loss of control and sexual depravity among women, it was argued that it placed them in a detached state in which they were unable to experience pain and thus retain their link to the real world of feeling and experiencing childbirth. Dr Webster presented several cases of insanity supervening after the use of chloroform in the

Journal of Psychological Medicine in 1850; all the women became maniacal and remained insane for many months, though only small doses were administered.[105] Following the boost to chloroform's popularity after it was administered to Queen Victoria in 1853 at the birth of her eighth child, it was increasingly advocated as a form of pain relief in childbirth, offering escape not only from agony, but also, many argued, from insanity produced by intense pain and shock.[106] Charles Kidd, Physician to the Metropolitan Dispensary in London, praised chloroform's good results, citing a case reported by James Simpson, who had pioneered its use in obstetrics in 1847, of a woman who had been attacked by puerperal mania in each of her five confinements. On the occasion of her sixth confinement, chloroform was administered for the first time, and she showed no signs of insanity.

> The poor lady herself said that the extreme intensity of her suffering in the first labours drove her out of her mind, and she attributed, perhaps quite correctly, *her escape on this occasion from the horrors of a lunatic asylum* to the use of chloroform during her labour.[107]

In one case of severe puerperal mania occurring in 1851 treatment first consisted of dashing buckets of cold water over the patient's head so that she could be brought to her bed. Thereafter she was kept under 'the benign and soporific influence' of chloroform for over four hours, after which her reason was 'completely restored'.[108] While midwifery practitioners retained an edge administering chloroform as a preventative during delivery, its more general use to treat mental disorders after 1850 detracted from their monopoly, although they often had the advantage of being first at the scene in cases of puerperal insanity if they had delivered the woman and many would already be trusted family attendants.

Alienists had, however, one very different claim to expertise and special knowledge: to be able to identify mental disorder in the face of the patient. From the early nineteenth century onwards illustrations and later photographs were used to demonstrate and communicate about mental illness, and also to show the progression of patients towards a cure. Physiognomy (the science of reading the face) commanded great authority during much of the mid-nineteenth century, extending beyond internal discussions within psychiatry to reach informed lay audiences. Primarily, however, the use of visual images enabled doctors to communicate with each other through plates published in journals and textbooks, giving practitioners access to 'virtual' patients, whose illnesses could be diagnosed by examining their faces.[109]

Interest in puerperal mania dovetailed closely with the development of methods of illustration, and signs of the disorder were claimed to be particularly legible on both the body and the face of the patient, the body through its emaciation and debility, the face through its alarming countenance and exaggerated expressions. It was not without reason that puerperal mania was the second form of mental disorder to be delineated in plates in Alexander Morison's *The Physiognomy of Mental Diseases*, published in 1838.[110] Morison's lithographic techniques 'permitted a more impressionistic sense of the mobility of the features' than earlier engravings with their sharp, clearly delineated features.[111] Morison, who instituted the first formal lectures on psychiatry in London and Edinburgh in 1823, and published widely on mental disease, included a number of studies of puerperal insanity in his publications, alongside advice on how to manage such patients. He commented that '[i]n Puerperal Mania the physiognomy is more variable, the changes are more frequent and more sudden, and the appearance of exhaustion is greater; the cure is also more frequently effected than in other varieties of Mania', and this gave him the opportunity to outline and illustrate the route to recovery.[112]

Morison's claim for the special status of puerperal insanity as a vivid example of how mental disease was stamped on the face of its victims would be shared by others interested in presenting cases through illustrations, as too were the clearly defined signs of the route to a cure.[113] Limits on the effectiveness of images, however, are revealed in the descriptions attached to Morison's illustrations, as shown in the case of E.I.; the text barely described expression and appearance, devoting itself to the story of her treatment and recovery. In general the use of images should be put in perspective. Most physicians did not have the means or impulse to provide pictures of their patients and, as we will see in chapters 4 and 5, verbal descriptions were chiefly relied on to provide evocative impressions of the physical and mental state of patients.

Locating puerperal insanity

Despite the powerful use of illustrations to demonstrate the route to a cure, the mildness and relative ease of administration of many of the therapies used to treat puerperal insanity served to undermine arguments outlining the necessity for removing patients to an asylum. Beef

Figures 2.1–3   Morison's Three Stages of Puerperal Insanity.
Figure 2.1   **E.I. aged 33** 'This female, who had no hereditary disposition to insanity, was seized with Puerperal Mania three days after the birth of her first child; she is here represented eight weeks after the commencement of the disorder – her face pale, and her eyes and mouth shut; at times she is very silent, at other times she is very noisy, and screams; she attempted to jump out at a window, is disposed to tear her clothes, and frequently drops on her knees; her conversation is incoherent, sometimes she says that she is strange, that she is mad, that she shall destroy her child, or cut her own throat; restraint is found necessary.'
(Source: Wellcome Trust Library)

Figure 2.2   **E.I. taken seven months after her disorder commenced**
'[T]aken seven months after her disorder commenced. Gentle laxatives, nourishing diet, fresh air and exercise, effected some improvement; she, however, required occasional restraint, on account of a disposition to tear her clothes during the whole interval. Premature communication with her friends was prejudicial, and was succeeded by greater violence, her conversation became incoherent, and she spat at those around her.'

Figure 2.3   **E.I. restored to reason**
'[R]estored to reason. In this case a blister applied to the nape of her neck, and a discharge kept up by the application of Savine Ointment, appeared to expedite the recovery, which was completed by the use of Sulphate of Quinine, in about nine months from the commencement of the disorder.'[114]

tea could after all be administered just as easily at home as in an asylum. The events surrounding a woman's decline into insanity, meanwhile, were linked closely to her recent delivery and in the first instance were physically located at home, in the lying-in room. However, once diagnosed, decisions needed to be made about what to do with the patient, whether she should be moved and how she should be treated. Although considerable numbers of patients suffering from puerperal insanity were placed in asylums, many others remained under domestic management. This could be by default, when the woman and her attendants – medical or otherwise – simply struggled on until she recovered. Or, chiefly in the case of better-off families who could afford to pay for the help of nurses and private medical attendance, a considered decision may have been reached to keep the patient at home, or to remove her to a private residence or a private asylum.

Cases recorded by midwifery practitioners mainly discuss treatment in a domestic setting, while those written up by alienists describe women who were brought at some stage to an asylum, sometimes after being ill for some time. This was by no means a rigid division, and occasionally obstetric physicians were invited to visit asylum patients, while some alienists, like Morison, treated women at home. Those treated privately by obstetric practitioners tended to be well-to-do, those in the asylum working-class or pauper patients, but the dividing line was by no means hard and fast. A great many wealthy women were cared for in private institutions, while the number of middle-class patients in asylums increased during the latter half of the nineteenth century.[115] It is not clear whether being assisted by a doctor rather than a midwife at the delivery would slow down or speed up the process of referral to an asylum; that would depend on the family circumstances, the socio-economic position of the patient and the confidence and expertise of the practitioner involved, but it is likely that poorer women attended by midwives would have fewer buffers delaying their removal to an institution of some kind, be it a workhouse or pauper asylum.

Many women giving birth in the workhouse remained there even when they became disturbed, unless they became particularly disorderly and unmanageable. Of the 357 women admitted to Devon asylums with puerperal insanity between the 1860s and 1920s, thirty were brought from local workhouses. Many of these were noted to be violent and many were single women lacking in family support.[116] Elizabeth A., accompanied by several children, arrived on several occasions at the Plympton St Mary Workhouse in Devon in desperate circumstances between 1868 and 1872. Moving in and out of the workhouse, ill and destitute, she was finally removed to the Exminster Asylum in 1872, when, following the birth of a baby, she became very disruptive, calling out the names of imaginary people and trying to burn herself in the fire. She remained there until her death in 1894.[117]

Although reports of cases of puerperal insanity occurring as part and parcel of general practice are rare, particularly given the expansion of general practitioners' involvement in midwifery work during the nineteenth century, this is probably due to the fact that few full records of such practices survive. General practitioners, however, appear to have been aware of the condition, wrote up cases and their responses to the disorder, and had these accounts published in medical journals. Alexander Morison commented that, as it was not advisable to move patients recently delivered and in a state of exhaustion from their homes, puerperal mania 'frequently fell under the observation of general practitioners of medicine'.[118] General practitioners appear to have been neither surprised nor baffled by the disorder, and took it on board as they would other complications of childbirth.

In 1846 a case of puerperal mania occurring in Pwllheli, North Wales was reported to the *Medical Times* by Howell Evans, the general practitioner who attended. The young woman, single and a victim of seduction, was raving and showed great antipathy towards her child. She was given opiates, 'freely' purged, and her head shaved and wetted with spirit lotion. Evans related that she improved rapidly, became affectionate towards the child whom she began to breastfeed and was able to sit up at the end of a fortnight.[119] It was general practitioners or surgeons who often became key witnesses in infanticide trials, and again they seem to have been knowledgeable about the disorder, although they presumably encountered it rarely.[120]

A handful of surgeons and general practitioners kept extensive records of their practices, including John Thomson, a surgeon working in Kilmarnock in Scotland, who summed up his midwifery work extending over a fifteen-year period in the *Glasgow Medical Journal* in 1855.[121] He recorded 3,300 cases of 'obstetricy', and a table breaking down his 'preternatural and instrumental cases' included just two cases of puerperal mania. One of the women became disturbed several days after experiencing severe bleeding and delivering a stillborn child. Thomson without hesitation labelled her condition 'puerperal mania' and treated it with purgatives, opiates, medicines to induce vomiting and cold douches. He soon decided that such remedies were not proving efficacious and abandoned them, recommending in their place a 'nourishing diet and kind and gentle treatment'. The woman recovered, though she remained pale and emaciated, and occasionally her mind wandered.[122] Thomson presumably had little expert help to call on, but he did not seem to require it. He managed the case calmly, trying different regimes until he hit on one that produced an improvement. He seems to have regarded puerperal mania as one of the 'preternatural' occurrences he would exceptionally encounter in the course of his midwifery work.

In 1847 Thomas Salter recounted in the *Provincial Medical and Surgical Journal* the dreadful story of one woman's experiences of puerperal mania.[123] Her first attack of mental disturbance came on in the seventh month of pregnancy, but after a fortnight the woman was delivered of a healthy child and recovered. In the second pregnancy the woman became ill during the sixth month, was extremely restless and sleepless, and showed such 'a degree of violence of manner, as often to require four or five persons in constant attendance upon her night and day'.[124] The woman continued in this state for three weeks, before the medical attendant decided to bring on labour by breaking her waters. Five days later the woman gave birth to a stillborn child, and her violent insanity, which, it was claimed, had threatened her life, ceased two days later. The woman subsequently moved to the North of England, and at this point Thomas Salter took over as her medical attendant. She was then pregnant for the third time, and this time her unsoundness of mind showed itself after the fourth month. She was given sedatives, but was unable to sleep and again required several people to keep her in bed; 'incessantly talking, either religiously or quite the reverse; she had occasional fits of screaming, so loud as to be heard by the neighbours living at a considerable distance from her own residence. She continued in this state, and without any sleep, day after day.'[125] Salter decided to follow the example of his predecessor, managing with some difficulty to rupture her membranes and to administer ergot of rye in order to procure an abortion, and finally nine days after the commencement of mental derangement and fifty hours after puncturing the membranes the child was aborted. Her recovery this time was more gradual, but Salter reported that a few years later she was in good health; she had one further pregnancy, which ended in a miscarriage without 'any unusual circumstances'.[126]

Alienists may have argued that an unfortunate case like this would have been better managed in the asylum, and some women who became insane while pregnant did remain under asylum treatment until after their delivery and restoration to sanity. There is no comment in this case on the responses of the family to this extreme method of saving the mother and restoring peace to the household. Indeed, the clinical way in which it is reported makes this account deeply disturbing, but it also serves to show that doctors in general practice were not daunted in even the most challenging of circumstances.[127]

Bucknill and Tuke remarked that the small number of women treated for puerperal insanity in lying-in hospitals was accounted for by 'the very favorable [*sic*] circumstances (such as quiet, good nursing, and sufficient nourishment), which surround the hospital patient, as compared with those of a patient of the same destitute class at her own home'.[128] It is also likely, as patients were generally discharged from maternity hospitals about two weeks after delivery, that many cases of puerperal insanity were missed. Most lying-in hospitals, however, recorded a few cases. Patients in the General Lying-in Hospital, London, were noted as suffering from bouts of 'nervous excitement' or melancholy. One such was Frances Butler who, after 'talking wildly', became silent, indifferent to her infant and fearful of becoming permanently melancholy; another was Elizabeth Nicholls, who, after giving birth to a premature baby who died after two days, was alternately excited, regretful at causing so much trouble, then abusive. While Frances Butler recovered and was discharged within a few days, Elizabeth Nicholls was still 'improving slowly' in the hospital three weeks after her delivery.[129]

Women displaying signs of mental disorder in lying-in institutions were often dispatched quickly to local asylums or the workhouse. Elizabeth Redfern, a patient at the General Lying-in, was returned to the workhouse from which she was admitted when she became excited and delusional after the delivery of her eighth child.[130] The Glasgow Maternity Hospital recorded only very occasional cases of puerperal mania in their patient registers, including a case in 1855, a first pregnancy with a normal, though lengthy delivery: 'a case of puerperal mania – the patient died from exhaustion on the 15th day after delivery'. Another woman '[t]ook mania after leaving Hospital' in 1866, and in 1870 another patient was removed to the Poor House after she became disturbed.[131]

Thomas More Madden, Physician to the Rotunda Lying-in Hospital, Dublin, reported, however, that he had extensive opportunities to study puerperal mania. Indeed, Madden stressed that doctors were far more likely to meet the condition in hospital obstetric practice than privately, where 'patients were generally in better circumstances and social condition, having less mental anxiety and physical privation'.[132] He added that an important aspect of the physician's brief in such institutions was to guard against the dangers of the woman harming her infant.

At the Rotunda, if early recovery was not likely, patients were dispatched to an asylum, such as a woman delivered in No. 7 ward in June 1868 who developed a deep dislike of her child and who was loquacious and excited.[133] Another case, Mrs B., confined in 1869, was violent, delusional, harmful to her infant and suicidal, and was saved from jumping from the window to the paved area below only when the ward maid managed to grab her hair. She was put into a strait waistcoat and packed off the same day to Richmond Asylum.[134] Other cases, less violent and troublesome, were retained and treated at the Rotunda until they were discharged, together with their infants. Out of 26 cases of puerperal mania following on from delivery at the Rotunda between 1847 and 1854 (out of 13,748 deliveries, fewer than Madden implied), 18 recovered, five were removed and three died.[135]

The task of the medical attendants became pre-emptive in the sense that they were to guard against and if possible prevent a recurrence of the disorder in future confinements. The physician who had attended the woman in childbirth also had a vested interest in treating the condition; it was he who witnessed the slippage of his patient into insanity and it was his task to cure her and ensure that insanity would not recur in future confinements, at the same time taking care to preserve the custom of the family. Prevention in effect became part of the treatment regime, sometimes involving medical interventions such as purging around the time of delivery or simply avoiding excitement. F.W. Mackenzie, who as a dispensary physician had a great deal of contact with poor women as well as his private patients, warned of 'irritative disorders' of the body or particular organs, and 'all painful states of mind should, if possible, be prevented: distress, anxiety, grief' as well as 'fright, agitation, shock, or alarm'.[136] Doctors would emphasise their tact and skill in treating patients and counselling their families, being attentive at the same time to their reputations, fees and clientele.

Midwifery practitioners emphasised the value of treating women in a domestic setting. Domestic management was advocated because of the special therapeutic regime required of patient treatment and close watching, but also because of the 'specificity' of the condition – it not being like other forms of madness – and the generally good prognosis.

Treatment should not, it was argued, be linked to incarceration, but to separation from the specific domestic circumstances that so distressed the patient. Obstetric physicians were more or less unanimous in expressing their opposition to asylum treatment, encouraging instead private seclusion and proper nursing. William Tyler Smith opposed placement in institutions on the grounds that women suffering from puerperal mania were not like other insane patients and should be protected from the stigma of the asylum. Such cases should always be treated as if they would recover fully, Smith emphasised, as it was temporary conditions that produced the insanity. Smith believed 'that great mischief is frequently done by placing puerperal patients in public and private lunatic asylums'.[137] He cited the example of Bethlem, where some of the women incarcerated had killed their children while suffering from puerperal mania, but even then the asylum was rejected:

> Some of these poor creatures have passed the childbearing age, are perfectly clear in intellect, conscious of what they have done, and suffer intense misery therefrom, while they have their desolation enhanced by mixing constantly with confirmed lunatics.[138]

James Reid maintained that 'general opinion' was against the removal of patients labouring under puerperal insanity to the asylum, unless the case lapsed into a 'chronic and lingering form'. Rather,

> If change of scene be deemed requisite, it is better that the patient should be removed at first to a quiet country village, or to the seaside, under the care of an experienced nurse, but the frequent visits of the medical attendant will here be advisable ... The friends of the patient should also pay occasional visits, to examine into the domestic arrangements and comforts of the place, without, however, seeing the patient herself ...[139]

In his *Principles and Practice of Obstetric Medicine and Surgery*, Francis Ramsbotham considered the propriety of removing the patient from home. He recommended removal to another residence, particularly from 'confined to a purer air, and from a noisy to a quiet situation', yet not to placement amongst 'permanent maniacs, because of the great hope we entertain of a restoration, the uncertainty how long the affection may continue, and the chance of her recovery being sudden'. The best option was to have a whole or part of a house devoted to the patient; failing that, lodgings in a smaller, presumably private, establishment were recommended.[140]

Many patients could not afford the luxury of secluded care with suitable nurses, and some physicians recommended that if it proved impossible to remove the patient from her home, then her husband and family should move out, leaving her to rest in quiet. Gooch recommended the removal of husbands and relations, particularly if the disorder threatened to be 'lasting'.[141] William Tyler Smith argued that philanthropy should offer a remedy for treating puerperal insanity in a humane, home-based setting when family means were limited,[142] while W.S. Playfair explained as late as 1878 that only the worst cases found their way into asylums, 'a step which everyone would wish to avoid if possible'.[143]

The infant in all of these discussions gets short shrift; it seems to be simply assumed that someone will take care of the newborn in the mother's absence. In poorer families infants were boarded out under Poor Law provisions to be cared for by other women, but many infants were kept at home with family members, or, in richer households, a nurse was employed. The absence of debates on the impact of separation of the infant from the mother is striking, given the emphasis on the importance of mothering in general and breastfeeding in particular.[144] Yet it was frequently emphasised in the medical literature and asylum case notes that the ultimate benefit of separation was the restoration of the healthy mother to her family.

Alienists were more guarded in making recommendations about where treatment should take place, weighing up good rates of recovery and cure from puerperal insanity against the need to incarcerate such women in the first place. Overall many were of the opinion that the asylum was often no place for the puerperal maniac. Early in his career, John Conolly was remarkably forthright in his criticism of the asylum with its associated privations as the appropriate place of cure, though, in line with his broader shift to the advocacy of asylums after his appointment at Hanwell in 1839, he later argued that specialists in mental alienation were best equipped to treat puerperal insanity.[145] In the 1830s, however, he spoke powerfully against removal to an asylum, especially in cases associated with debility: 'what can be more barbarous than to subject the morbidly susceptible system to new and painful impressions?'[146] He argued against the axiom that 'mad people never get well at home', particularly in the case of puerperal insanity, even in unresponsive cases:

> To separate her from her infant and her family, and to place her amongst strangers, is to debar her from almost every hope of being

soothed and calmed ... If the disorder continues long, if home and its associations are evidently sources of irritation, <u>then</u> a change of place and of faces may be advisable; but not to a Lunatic asylum.[147]

Conolly may have irritated some of his colleagues with such powerful assertions, and he also flew in the face of conventional ideas on treating puerperal insanity by stressing the potentially positive role of the family in the cure of the patient.[148] But the later shift of such a key figure to support of the asylum, under humane conditions, as the best place to treat puerperal insanity must subsequently have bolstered the claims of alienists.

Thomas Clouston, reporting on the activities of the Cumberland and Westmoreland Lunatic Asylum, which he superintended in the 1860s, remarked that cases of insanity occurring after childbirth helped 'keep up the standard of curability', at the same time suggesting that when symptoms were mild and manageable, there was really no need for such women to be in the asylum at all. Admission of cases of puerperal insanity under Clouston were substantially higher than at most other contemporary asylums, peaking at almost 22 per cent of female admissions in 1869; he was at a loss to explain this phenomenon, but surmised that he may have taken more care to ascertain particular forms of insanity.[149] Tuke, writing on cases treated in the Royal Edinburgh Asylum in the mid-nineteenth century, referred to puerperal mania as '*the* most curable form of insanity', to be placed under asylum treatment 'as a *dernier ressort*', but at the same time urged early removal in serious cases.[150] While alienists professed that milder cases need not be admitted to asylums, at the same time emphasis was placed on treating puerperal insanity quickly, catching cases before they became serious and entrenched.

Some features of the asylum were duplicated in domestic settings, and even strengthened. A form of moral management evolved which stressed the authority of the physician, the importance of attendants and the avoidance of excitement. The need for close watching was stressed, on a one-to-one, or even two-to-one, basis, with nurses working shifts. For although those suffering from puerperal mania were regarded as 'helpless', they were also believed to be endlessly cunning, testified to by their ceaseless efforts to trick their attendants into leaving them, giving them the opportunity to do away with themselves or their infants.

It is scarcely necessary that I should insist on the patient's never being left a moment alone in any of the forms of mental aberra-

tion, nor in the removal from her reach of whatever can be can-
verted [*sic*] into instruments of self-injury; such as knives, cords,
garters, or any articles of dress by which strangulation could
possibly be effected. The door should be kept locked, and the
windows tightly nailed down, or so secured as to be only capable of
being opened to the extent of admitting the requisite quantity of
fresh air. If the case is likely to be of long standing, her nurse and
other domestic servants should be removed; two females accustomed to the charge of insane patients must be substituted; and
they should take their rest alternately.[151]

Seclusion was to be matched by restraint and 'general moral management'. A particular type of nurse should be sought out, which
reaffirmed the patient's isolation from loved ones:

the *attendants* should be *peculiarly adapted* for their duty; the usual
domestics, and even the experienced monthly nurse, are not so
valuable as one who is accustomed to the care of this class of
patients, and it is advisable, therefore, to replace them. A vigilant,
firm, though kind superintendence, soothing violence, encouraging
and cheering despondency, soon produces its effect; and a nurse
who is skilful and experienced in these cases, will have much more
moral control ...[152]

In 1848 Arthur M'Clintock and Samuel Hardy, who had both worked
at the Lying-in Hospital in Dublin, passed on their techniques of
patient control, offering the following advice if the woman proved
refractory, when 'the contest will be whether she or the doctor is to
have the ascendancy':

Under these circumstances it is requisite for the physician to
exercise much tact and resolution in his language and conduct
towards the patient. If he does not succeed in enforcing his
directions by mild expostulation, he must shew himself to be
determined, and, without harshness, insist on his orders being
obeyed; for if, through vacillation, or want of resolution, he now
fails to establish his authority, he will lose all control or restraint
over the patient; whereas, on the other hand, if he carry his
point, and bring her into compliance, it will have a lasting effect,
and she will probably stand in awe of him during the remainder
of her illness.[153]

The role of family members was typically minimised to inspecting the arrangements concerning treatment. Husbands, often the butt of hatred on the part of the women sufferers, were decisively excluded from the therapeutic regime. Even in wealthy households, wives were removed from their husbands' presence and authority, taken from their homes and separated from 'dearest' connections, with visits severely restricted. The authority of the doctor in attendance, who replaced the husband, and insisted on removing friends, relatives, familiar nurses and servants, reigned supreme, and gained the time and space needed to effect a cure: 'It is necessary to gain, by firmness, a mastery over such patients. They recover more quickly when not allowed to see their husbands, infants or immediate relatives, and they are most easily managed by nurses whom they have not previously known.'[154]

* * *

The rewards for treating puerperal insanity were considerable. As alienists and obstetricians laid claim to professional expertise in treating the disorder, they could point to good, often excellent, rates of cure, a reflection of the importance of establishing who treated whom and where. Haslam had reported rates of cure of 50–85 per cent at Bethlem in the late eighteenth century, with most patients recovering by the fourth month.[155] Of the 111 cases of puerperal insanity admitted to Bethlem between 1842 and 1847, 69 (62 per cent) were discharged cured.[156] Gooch in private practice and Johnston and Sinclair at the Rotunda Hospital claimed cure rates of almost 70 per cent.[157] Tuke reported 58 out of 73 patients (79 per cent) as cured within six months at the Royal Edinburgh Asylum. Expressing a widely held view, Tuke emphasised that the chances of cure diminished rapidly as time passed: 'if the mania is prolonged more than a month, or at the outside six weeks after confinement, the probabilities of ultimate cure are very faint.'[158] Insanity of lactation was in general considered more difficult to treat, as were cases of melancholia, which set in later than mania, often several months after delivery.

More than many other mental disorders, puerperal insanity was seen as fundamentally disrupting the household, a theme to be explored in chapter 3. The doctor able to cure this disorder was also healing the family, putting the mother back with her child, the wife back with her husband, and the woman of the house back at the centre of domestic activities. The authority of competing groups of specialists to treat women suffering from puerperal insanity was contested until the close of the century, and revealed too, at least for this condition, the significance of treatment outside the asylum. Obstetric doctors argued that to give women the best chance of a cure, as well as preserving reputations – their own as well as their patients' – they should be kept under their close supervision in a domestic setting. In this way the nature of the cure taking place would be masked, the patient protected from the stigma of the asylum. Ramsbotham was far from alone in emphasising the need for great diplomacy in alerting the family to the nature of the disorder, which was such a dreadful calamity following delivery, and husbands and families were often reluctant to hand over their wives, sisters and daughters, who had recently become mothers, to an asylum.[159] It was a particularly sad and shocking conclusion to what should have been a happy event. Such patients were kept out of sight and under the authority of the obstetric doctor. For alienists, their uncertainty about whether puerperal insanity should be treated in the asylum was replaced increasingly by certainty that it should be as the century progressed, and they – as so well demonstrated by Conolly's change of heart – argued ever more resolutely for their authority to treat such cases. Women suffering from puerperal insanity tended, as we will see in chapters 4 and 5, to be admitted to the asylum once the condition had fully manifested itself and perhaps reached an alarming state. The more expectant, watchful approach of obstetricians, who perceived the condition not only as part of their remit, but also as a potential aspect of the cycle of giving birth and recovery, was lost. The subsuming of such cases into asylum regimes along with other forms of mental illness may help explain why, towards the end of the century, puerperal insanity came to be treated less and less as something different, special, requiring specific treatment in a specific environment; this will be further explored in the final chapter. Here the overall mildness of approach and the often sensitive and supportive therapies adopted have been emphasised, but it is worth emphasising too that the proponents of this approach, alienists and obstetric physicians alike, were responsible for firming up the idea that women were highly vulnerable and their bodies extremely capable of producing mental disorder. Even Gooch who, as we will see in chapter 3, embodied the sympathetic approach to treatment, was culpable in stressing that maternity would be a route of physical and mental breakdown for many women.

3
Disordered Households: Puerperal Insanity and the Bourgeois Home

> Nervous irritation is very common after delivery, more especially among fashionable ladies, and this may exist in any degree between mere peevishness and downright madness. Some women, though naturally amiable and good tempered, are so irritable after delivery that their husbands cannot enter their bed-rooms without getting a certain lecture; others are thoroughly mad.[1]

Puerperal insanity was unsurpassed in its ability to challenge the hegemony of domestic ideology. The household was disrupted, turned upside down, social mores and notions of domestic order faced an open and forceful challenge, and the place of mothers within the home was abandoned as they took on an unaccountable manner so unlike their usual selves. Rude, bossy and domineering, irritable and bad-tempered, annoyed with their family and servants, whom they slapped and insulted, violent, loud, dirty and careless of their appearance, women became alarming spectres in their own homes, distorted mirror images of their 'true selves'. Self-sacrifice, maternal love, tenderness and moral influence were lost, aggressiveness and blatant sexuality unleashed. Women suffering from the more insidious and stealthy melancholia, meanwhile, became silent, detached, self-absorbed, often ridden with guilt about real and imagined failures and crimes, unaware and unable to care about how their families and homes fared. Women who had recently become mothers should, according to nineteenth-century obstetric protocols, rest and eat in order to recover from the delivery and prepare for their new role, but instead they refused food and were unable or unwilling to sleep, ignored their infant or turned against it. Women's most important contribution to the consolidation of bourgeois wealth and power, as 'superintendents of the domestic sphere', incarnations of virtue, offering 'emotional labor motivated (and guaranteed) by maternal instinct' was not merely destabilised, but totally derailed; the aggression it was feared that women harboured broke forth with almost unimaginable force.[2]

In a book dominated to a large extent by asylum records describing the experiences of poor patients, which will be explored in chapters 4 and 5, this chapter will focus on upper- and middle-class sufferers, particularly the mental and familial turmoil of Sara Coleridge and Isabella Thackeray. These two women fell prey to mental disorders following childbirth in 1832 and 1840 respectively, although these took very different forms, prompted very different responses and had very different outcomes. One of the features of puerperal insanity was that it seemed capable of striking at the heart of wealthy families, whose womenfolk lived lives of relative luxury and comfort, just as much as at poorer homes. Obstetric physicians concentrated their attentions on just this type of setting, while private asylum keepers drew patients considered too ill to function at home into their institutions where, in the best and most expensive, the conditions of the bourgeois home were closely replicated. As well as focusing on the emphasis placed on the disruption of the domestic sphere, this chapter will explore how physicians worked not only to cure the patient, but also to seal, or at least conceal, the cracks that opened in the Victorian household, particularly drawing on evidence taken from Robert Gooch's case histories. This repair work would also cement a bond between middle-class families, women patients and their physicians.

It was the disruption of domestic life in the bourgeois home that would first attract the attention of obstetric commentators, though this would subsequently spill over into discussions about working-class patients as they too turned their backs on their homes and families. At a time when middle-class domestic values were being transplanted onto poor women, through the activities of home visiting, missions, Bible Nursing and sanitary reform,[3] so too would bourgeois notions of domesticity be transposed to asylum care, 'to transform the company of the deranged into at least a facsimile of bourgeois family life'.[4] Mid-nineteenth-century asylums, as we will see in chapter 4, would come to stress the importance of replicating a domestic environment and encourage the practice of domestic functions and virtues as one of the fundamentals of treatment. Great emphasis was placed on curing women whatever their social class to enable them once again to take up their household duties and roles as mothers.

Disordered households

Women and their families described mental disorders following delivery in letters and diaries, as part of their dialogue on the preparations for and discomforts and dangers associated with pregnancy and childbirth.[5] In the early modern period, as we saw in chapter 1, mental disorders following birth blurred into fear, dread, maternal imaginings and grief at the loss of children. The upset occasioned also greatly disrupted family life. Elizabeth Walker's eighth child was stillborn and afterwards she suffered depression for three months, Katherine Stubbes told her husband, neighbours and friends that her forthcoming child 'woulde bee her death', while Elizabeth Joceline secretly purchased a new winding sheet, not expecting to survive.[6]

Aristocratic families were far from immune from mental disturbances during pregnancy and following childbirth despite availing themselves of the best attendance and being cosseted by servants and nurses. Lady Cowper fell into a deep depression following the birth of her second child in 1738, accusing her husband of mistreating her. Lord Cowper believed her to be suffering from 'the vapours' and recommended the invigorating exercise of horse riding as a cure. The distinguished London physician Richard Mead was consulted who recommended a regime of bleeding, vomiting and purging 'at such intervals as her strength will bear, (and her Ladyship is not at all weak) to bring her spirits into a right course'. This approach, which would be scorned by many later authorities on puerperal insanity, failed to help. The marriage had been a love match, but broke down following Lady Cowper's illness. With Lord Cowper's ideals of domesticity shattered, he distanced himself from his children, and became mournful, fat and lazy.[7]

Lady Emily Lennox married Lord Kildare, later Duke of Leinster, shortly after her fifteenth birthday and in 1748, at the age of 17, give birth to her first child. Her childbearing career would not end until 1778 when she was 47, by which time she was married to her children's tutor; she had delivered twenty-two children and suffered several miscarriages. Almost half of Emily's children failed to survive infancy, and in 1755, following the death of her newborn infant Caroline, she told her husband that she would not have thought it possible to grieve so for 'an infant that I cou'd know nothing of'.[8]

Emily found her children an enormous blessing and spent a great deal of her time in their company, becoming an expert and something of a 'connoisseur of childbirth',[9] but also felt the strains of a life so dominated by childbearing and raising and looked forward to her menstrual periods, 'the French lady's visit', with great eagerness.[10] In 1760 her sister, Lady Caroline, wrote to her: 'I do feel vastly sorry you should lead so uncomfortable a life as continually breeding and lying in makes you do, and I don't wonder it wears your spirits.'[11] Aside from the physical burden of so many pregnancies and births, Emily was low spirited and depressed following her deliveries. In September 1775, her sister Louisa wrote to her following the birth of a daughter, Cecilia, 'I am very sorry, my sweet Siss, that your nerves were so much affected in your lying-in; 'tis really a miserable disorder that I would give anything you could get rid of'.[12] Following Emily's final confinement Louisa received a good account of her delivery after just a few hours of labour. However, 'I won't expect the next letters to bring quite such good accounts, as after a week or ten days your poor spirits always used to flag'. She urged Emily to cry, it being 'rather of use to you than harm'.[13]

In a captivating account of the childbirth experiences of Georgian commercial, professional and gentry families in Lancashire and Yorkshire, Amanda Vickery cites several cases of melancholia following childbirth. Jane Scrimshawe endured 'low spirits' for at least four months after her delivery and Elizabeth Parker suffered for six months in 1756. Elizabeth Addison of Liverpool regained her general health quickly after her delivery in 1816, but 'her nervous head will not allow me to take very great liberties with it', while Ellen Stock went to convalesce at the seaside in Southport, but 'laboured under so great a depression of spirits that my recovery was slow'.[14] The death of a close family member, particularly a husband, placed the household in great jeopardy, financial as well as emotional, and Vickery describes how families and friends rallied round widowed mothers-to-be prostrated with grief at the loss of their husband.[15]

In her study of aristocratic childbirth between 1760 and 1860 Judith Schneid Lewis greatly plays down the extent of nervous complaints following childbirth, claiming that the medical profession did not recognise such disorders. However, she gives several examples of women disturbed and depressed following childbirth, which she suggests could be related to anxiety about having too many children, or too many of the wrong sex, or having children born too close together.[16] Lady Londonderry was depressed after giving birth to a second girl in 1823 after enduring a long and difficult labour, while Lady Verulam felt her husband was not as attentive as he should be after the birth of her ninth child in 1822, when he went out shooting with his friends. He was 'annoyed to find his wife nervous and irritable' when he returned home after his jaunt, particularly as she continued so for several days.[17] Her disgust at the neglect and lack of sympathy shown by her husband reputedly manifested itself in a horrific way when her daughter, Lady Catherine Barham, delivered a stillborn child in 1835, and it was reported that 'Lady Verulam wanted to keep it to show Mr. Barham, & so pickled it and salted it and hung it up by the hindheels in the Ice House with the Venison'.[18] Lady Emmeline Stuart-Wortley suffered from depression following her confinement in 1835, and the household fell into disarray as a consequence. Her husband reported to a friend:

> he would not have another child for £1,000 for that her Ladyship takes no part of the management on her own shoulders, and that he has to make its bed, and to cook its victuals with his own hand – & that Emmeline wd let it starve and sleep on the floor if he were to be accidentally out of the way.[19]

Even royalty would not escape. Queen Victoria was troubled by 'lowness' that 'made me quite miserable' after the birth of her second child, Edward, Prince of Wales, in 1841. She lost interest in political affairs and an air of gloom descended on the Royal Household.[20] Victoria found childbearing hard to cope with, and this strongly influenced her decision to have chloroform administered at the birth of her eighth child, Prince Leopold, in 1853, the effect of which, she wrote, was 'soothing, quieting and delightful beyond measure'.[21] Even after this birth, though her delivery had been less painful and exhausting, she suffered her usual misery and when it was discovered that the child had haemophilia, her depression deepened.[22] Childbearing, she later concluded, 'is indeed too hard and dreadful'; 'men ought to have an adoration for one, and indeed to make up, for what they after all, they alone, are the cause of'.[23] As late as 1859 Queen Victoria was reflecting on the birth of 'Bertie'; in a letter to her eldest daughter, the Princess Royal, she wrote of the severe suffering she had experienced almost twenty years previously, as well as commiserating on the 'cruel sufferings' of her daughter during her first delivery.[24]

One unique scandal demonstrated the extent of the disruption that could be unleashed on aristocratic families by puerperal insanity, or, according to many who observed the case unfolding, the feigning of insanity. The remarkable case centred on the alleged mental stability of the vibrant and independent Harriett Moncreiffe who, following her marriage in 1866 to the much older Warwickshire landowner and avid huntsman Sir Charles Mordaunt, gave birth to her first child Violet in 1869.[25] Harriett had insisted on maintaining her independent lifestyle and had also surrounded herself with admirers, including Edward, the Prince of Wales, and there were many rumours of illicit liaisons. According to Harriett's confession made shortly after her daughter's birth, the child was suffering from a venereal disorder and there were several contenders for her parentage. The subsequent trial for adultery, which attracted immense public interest and brought in some of the most prominent medical men to testify, would also see the appearance of the Prince of Wales (who had caused his mother Queen Victoria such anguish at his birth in 1841) in the witness box. The extent of the furore and publicity surrounding the trial was even reflected in asylum admissions, with one woman admitted to St Luke's Hospital in February 1870, after giving birth to another man's child while her husband was at sea, enquiring 'Am I Lady Mordaunt?'[26]

While the *Medical Times & Gazette* suggested that many believed Harriett Mordaunt was shamming her mental disorder, evidence to dispute this was based largely on her husband's correspondence shortly after her delivery and confession, in which he referred to her as hysterical, nervous, refusing food and 'quite still, without ever speaking and without understanding anything'.[27] Sir James Simpson, one of the medical experts giving evidence, drew attention to her congested bowels and delusional state, which explained her wild confessions. Her dislike of the child and her desire to poison herself and the infant, as well as her apathy towards her husband, were also read as signs of puerperal melancholia, and it was stressed that it was this, rather than mania, she was suffering from.[28] In 1871 Harriett Mordaunt was brought to Manor House Asylum, Chiswick, where, because of continuing anxiety about claims that she was feigning insanity, she was again examined by several doctors, including the Medical Superintendent Thomas Harrington Tuke and the Physician to the Royal Household, Dr Gull, who confirmed that Harriett was suffering from puerperal insanity. She was reported to be happy and gentle but incapable of doing anything for herself and very childlike and demented.[29] Insane or not, the consequences for her were dire: she remained in a private asylum until her death.

'Confinement' took on a poignant meaning when connected with puerperal mania. The happy event of giving birth was tainted by the illness of the mother and the disruption of family life. Though childbearing women were depicted as vulnerable to a range of mental and physical challenges and disorders, and childbirth expected to be a dangerous passage, puerperal insanity was seen as a cruel twist of fate and in many women was unexpected and unaccountable given their previous medical, moral and family history. In some cases an hereditary connection was discovered, but for many it was seen as an unlucky combination of events and attributed to sleepless nights, agitation, difficulties feeding or worry compounding the tribulations of the delivery, or even the 'interruption of mental tranquillity required during the susceptible puerperal state', including the admission of boisterous relatives or nurses into the lying-in room.[30]

Dr Robert Gooch was the first to stress the link between puerperal insanity and domestic disruption, which he discussed at length in his case histories. While he claimed to have learnt much about midwifery and its associated disorders among poor patients in the wards of London lying-in hospitals, his case studies of puerperal insanity were devoted single-mindedly to the types of patient who dominated his private practice, well-to-do upper- and middle-ranking families whose domestic lives were falling into disarray. In one of Gooch's earliest cases,

> A pale, delicate lady, nursing an infant four months old, told me that she scarcely knew what was the matter with her: her sight was so impaired that she could not read; her powers of attention were so much impaired that her household accounts were burthensome to her; that she often rang for the footman, and when he came she had forgotten what she had rang for. She said she had a good husband, sweet children, ample property, everything to make her happy, yet she felt no interest in life. She added, that if this went on thus she should lose her senses. She had lost flesh, and had little milk ... She next began to accuse her friends, especially her husband, whom she charged with infidelity, and an intention to poison her; and it became necessary to separate her from her family, and place her in that state of seclusion and control usually employed under such circumstances. She continued in this state many months, but ultimately recovered, and has had a child since without a recurrence of the disease.[31]

Though Gooch would give an almost humorous spin to some of his cases, he highlighted a phenomenon about which he and other authors expressed grave concern. Gooch's ladies attempted to keep a grip on the household and their duties, but finally, inevitably, domestic order was replaced by misrule:

> One of my patients, almost from the day of her delivery, was observed to have restless nights, a quick pulse, and an irritable temper, compared with her natural one. She scolded the nurse about the merest trifles: one minute sent for the child to suckle, and the next ordered it angrily to be taken away. She would superintend her housekeeping, though she was entreated not to do it, and sent for the cook up into her chamber several times a day to enquire into the consumption of the family, and to give directions about its regulation. She talked almost incessantly, with disproportionate earnestness; and complained that her husband was not attentive to her: at length she accused him of incredible things, and soon after became so violent as to require confinement.[32]

The link with disorder, households in chaos and inappropriate behaviour would be remarked on and developed in many accounts of puerperal insanity. Gooch and other commentators recognised sleeplessness as an early symptom of puerperal insanity, but so too was a deviation in the patient's language or behaviour, especially if she became quarrelsome, critical, peevish, fretful, sullen or discontented.[33] Such infringements of codes of conduct and duty were reckoned to be as telling as the more violent outbursts that could typify puerperal insanity. Disrupted households would not only result from puerperal insanity, however; they could actually cause it.

Robert West, a rural general practitioner, recognised the extent of Mrs V's disturbance, a farmer's wife he had attended in labour in 1854, when five days after delivery she 'actually got out of bed, proceeded down stairs and into the yard, with nothing on, but her night-clothes, and without shoes; and had helped herself to a copious draught of cold water at the pump!'[34] West described the case in a letter to his midwifery lecturer, Dr Ryan. The letter was later published in the *Association Medical Journal* and attributed Mrs V's attack of puerperal mania principally to the circumstances of the lying-in.

> Her mother, who was to nurse her, did not arrive until eleven o'clock in the evening, and then found no one in the house with the patient but a little girl. The fires up stairs and down were all out. For the four following days, the mother had all the work of the

house to do, cows to milk, calves to feed, and I don't know what, besides attending to her nurse-lings; Mrs. V. all the time lamenting that her mother had much to do. A servant had been hired, but had failed to come, which was another source of mental disquietude.[35]

On the third day Mrs V. sat up for an hour, and on the fourth four hours, 'being all the time surrounded by gossiping friends, and being driven to bed at last by the pain in her head ... It was said that she drank a quantity of rum on one of these days.'[36] Mrs V. died shortly afterwards. West was 'sadly troubled' about the case, but blamed the woman's madness and death on the lack of care devoted to her and the disturbance of the household. This, together with her lack of sleep, and the heat and discomfort of her cottage, had confounded his efforts to cure her with camphor, sedatives, blisters and leeches to her head. He was also openly worried about the effect this would have on his practice, but his strategy seems to have been to provide in print a detailed narrative of his sensible management of the case compared with the family's mismanagement:

A death in childbed makes a great hubbub in the country; and as this was only my third midwifery case since I commenced in practice in this village, about nine weeks ago, I fear it will injure me; though every one blames the poor husband for not having sooner fetched 'the doctor'. For my own part, I don't know what to blame, and I certainly don't like to blame myself.[37]

It was often claimed that the calamity of puerperal insanity was triggered by a household mishap or more serious disaster, with disruptive visitors, husbands, mothers and nurses being implicated. W.E. Image, a surgeon attending a woman during her first, very lengthy labour, described how, while she was still weak, she received a letter from her husband, which informed her that he had been declared bankrupt. She suffered a fit and when she regained consciousness he described how she had taken on a strange countenance, 'expressive, quick, sullen, peculiar, but yet without intelligence'. Her milk and lochia ceased to flow and the woman died thirteen days later, which Image attributed to the mental shock she had received so soon after an exhausting delivery.[38] James Simpson would also later comment on this case and cited an even more remarkable incidence, where a bout of insanity was triggered by the 'cruel gift' of a box of mice, given by an 'evil-minded relative' to a lady convalescing after childbirth who had a 'constitutional horror' of the creatures.[39]

Where the delivery had been attended by an accoucheur there was a certain amount of apprehension on his part about declaring what the woman was actually suffering from, as he had assured her and her family of the best available medical attendance as well as protection from the rigours and risks of giving birth. Yet it was not without sincerity that doctors described how the condition marred a scene of potential happiness to the mother and her family. Fleetwood Churchill described the condition as 'a very distressing malady in itself, but doubly so from occurring at a moment ordinarily so joyful'.[40] Francis Ramsbotham urged great care in monitoring symptoms that might well show themselves during pregnancy as well as tact in revealing to families the 'anticipation of such a calamity supervening on delivery'.[41] John Conolly found it a particularly difficult disorder for a medical practitioner to deal with as the household dissolved into chaos.

> Few cases can be supposed to occasion more distress in a family than the unexpected appearance of insanity in a young woman, just when she, her husband, her relatives, and her friends, are full of natural joy on account of her having safely become a mother. A more than ordinary weight of anxiety is consequently thrown upon the practitioner, who is earnestly besought to answer interrogatories for which his practical experience has scarcely prepared him. He is questioned in whispers at the door of the bed-room, intercepted on the stairs, and cross-examined by an anxious assembly in the drawing-room... he is hoping that the anxious part of the case is at an end, and he has to meet new difficulties, against which he finds little aid in a house now thrown into confusion and with monthly nurses seeming, in these new circumstances, without resources, and full of alarm. It is well if these perplexities do not drive him to hasty and injudicious proceedings.[42]

The question of preventing future attacks preoccupied obstetric physicians, keen to preserve a family's custom. The physician would not merely treat the patient, but would vouch to repair the household and put domestic wrongs to right, and this, as we have seen in chapter 2, would form the cornerstone of claims to cure between competing groups of practitioners. Gooch wrote in detail about one case he was called to attend without being aware of a previous instance of puerperal mania until informed by the nurse during the delivery. Gooch was very concerned as, he reported, nearly all the woman's relatives were mad, or had died mad, and she had married a gentleman whose family was 'equally mad'. During the previous labour Gooch heard that her friends thought it a time of merriment, 'footing it about the house, which resembled a rabbit warren'.[43] Gooch determined that this time it would be different and ordered that the house be kept quiet. However, after ten days of good progress, a fire broke out nearby, which the woman witnessed, and consequently she began to talk and look 'oddly'. Gooch decided to sleep in the house, and at 2.00 am was summoned to the woman's room, where 'clasping her hands, with a whining methodistical tone, she exclaimed ... "I thought a glorious light issued from my temples, and that I was the Virgin Mary"'.[44] After three weeks, with careful attendance to her bowels, the woman recovered her senses. During her next pregnancy, Gooch became anxious about her state three weeks before the delivery and prescribed laxatives. These were continued after the delivery; the woman did well and suffered no 'maniacal affection'.[45]

Gooch's advice to midwifery attendants was to be aware of any pre-disposition to insanity and to endeavour to prevent it by purging the woman before and after delivery. The physician should be alert and watch for potential complications, and act quickly if any traces of disorder began to reveal themselves. The physician's role was not only to cure the woman, but also to restore a semblance of normality to the home and to retain his own authority, which his patient, due to the nature of the symptoms, would challenge at every turn.

Cures could be straightforward to the point of banality or complex, taking many months of patience. All cures involved the re-establishment of the fundamentals of self-control, in some cases marked by the patient ceasing to flout bourgeois standards of respectability by defecating when and where she chose:

> A lady, who had for some time been maniacal, was confined by a strait waistcoat, and had been accustomed to evacuate her bowels in the bed; an active purgative was given to her: soon after taking it she said, 'Nurse, let me go to the water closet;' and the nurse, astonished at this rational request, unloosened the straight waistcoat, and led her to the water closet adjoining, where the bowels were relieved of a prodigious load of feculent matter: she very soon recovered perfectly.[46]

Gooch was prepared to take a waiting stance and warned repeatedly against rushing patients into an incomplete recovery. Stressing the importance of rest, quiet and separation from 'loved' ones, Gooch aimed to protect the patient from herself, but also to shield her from her responsibilities until she recovered. He also stressed the importance of 'moral treatment', but argued that it could have little effect in the early stages of the condition. His patients, meanwhile, were to be distracted with care from their 'strange ideas':

> you must not combat these fancies by argument, for they will defend their absurdities stoutly, and your attempt to correct, will serve only to confirm them. You must draw their minds from their morbid fancies by engaging them on some other subject. I would rather allow a patient to think her legs were made of straw, and her body of glass, than dispute either proposition.[47]

In a case that attracted a good deal of attention and commentary, Gooch revealed the extent to which cooperation, even trust, could be built between patient and physician in cases where recovery took several months.[48] According to Gooch, a near relation having a 'frightful accident' precipitated the woman's severe melancholia. This was followed by sleepless nights. The woman, recently delivered of her second child, feared that she would lose her reason. She nursed the child for three to four months 'without feeding it', was thin, confused and low-spirited. She was persuaded to wean the child, dosed with light tonics and gentle laxatives and sent to the seaside. Yet the woman worsened and became increasingly violent. Finally, she was moved to the countryside 'under the care of an experienced attendant, separated entirely from her husband, children and friends, placed in a neat cottage surrounded by agreeable country'.[49] Yet she remained preoccupied and disillusioned, believing that she was to be executed, that she had caused the death of her husband and children and that spirits haunted her. An ill-advised visit from her husband, according to Gooch, had the potential to make matters worse. The woman initially believed that he had returned from the dead to haunt her and took on a 'hideous countenance', yet he persisted in talking to her and gradually she became calm and interested in what he was saying, and the cure came by 'magic'.[50] Gooch explained away the need to keep the husband at a distance by claiming that the patient had been approaching convalescence 'in which the bodily disease is loosening its hold over the mental faculties, and in which the latter are capable of being drawn out of the former by judicious appeals to the mind'.[51]

Gooch responded to his patients' cases with a decorum which was to become the hallmark of his practice and which earned him the praise of his contemporaries. Henry Southey wrote of Gooch's well-adapted sickroom manner, his kind-heartedness and ability to connect with the experiences of others; he 'rarely failed to attach his patients strongly'.[52] Gooch's extensive experience of illness – he died of tuberculosis in 1830 aged only forty-five – and his melancholy disposition, self-doubt and fear of death, seem to have imbued him with a sympathetic approach towards his patients.[53] Even when reporting cases where the woman reduced the household to turmoil, Gooch pointed to her naturally good demeanour and pleasant character. The physician's role centred on efforts to treat and cure the woman, but he should also act as a friend and diplomat.

Gooch, however, expressed anxiety about some aspects of his patients' behaviour, including over-cultivation and bookishness, citing the case of a woman suffering from mania who was 'clever, susceptible and given to books', and another lady fond of music, poetry and painting, who had amused herself during labour reciting poetry.[54] Thomas Trotter had expressed concern as early as 1808 that, as part of a wider shift from a more 'savage state', fashionable manners were enfeebling young women, making them less able to cope with the strains of childbearing and motherhood, which was further aggravated by their insistence on studying: 'Fashionable manners have shamefully mistaken the purposes of nature', while the modern system of education induced 'debility of body ... so as to make feeble woman rather a subject for medical disquisition than the healthful companion of our cares'.[55] The obstetric physician Samuel Ashwell warned too of the particular liability to puerperal and lactational insanity of women 'whose minds have early and long been cultivated at the expense of their physical strength'.[56]

Other doctors apparently lacked the patience and diplomacy to deal with the havoc wreaked by their patients, particularly in households where it was not only the patient who was disruptive. Dr Hastings, an Oxfordshire surgeon, who attended Mrs L in childbed in 1849, was somewhat perturbed when thirteen days after her confinement he was summoned at about midnight:

> I found her in bed ... she looked *very wild*, and remained perfectly quiet; was rather inclined to be sulky, and would only answer my questions occasionally, and even then only in a faint whisper, using that detestable monosyllable Yes, or No.

The husband and nurse reported how Mrs L, who was 'of a nervous and irritable temperament',

> exhibited a great degree of excitement and performed some strange freaks, such as breaking favourite china tossing it into the fire, in fact, she had quite a propensity to throw everything which she could seize into this destructive element ...[57]

Hastings 'determined' to bleed the woman, which he did vigorously. He also sweated her, dosed her with purgatives and opium, applied vinegar and water to her head, ordered her breasts to be rubbed with oil, and arrowroot and gruel to be given as nourishment. The woman began to improve under this rigorous regime; she asked for the child and her 'hatred' of both her husband and mother began to diminish. However, Hastings' efforts were to be foiled by the husband, who, elated at his wife's recovery, 'indulged too freely in other exhilarating courses, which made him observe late hours; and when he came home, his naturally boisterous conduct thwarted all my endeavours towards his wife's mental tranquillity'. His wife became exceedingly nervous again, heard noises in her head, violently rejected her baby and her hatred of her mother returned. Hastings tried several other remedies. Although Mrs L was no longer violent, she continued to talk a great deal of nonsense, declaring she was going to dine with the Queen, and needed morphia to sleep. 'The disease becoming more confirmed, I advised her friends to send her to an asylum', and she was brought to an asylum at Headington, near Oxford.[58]

Removal to an asylum meant for the private practitioner the abandonment of the case. Women tended to be removed to private institutions when they became too violent, inconvenient or embarrassing to treat at home.[59] Many families resisted the asylum solution, but for others the removal of the patient appears to have offered relief and the household functioned much better without her. Frequently, a stay at a private residence cared for by nurses supervised by a medical man, or at a medical man's own home, supervened between home and asylum admission, or followed release from a private institution. Ellen Johnston first became melancholy and then developed 'violent hysterical mania' about four months after the delivery of her fourth child. Her screaming so disturbed her neighbours that she was eventually brought to Chiswick Asylum in July 1870. Here the home environment was replicated as closely as possible; the patients lived in comfort with numerous attendants and servants, and were urged to read, take walks, converse with other patients and play the piano. After several months, Mrs Johnston's worst symptoms had abated and the Medical Superintendent, Dr Tuke, found her a private residence and a companion to look after her, though she was not considered fit to return home.[60] As we will see in chapter 4, for poor patients the ability to work as well as general calmness and decorous behaviour indicated recovery. For wealthy clients, the ability to converse, sew, shop and play the piano – though not the same tune over and over again – were held up as signs of cure.

Mrs Harriet Chaplin was admitted to Ticehurst House Hospital in June 1861, her naturally weak constitution breaking down after the birth of four children in five years. Though a 'ladylike person', she revealed the extent of her unfeminine behaviour in her 'delusions, waywardness and obstinacy of temper, by paroxysms of anger, want of natural affection and suspicion of all about her'. She claimed to converse with God and spoke of the 'happiness she feels at death approaching and waiting for wings to fly to heaven'.[61] By November she was starting to behave with 'propriety', and by February 1862 had freed herself from the influence of voices and was starting to conduct herself properly: 'can converse rationally – she is able to give orders properly in shops and make purchases'.[62] Discharge followed the reassertion of an ability to take up a place in the household, even if this was a diminished and tenuous place.

For other women, that place was never to be regained. Mrs Eliza Gipps, the wife of a country squire, developed symptoms of insanity after her first 'difficult and dangerous' confinement at the age of forty and was brought to Ticehurst Asylum in 1860. Mrs Gipps, described before her illness as active and intelligent, had been treated at an asylum near Edinburgh, and was afterwards placed with a medical man and his family. She was then considered fit to return home under the care of an attendant. At first she was able to associate with her husband, child and friends, but became increasingly obstinate and dirty in her habits, 'voiding her urine about the house and soiling her linen with faeces', though she was not violent or suicidal. She also developed terrible delusions:

> She believes that there is still the same connection between herself & child as existed when in the womb, that he is influenced by her own state of health, by the food she takes, & by the action of her own bodily functions. For instance if the child is away from her she will eat immoderately, that he may be supported through her

during his absence – she also believes that when obeying the call of nature she is passing out her own & the child's life & consequently restrains the action of the bowels as long as she possibly can do so – She states that her attendants and servants have the power of taking her mind from her, that they can produce internal pain with her at pleasure that they are the cause of her hair falling off, of weakness in her back, deformity of her toes, &c, &c.[63]

By May, Eliza Gipps was more tractable and had abandoned her dirty habits, chatted to the other ladies and employed herself in needlework, chess and reading, but she remained deeply distressed about her separation from her son, and kept portions of her food aside for him. However, in the autumn she became more troublesome, refusing food, attempting to retain her faeces and needing two attendants at night to prevent her leaving her bed. She was 'uncomfortable' and 'obstinate' with her husband when he visited '& she cried a great deal & wished to return with him to see her boy – she shewed little affectionate regard for her husband'.[64] Mrs Gipps was to remain in Ticehurst until her death from bronchitis nine years later, her state of mind deteriorating as she became apathetic and sometimes 'obstinate and irritable', but through all of this retaining a heartbreaking affection for her son, talking about him and expressing a desire to go home. On New Year's Eve in 1867 she refused to go to bed 'under the impression that her carriage was coming to take her home to her "child"', but she was never to see him again after she entered Ticehurst.[65] Nor was her son ever to know his mother, which reveals how domestic tragedies could accompany puerperal insanity and go much deeper than temporary disruption of the household.

Two flights into mental disorder: Isabella Thackeray and Sara Coleridge

The comfort and well-being of the families and households of Sara Coleridge and Isabella Thackeray were threatened when they developed mental disorders shortly after giving birth. Both fell ill in what could be termed the 'golden age' of puerperal insanity as it was becoming rooted in medical literature and practice, Sara Coleridge in 1832 and Isabella Thackeray in 1840. Isabella was married to the author and illustrator William Makepeace Thackeray. Sara was married to her cousin, Henry Nelson Coleridge; she was a prolific writer, translator, poet and author of pedagogical texts in her own right, and the daughter of the Romantic poet Samuel Taylor Coleridge. Both were raised in impaired families, in genteel poverty and dependent on the goodwill of others, though the financial struggles of the Thackeray family were far more pressing. Isabella was brought up by her mother in a succession of lodging houses after her father's early death, while Sara, her mother more or less abandoned by Coleridge as his drug addiction deepened, lived for many years in the Lake District with her uncle, the poet Robert Southey. For Sara, her mother, who lived for many years in the family home helping to run the household, would be part of her salvation. Isabella Thackeray's mother, according to William Makepeace Thackeray, was a frightful, selfish and unaffectionate woman, who failed to offer any support and this, together with his mother's resentment and comments on Isabella's domestic incompetence, appears to have exacerbated her illness. Sara and Isabella were both regarded as delicate, their pregnancies greeted with apprehension, and both were considered incapable of running a household. They were prime candidates for mental disorders associated with childbirth, for not only did puerperal insanity potentially destroy the household, but, as we will see in chapter 5, poverty, stress, disappointment and difficult home circumstances were held directly accountable for the onset of puerperal insanity in its varied forms.

Though the two women's experiences of illness were very different, both are sad and disturbing. Isabella's case was described as one of melancholia with episodes of mania. Sara defined her case herself as a direct result of childbirth and her associated nervousness and weak physical state. The stories of their illnesses were also narrated very differently. Aside from one letter to her mother-in-law in which Isabella wrote of her sense of encroaching mental derangement, her husband related Isabella's illness in correspondence with his mother and acquaintances. Sara kept her own counsel and narrated a detailed account of her illness, recorded in a diary between 1830 and 1838 and in letters to her children's nurse, family and friends. It could be questioned whether Sara's depression combined with feelings of hysteria really was 'puerperal insanity'. The definition of puerperal insanity was broad, but Sara's disorder was only rarely named as such and its duration was extraordinary, as was her ability to continue to work while ill. Sara and her doctors, however, were convinced that her mental disorder was triggered by the strain of childbirth. Sara also negotiated her condition in a very different way from Isabella. Isabella rapidly lost all control over her destiny and her links with her children and household. Sara was able to keep the home running, largely by delegating her tasks to others, to continue with her intellectual work and to present historians with the difficult task of understanding her illness and suffering alongside her productivity as a writer and ambivalence about her roles as mother and housekeeper.

Isabella Thackeray

Isabella Shawe was just 18 when she married William Makepeace Thackeray and within four years of marriage she had given birth to three children. Isabella was described as delicate and inexperienced, light-headed, though not light-hearted, silly but not amusing, lacking in passion and inept in household matters. William had intended to marry into money, but married into penury instead and their living was to be a precarious one. Even before Isabella became pregnant, several precedents to nervous collapse, as described by doctors at the time, were firmly in place: the youthful marriage of a woman untutored in household affairs, poverty, domestic strain and marital disappointment, probably on both sides. Thackeray depicted Isabella in his sketches as a child-sized woman, cool towards him and helpless, and complained that she often interrupted his work.[66]

Isabella's first child was born in June 1837 in a horrible scenario with both grandmothers present at the birth and squabbling over the choice of attendant. Thackeray's mother was a recent convert to homeopathy and insisted on calling in a homeopathic practitioner. After absenting himself from the scene to attend a business appointment, Thackeray returned and summoned another doctor who finally safely delivered the child. Isabella and the baby, Annie, recovered, but Isabella was unable to feed her daughter and took a long time to regain her strength. Isabella was despondent after the birth – as her mother had reputedly been after her deliveries – but when she did recover all was assumed to be well.[67] Just one year later a second child, Jane, was born. This time 'the very pink of accoucheurs', Sir Charles Herbert, attended. Thackeray wrote to his mother-in-law that Isabella produced children 'with a remarkable facility':

> She is as happy and as comfortable as any woman can be …We have been [none] the worse I assure you for being alone: the last time there were too many cooks for our broth, all excellent ones: but I make a vow that for the next 15 confinements there shall not be more than <u>one</u>.[68]

The baby, however, became ill and in March 1839 died at the age of eight months.[69] Before the end of the year Isabella was pregnant again.

As her confinement approached William described her as depressed, remembering the difficulties surrounding Annie's birth and the death of Jane, and she was despondent too about the state of the household and had been criticised by her mother-in-law for her ineptness. The baby, a girl called Harriet, was born in late May and Isabella appeared to make a good recovery, with the care of Nurse Brodie who was to be a pillar of strength. A few weeks later Thackeray wrote to his mother that 'the two small patients are getting on very well'.[70]

However, in August 1840 Isabella wrote to her mother-in-law:

> I feel myself excited, my strength is not great and my head flies away with me as if it were a balloon. This is mere weakness and a walk will set me right but in case there should be incoherence in my letter you will know what to attribute it to ... I think my fears imaginary and exaggerated and that I am a coward by nature.[71]

This was to be Isabella's only letter describing her illness. After a short trip to Belgium in August, Thackeray returned to find Isabella in an 'extraordinary state of languor and depression'.[72] He was advised to take her to the seaside for a break, but, though Isabella's health improved, she remained 'very low', breastfeeding 'pull[ed] her down' and Thackeray observed to his mother, 'For the last 4 days I have not been able to write one line in consequence of her. I must now work double time.'[73] In his next letter to his mother, Thackeray complained of her gossiping 'about her faults not doing her duty & so on', so that she now fancied herself 'a perfect demon of wickedness – God abandoned & the juice knows what: so that all the good of your reproof was that she became perfectly miserable, & did her duty less than ever'.[74]

Shortly after, the Thackerays set off on an ill-fated journey to visit Isabella's mother in Ireland. During the voyage Isabella made several attempts on her life, one night throwing herself into the sea; '[m]y dear wife's melancholy augmented to absolute insanity during the voyage, and I had to watch her for 3 nights (when she was positively making attempts to destroy herself,) and brought her here quite demented'. Thackeray tethered himself to Isabella with a ribbon while they slept, so fearful was he that she would seek to harm herself again: 'You may fancy what rest I had.'[75] Thackeray also realised that he had underestimated the seriousness of his wife's condition. The doctors they consulted recommended 'calmers, restoratives, plenty of food & quiet & very little diet',[76] but over the next weeks Isabella was 'devoured by gloom', 'clouded and rambling'. Her mother meanwhile refused to take Isabella into her own home, claiming that she herself was too ill with her nerves to receive them, and did little to care for her; the atmosphere between Thackeray and his 'demented' mother-in-law was one of mutual loathing. Thackeray also described how Isabella did not 'care in the least for me or her children' as well as her feelings of unworthiness: she 'won't sit still, wont employ herself, wont do anything that she is asked & vice versâ'.[77]

The search for a place of refuge and cure for Isabella commenced. When the Thackerays departed for France, Isabella's mother advised William to put her daughter in a madhouse, but Thackeray tried over the next few years to find a cure both on the continent and in Britain. Isabella, however, was never to recover, although at times she showed signs of improvement and in the first years of her illness lived for short periods with her family. She was brought in November 1840 to Jean-Etienne-Dominique Esquirol's 'famous Maison de Santé' at Ivry, but by January 1841 her condition had become chronic. This would have alarmed many doctors as a sign of the illness deepening, but Thackeray appeared to be relieved:

> at first she was in a fever and violent, then she was indifferent, now she is melancholy & silent and we are glad of it ... She knows everybody and recollects things but in a stunned confused sort of way. She kissed me at first very warmly and with tears in her eyes, then she went away from me, as if she felt she was unworthy of having such a God of a husband. God help her.[78]

In April 1841 Thackeray removed Isabella from Ivry, as the doctors had confessed that they could do nothing more for her. Thackeray decided to try an alternative cure, took Isabella for a walk and fed her dinner and champagne 'and actually for the first time these six months the poor little woman flung herself into my arms with all her heart and gave me a kiss'.[79] For the next six weeks Thackeray was Isabella's only attendant but 'almost broke down under her slavery'.[80] He hired a nurse, still hoping she would recover outside an asylum:

> My wife will get well, I hope and believe; perhaps not for a year, perhaps in a month. There is nothing the matter with her except perfect indifference, silence and sluggishness. She cares for nothing, except for me a little ... She is not unhappy and looks fresh, smiling and about sixteen years old. To-day is her little baby's birthday. She kissed the child when I told her of the circumstance, but does not care for it.[81]

Thackeray left Isabella for several weeks, and returned to find her a little better. Under the advice of a hydropathic physician, in August 1841 he brought her to a spa at Boppard on the Rhine. William had 'a strong hope that under this strange regimen my dear little patient will recover her reason'.[82] Subject to a rigorous regime, which involved sweating in blankets, douches and ice-cold baths, Isabella at first seemed to improve, becoming at one point 'extraordinarily better', but then became 'excessively violent & passionate wh. the Doctors say is a good sign'.[83] In March 1842, Isabella was placed under the care of Dr Puzin at Chaillot and was 'much better since I parted from her, wh. I was obliged to do ... for she was past my management'.[84] Thackeray wrote of his despair to his friend Edward Fitzgerald. Isabella again seemed to be better, she was 'perfectly happy, obedient and reasonable', but he missed her, and the children 'are not half the children without their mother – A man's grief is very selfish certainly, and it's our comforts we mourn'.[85] In September 1842 Thackeray was looking for a private asylum in London where he could place Isabella. Accompanied by Bryan Proctor, a Lunacy Commissioner, he visited 'his favourite place which makes me quite sick to think of even now. He shook his head about other places.'[86]

Occasional improvements in Isabella's state would be followed by relapses, but hope of her recovery faded as the years passed. In November 1845 Thackeray took Isabella away from Dr Puzin and placed her in the care of Mrs Bakewell, an 'excellent worthy woman', and her daughter Mrs Gloyne at Camberwell.[87] In 1846 Thackeray commented that 'the poor little woman gets no better and plays the nastiest pranks more frequently than ever' and in 1848 declared that she was 'dead to us all' and 'cares for none of us now'.[88] Isabella remained with Mrs Bakewell until her death in 1893, after 53 years of illness.[89] Though Thackeray found the idea of placing Isabella in any form of asylum repulsive and inappropriate, he mentions her little in his correspondence after 1845; she became a distant figure, out of sight, lost to the household but also no longer troubling it.[90]

Sara Coleridge

In June 1832 Sara Coleridge gave birth to her second child, Edith, and two months later slipped into a deep depression. Towards the end of August she had complained in her diary of a slight bowel complaint, bad pains and diarrhoea. A local surgeon, Dr Evans, attended her and prescribed aromatic draughts, a careful diet and castor oil. She was also prescribed ten drops of laudanum by 'Henry'.[91] By 12 September Sara felt herself increasingly debilitated, 'weak and miserably nervous and fluttered'.[92] A week later, she recorded in her diary that she was 'going on very sadly ... Disordered bile accompanied with derangement of the nervous system is my complaint. Stomach and bowels out of order– great weakness – nervousness – shivering and glowings, etc.'[93] Sara had trouble sleeping and started to take opiates regularly to induce sleep. By December, after a brief stay in Brighton, followed first by improvement, then relapse, Sara described herself as 'hysterical', 'low' and 'in despair'.[94] Her journey had commenced into a depression that would not lift for many years and which would be compounded by her attachment to opiates and further pregnancies.

Although not labelled as such by Sara except implicitly in references to 'nervousness' associated with childbirth, she appears to have fallen victim to a form of puerperal melancholia. Her mother confirmed the diagnosis in October 1832 when she wrote: 'Our poor Sara is reduced to a very sad state of stomach & nerves by over-nursing; and her disease, which, by the Medical-man is called Puerperal is of the most distressing kind.'[95] Sara's impressions of her illness evolved into a gruelling narrative of depression and confusion, as the illness became indistinguishable from the effects of the narcotic drugs with which it was treated and to which Sara clung. Sara negotiated her way through her illness in a way that often gave her freedom to work on her literary projects and to absent herself from the household, and it could be questioned how much she actually learnt about her disorder from reading medical texts and from her circle of acquaintances. In any case, Sara responded and even 'managed' her illness in a very different way from Isabella Thackeray.

Sara's first pregnancy, shortly after her marriage, had been greeted with consternation by family and friends alike who thought she was too frail to withstand the rigours of childbirth. Her mother came to Hampstead for the confinement and helped take care of the infant, Herbert, after the birth. Even 'poor father at Highgate', with whom Sara had little contact, 'has been very nervous about her'.[96] Despite her frailty, Sara would go through three more deliveries and several miscarriages during the next ten years. Each pregnancy caused her great anguish and physical and mental collapse, but she was to have no respite until her husband became severely ill in 1841.

Nursing and caring for Herbert wore Sara down, and, as we have seen, her physical health deteriorated and spirits plummeted following the birth of Edith. The diary in which she recorded her struggle with encroaching depression is a miserable read. What started out as a baby diary after the birth of Herbert in 1830, intended to record her children's progress and physical development, as well as an astonishing array of colds, 'feverlets', skin and bowel disorders, became a narration of her own mental and physical condition. On many days Sara noted down 'nervous flutterings', 'no tears though low enough' or a 'yawny hollow' feeling, and she also recorded lapsing in and out of 'nervous terror' and 'wretched spirits'. Throughout June 1833 Sara described herself as 'fatigued, exhausted and heavy', she had a sore throat and mouth, her back ached. She briefly improved and became 'less nervous' and 'without tears', but on 20 June was 'low', 'upset and hysterical', had 'nervous twitchings in my arms'.[97] In early autumn she recorded more frequent episodes of weakness and hysteria and was discouraged and fearful about what she referred to as 'the stoppage'. Sara was once again pregnant. In November 1833 she wrote:

> O woe is me! O misery! The sensations of utter helplessness and heart sinking wretchedness with slight intervals of mitigation have been more over powering than ever since I wrote last. And then she wept again, because she had no more to say of that perpetual weight which on her spirits lay.[98]

In January 1834 Sara gave birth to twins, a boy and girl, both of whom, Sara wrote in her diary, 'gave up their little feeble lives' a few days later.[99] The death of three children shortly after birth and several miscarriages during the years covered by the diary can probably be attributed to Sara's large intake of morphine and other drugs. Sara was to remain in this melancholy state, with ups and downs and episodes of relatively good health, until her husband's death in 1843. Sara had but a few years of improved health and spirits before she developed breast cancer in 1848; she died shortly after visiting the Great Exhibition in 1852 at the age of 49.[100]

Sara's account of her illness is particularly intriguing because of the way she interacted and allied herself with the physicians who treated her. Her diary reveals that she was not treated by doctors with any particular expertise in midwifery or nervous complaints. Mr Haines and his partner Mr Evans, who both practised near her home in Hampstead, attended her. Sara also saw Mr Gillman, her father's doctor,[101] very regularly after he attended her first confinement, and occasionally consulted her uncle, Henry Southey. In 1833 she had a brief and somewhat distressing interaction with the establishment figure and wealthiest physician in London, Sir Henry Halford, President of the Royal College of Physicians. Regarded by colleagues as 'more a courtier than a doctor' with his 'highly developed, professional bedside manner',[102] Halford failed Sara badly. Although a Commissioner in Lunacy, he appeared to know little about depression following childbirth and openly derided midwifery as a low branch of practice.[103] He promised to 'cure' Sara, but merely prescribed more drugs. His bedside manner did not match expectations, and in May 1833 Sara declared herself 'alarmed and dejected by his manner. He considers mine an obstinate case.'[104]

Sara also seems to have fared badly in terms of therapeutics, but this was at least in part through her own volition. She remained mainly at home, in the company of her family, although her husband, a barrister, spent a good deal of time away during the week working in the law courts and living in lodgings. Sara put into effect the advice of experts on puerperal insanity by 'escaping' from the family home, letting the responsibilities of the household fall on her mother and leaving the children with their nurse, Ann Parrott, on a number of occasions. But she never freed herself from home and family for substantial periods of time, partly because she felt so committed to her own methods of teaching and raising her children.

Sara continued to dose herself with opiates, taken for insomnia since 1825. As mentioned in chapter 2, many later medical authorities on mental disorders following childbirth denounced the use of narcotics; however, Gooch and Prichard, Sara's contemporaries, supported their use. Prichard recommended fairly hefty doses, 10 grains of Dover's powder at night, a grain and a half of solid opium, or 30 drops of tincture, similar amounts to those taken by Sara on many occasions.[105] Sara made only occasional feeble attempts to give up morphine and laudanum, continuing to consume them on a regular basis.[106] As a result, she became very constipated, and her struggle to deal with this became a daily preoccupation, as she also took a variety of laxatives and recorded her motions meticulously.

Sara's relationships with her doctors are revealing. She demonstrated her knowledge in her interactions with them; she consulted a number of local practitioners and physicians, but never feared to express her own opinion. She classed some of her attendants as 'kind', while others disappointed her, either because they failed to make her feel better or, one suspects, because they tried too hard to wean her from her opiates. In January 1833, together with her 'kind' friend Mr G. (Gillman), it was agreed that her state, particularly the suspension of her 'monthlies', was

'very common after confinements connected with nervous derangement'.[107] Mr Gillman supplied Sara with opiates; she records sending for
them in great urgency at night. Later in the same month, Sara was seen
by Dr Nevinson, who gave sound advice: 'He says I shall surely recover
but not very soon – advises me to reduce the morphine <u>as soon as I can</u>
and to go on quietly making no painful exertions but attending to my
general health and not relying on medicine.' He prescribed a mild purga-
tive, but nothing else. 'He thinks the weakness I feel still more of today of
no great consequence.'[108] His advice was not followed, and Dr Nevinson
was never referred to again. Sara continued to take laudanum.

As well as giving us details of her illness and suffering, the diary
also shows that Sara was well informed about children's complaints
and their treatment, and about the diseases of women and the
workings of her own body, nervousness and moral insanity. She was
precise in monitoring her menstrual periods, knowing when to
expect her 'show' calculated to the day, even the hour. Their
absence made her quick to fear that she could once again be pregnant. Sara's essay on 'Nervousness', written in 1834 during a period
of relatively good health, was largely an effort to analyse her condition, to confront and understand the nervous affliction which she
saw as having both physical and psychological origins. It revealed
familiarity with recent medical literature and appeared to have been
influenced by the publications of several authorities on the diseases
of women, Robert Gooch, James Cowles Prichard and Marshall
Hall.[109] She used words and descriptions that indicated confidence in
explaining female disorders, showed familiarity with current medical
disputes and was able to link her condition with what she had read.
She was also well read on moral insanity and her essay was largely
an effort to differentiate between madness as a disease of reason and
hysteria as a disease of the emotions. She blamed her reproductive
organs as a contributing factor in her 'nervous derangement', and
'suspicious discharges', which she labelled 'whites' or 'albifluores',
confirmed for her 'irritation of the uterus, either as a cause or accompaniment of general nervous weakness'.[110] She also referred to
herself on many occasions as 'hysteric'.

In November 1832, not long after she became ill, she explained to
her husband that her 'nervous derangement' was 'a disease in the
mental powers – not madness – yet in some respects akin to it'.[111]
Dr James Cowles Prichard had described 'moral insanity' in 1833, just
one year before Sara wrote her essay, defining it as 'a morbid perver-
sion of the feelings, affections, habits, without any hallucination or

erroneous conviction impressed upon the understanding; it sometimes co-exists with an apparently unimpaired state of the intellectual faculties'.[112] It was a description that would have suited Sara, as it did not compromise reason or intellect.

This raises the question of how far it was possible for Sara to have 'learned' her illness and how far her condition was mediated by what she had read in medical texts.[113] Just as she had observed the opium usage of her father and a wide circle of his family and acquaintances, she also had a close acquaintance with mental disorders. Her aunt Edith, Robert Southey's wife, suffered many years of depression and had a further collapse in 1834 when her affliction 'assumed a decided form of madness in its most frightful manifestations'.[114] Edith was brought to the York Retreat for treatment and subsequently was cared for at home by Southey until her death.[115] Around the same period, Dorothy Wordsworth, a close family friend, was lapsing deeper into mental derangement and confusion: 'Her mind ... much shattered.'[116] Accounts of the illnesses of both women formed the subject of much correspondence within the Coleridge circle.

The foremost author on puerperal disorders, Robert Gooch, was also intimate with the family. Gooch had briefly treated Samuel Taylor Coleridge in 1812, when his health had taken a sudden turn for the worse. Though Coleridge placed his case in Gooch's hands 'without the least concealment' and Gooch professed 'strong hopes' of reducing his opiate addiction, if not emancipating him wholly from it, his attempt failed.[117] By the time Gooch was called in to treat Coleridge, though his own health was worsening, his private practice, based largely on midwifery work, was flourishing. Gooch also had close links with the Southeys, having studied medicine in Edinburgh with Henry Southey, who remained a close colleague.[118] It was Southey who wrote a moving epitaph in honour of Gooch in *The Lives of British Physicians* in 1830.[119] Gooch befriended Robert Southey after a visit to the Lake District in 1811, and Southey recruited Gooch as a contributor to the *Quarterly Review*, which he edited. Without doubt, Gooch was part of the Southey and Coleridge circle and his work known to them. He was a marvellous fit with the role of a 'Romantic' physician, well educated, attractive, sensual, sympathetic and also given to severe bouts of melancholy and self-doubt, carefully recording his own dreams and fears. By the time Sara fell ill in 1832, Gooch had been dead for two years, but Sara may have read his work (she certainly read the *Quarterly Review*), some of it published posthumously in 1831; she may even have met Gooch on his visits to the Lake District or in London.

Sara's interest in the concept of moral insanity, which did not compromise understanding or intellectual faculties, may have provided a framework for her escape from maternal duties to resume her writing. Even before her marriage there was widespread speculation in her circle about her unsuitability for marriage and motherhood, not only because of her frailty but also her lack of aptitude in organising a household. In 1825 Southey, in talking of Sara's accomplishments as a translator, suggested that these skills disqualified her 'not a little (in my judgement) ... for those duties which she will have to perform whenever she changes from the single to the married state'.[120]

The deepening of Sara's mental anxiety and her feelings of nervous derangement in 1832 persuaded her that it was time to wean Edith even though she was only two months old, and the diary takes on a particularly disturbing tone in the early autumn of 1832, recording the struggle to persuade Edith to suck from a bottle, Sara's sore breasts, Edith's bowel upsets and sickliness, Herbert teething and Sara desperately depressed.[121] There is a strong sense that Sara may have been negotiating her illness on her own terms, in a much more complex way than the 'unconscious form of feminist protest' suggested by Elaine Showalter,[122] partly in her choice of attendants and advice, but also to take time away from household duties, leaving them to her mother or the nurse. Her mother sympathised with Sara's change in circumstances. Her days spent with her uncle had been occupied by reading, writing, walking and mountaineering, while now she was mistress of a household and the mother of two young children with little time to study, 'transported from a *too* bustling family, to one of utter loneliness'.[123]

Days haunted by depression, as noted in the diary, coexisted at times with an active working life. In February 1833 Sara recorded being 'in a miserable state of irritation. I read all day.'[124] By the summer she was helping Henry with his Chancery papers and reading biographies of Shakespeare, Ben Jonson and Dryden.[125] Sara fell into the habit, according to acquaintances, of letting her mother look after the household so that she could devote herself to reading and writing, sorting out her father's papers and educating Herbert.[126] During one of her worst periods of 'nervous hysteria' in October 1836, Sara set off with her children and Ann Parrott for London after a stay with her relatives in Ottery St Mary in Devon. At the end of the first day, on reaching Ilchester in Somerset, she declared herself unfit to proceed – 'weak, tongue furred, stomach flatulent & sick ... and altogether so ill that if I put a further force upon my nerves, I know not what will happen ... If

I am now quiet I shall gradually recover but if I proceed I never shall.'[127] She sent the children back to Ottery St Mary and remained at Braine's Inn in Ilchester for a month. Her letters make it clear that she recovered enough to immerse herself in copious reading and writing with 'a clear head and vigorous hand'. However, whenever her husband tried to persuade her to return to London, her symptoms worsened. He pleaded and even commanded Sara to return, but she was determined to stay where she was.

> Say that I may rest here till my shattered nerves have recovered some degree of tone, and I shall be happy: but assuredly that will not be in ten days, nor perhaps in ten weeks. For the rest of my life I would keep my expenses within the closest bounds possible.[128]

In neither Sara nor Isabella's case were the doctors consulted able to cure their patients. Isabella soon lapsed into an irredeemable state of mental derangement and became estranged from her family, leaving Thackeray to raise his daughters with the help of his mother. Isabella was lost to her household. Sara's illness was marked by breaks from the responsibility of family and the negotiation of her treatment complicated by her knowledge of mental and female disorders. The household ticked over with the help of her mother and servants, while Sara reserved the right to educate her children herself and she certainly seems to have been in control when it came to organising medical attendance. Sara finally emerged from her depression in the early 1840s, with a desperately ill husband, but free from the burdens of childbearing.

Isabella and Sara were marked from the start of their marriages as ineffective housekeepers and constitutionally ill-equipped mothers. Isabella, in her husband's narration, appears powerless, depicted as his 'poor little wife'. Sara was aware of the disruption to the household caused by her illness, strove to minimise it by bringing in her mother and Ann Parrott to make up for her absences, and herself used the term 'poor' to describe those around her who suffered on her behalf – 'poor Henry' and 'her poor darletts'. Both women foiled the physicians' potential to put the household to rights.

* * *

As early as 1792 William Pargeter described a case where 'by management' a lady who developed '*Mania furibunda*' following a difficult parturition was happily reduced to '*Mania tranquilla*'. Before Pargeter's appearance the woman ill-treated and beat the servants and rejected medicine and food; after he won her 'good opinion' and dosed her with aperients he was able to restore 'a very valuable woman to the enjoyment of her family and friends'.[129] It is this theme of restoration that recurs and persists in the literature on puerperal insanity in the mid-nineteenth century. Disorder within the household was seen as both a cause and symptom of puerperal insanity, just as disordered households resulted from it, giving support to the assertion that 'the smooth surface of Victorian ideology, like that of an unhappy family keeping up appearances, is artificial and deceptive'.[130] The dominance of domestic hegemony and the sanctity of the home and mother were revealed again and again, as it was pointed out that this very sanctity had been violated by puerperal insanity.

Religious themes occur too, but more rarely. It was pointed out that in their normal state the women were good Christians, and many of their delusions, as we will see in chapter 4, revolved around the themes of God's punishment or of religious exaltation, which provided them too with a vocabulary to describe their distress and emotions. However, in many ways domestic ideology seems to have been the new religion. The physician's role was to re-establish order, and this would often be achieved – even though domestic situations were the root of the trouble in many cases – in a domestic setting, the patient's own home, a well-suited private house or an asylum, which increasingly came to replicate the home environment. Many physicians, as we have seen in chapter 2, recommended the removal of women from the immediate scene of their distress as one of the first principles of successful treatment. Others stressed the importance of the doctor establishing what appeared to be fearsome levels of authority over their patients. But most, like Gooch, recommended patience, mild remedies and common sense, allowing the women time to recover. Such an approach could be expected for paying patients tended carefully by their family doctor or a well-reputed obstetric physician like Gooch. However, this form of care also carried over into asylum practice, where great emphasis was placed on ensuring that patients were fully convalescent and not released too early to a family struggling in the absence of their wife and mother.

James Reid summed up in a nutshell the typical development and progression of puerperal mania, the form of preferred treatment and the best possible outcome:

> Mrs.__, aged 20, had been distressed by some family occurrences which caused her much anxiety. Four days after her first confinement, she became much excited, and at length exceedingly

violent, swearing and using most obscene language, although at other times a lady of most correct demeanour. As the usual treatment seemed to produce no good effect, and she had taken an inveterate dislike to her husband and child, it was thought advisable to remove her to a cottage in the country in the charge of two experienced nurses. In about five months she was restored to perfect health, and on asking to see her husband, he was immediately allowed to visit her. She has since this period borne several children, but although of a nervous temperament, has had no return of the complaint.[131]

Cure could take several months, but the hoped-for result was to restore the woman to her proper role and position in the household, which had been shaken but not destroyed, though neither Isabella Thackeray nor Sara Coleridge would be redeemed in this way. An alliance of sorts was struck between the patient and her physician, which for many women would involve a respite of sorts from the drudgery, boredom and frustration of her role as household manager, though the ultimate aim was restoration to that very position. Though there was a fundamental optimism about the curability of nervous disorders following childbirth, the physician was not in a position to relieve women permanently from their household duties – and for Isabella and Sara this may have been the only cure.

4
Incoherent, Violent and Thin: Patients and Puerperal Insanity in the Royal Edinburgh Asylum

In 1883 Thomas Clouston, Medical Superintendent of the Royal Edinburgh Asylum, found himself entranced reading back through the old case notes of the institution, remarking on the 'strange biographies', 'false beliefs' and 'strange conduct' of the patients.[1] These case histories prove equally captivating and illuminating for historians.[2] First-hand accounts of madness – diaries, letters and memoirs – are comparatively rare and difficult to work with, most being written after recovery or during 'lucid intervals'.[3] Very few refer to puerperal insanity.[4] Although asylum case notes have their limitations, they provide evocative and substantial evidence of the asylum experience, which would come to be the norm for growing numbers of women suffering from puerperal insanity. Drawing on the rich archives of the Royal Edinburgh Asylum, Morningside, this chapter focuses on the case histories of women treated there between the mid-1840s and the early 1870s, the period of Dr David Skae's superintendence.[5]

By the mid-nineteenth century in numerous asylums across Britain scores of women were admitted every year, suffering from puerperal insanity, and the asylum was increasingly promoted as the proper place to cure such cases. John B. Tuke, whose interest in puerperal insanity has been referred to in chapter 2, was Assistant Physician to the Royal Edinburgh Asylum between 1864 and 1865. During this short period he became intrigued by the incidence of puerperal insanity in the asylum over the previous decades and published two influential papers on the topic. He emphasised that cases reaching the asylum were 'of the most severe character, such as could not be treated at home';[6] and while he saw home care as a possibility in mild cases, Tuke strongly advocated asylum treatment. He also urged early removal to the asylum; provided cases of puerperal insanity were not permitted to languish, this was, he insisted, 'perhaps, *the* most curable form of insanity'.[7]

Manuscript case notes may well be an imperfect recording of actual practice and leave many opinions and events untold, as well as reflecting the preoccupations and preconceptions of the doctors who wrote them, but even so they tell us more than most other sources about how puerperal insanity was experienced by patients and physicians, and about the diagnosis and classification of the condition, its treatment and in many cases cure. Though the case notes largely represent the medical viewpoint and incorporate expectations about both the nature and symptoms of the disorder, they are also built on the reports of the family and friends of the patient, and read as individual accounts of personal suffering. Given the limited number of accounts scripted by patients themselves, case histories are often as close as historians can come to hearing the stories of poor asylum patients. They also show the views of an emerging psychiatric profession on a prevalent and troubling disorder, and reveal that doctors were prepared to engage with a broad causal framework for explaining the occurrence of puerperal insanity, an aspect that will be further developed in chapter 5, which draws on a wider selection of case histories, both published and unpublished, taken from asylums and obstetric practice.

The case of Ann Denholm who was admitted to the Morningside Asylum in September 1851 suffering from puerperal mania exemplifies the kind of information found in the case notes. She was single, aged 26, a domestic servant who lived in the parish of Liberton, and had just given birth to her first child:

> In disposition she is melancholy and reserved; but of quiet[,] sober and industrious habits. The causes assigned are regret at her misfortune in becoming a mother, followed by her recent confinement. This is her first attack and is now of sixteen days' duration, her delivery having taken place some three weeks ago.[8]

Ann Denholm was sleepless, depressed and gazed into space, believing that she had offended her Maker: 'She has been in heaven[,] thinks God will have nothing to do with her and that the world has come to an end'.[9] She had contemplated suicide and the murder of her friends. The case notes commented that, though in robust health, she seemed 'exhausted' and 'worn out' with the intensity of her grief. She would not answer questions, was untidy and self-absorbed.

On admission to the asylum, Ann Denholm was put to bed, given stimulants and purged. After a couple of weeks she was still unsettled, incoherent and unable to sit still or occupy herself. Her breasts required 'attention for some days', but otherwise her physical health was good. By the end of December, her restlessness alternated with fits of deep dejection. She blamed herself for all sorts of crimes and had particularly gloomy views 'on the destination of her soul'. She cried a good deal at times and was indifferent to her appearance and person. However, for the most part she was industrious and employed herself regularly in the laundry and washhouse. She was noted to be affectionate and kind. By the following March she had improved steadily, become more 'cheerful, sociable, active and sensible' and conducted herself with propriety. She was anxious to leave the asylum, and was allowed to visit friends in Liberton. On 14 April she was finally discharged 'cured' and 'kind, contented, mild and childlike'.[10]

Each case history is an individual story. However, many features repeat themselves and in some ways Ann Denholm was a 'typical' case. Single and a domestic servant, she represented one of the groups most often admitted to an asylum suffering from puerperal insanity. The timing of the onset of her illness, about a week after delivery, was common, as was the period required to restore her to sanity, some seven months. Her treatment was mild, confined to bed rest, purges to ensure the regularity of her bowel movements and stimulants to build her strength. Her self-neglect was commented on, but so too was her willingness to work. Her powerful delusions and sense of guilt gradually diminished, and she passed through phases of restlessness and gloom, before reaching a state marked by her good demeanour and positive, though 'childlike', attitude. Ann Denholm was a relatively trouble-free patient in that she does not seem to have been difficult to manage, and, unusually, her physical health was good. Case histories such as hers will form the core of this and the following chapter.

The Royal Edinburgh Asylum

The Royal Edinburgh Asylum, which became 'the largest, the busiest and, arguably, the most prestigious asylum in nineteenth century Scotland',[11] was opened in Morningside in 1813, the original building being known as East House.[12] The care of patients was initially entrusted to lay superintendents, but in 1839 Dr William McKinnon was appointed first resident Physician Superintendent.[13] In the early years only paying patients were accepted, but in 1843 West House opened to accommodate pauper patients and those paying low rates of board.[14] Edinburgh Asylum catered for a cross-section of patients compared with the large English pauper asylums, reflecting the make-up of Victorian Edinburgh with its mix of prestigious New Town, prosperous suburbs, appalling slums and poor environs. Private patients continued to be accommodated in the East House; paupers, who made up three-quarters of the asylum population, in the West House, which also admitted a small number of middle-class patients unable to afford the East House fees.

Medical education had long been one of Edinburgh's chief service industries when the Asylum was established; its medical faculty had been founded in 1726, the Royal Infirmary in 1729, and, developing as one of the main hubs of medical learning in Europe by the 1770s, it attracted large numbers of students through its emphasis on clinical teaching.[15] Dr Alexander Morison had introduced extramural lectures in psychiatry to Edinburgh in 1823, but David Skae was the first to offer clinical instruction to medical students, demonstrating with his West House patients.[16] Skae attracted a long succession of very able Assistant Physicians, who would go on to have impressive careers in psychiatry and who were responsible for compiling the case notes, a practice instituted by McKinnon.[17] The combined presence in Edinburgh of Skae and the 'mental physiologist', Thomas Laycock, author of *A Treatise on Nervous Diseases of Women* (1840), placed the city firmly at the forefront of psychological medicine.[18]

The first Medical Superintendent, William McKinnon, believed staunchly in moral therapy, opposed restraint, recommended wards and dormitories in preference to cells, and was enthusiastic about the curative value of gardening, music, games, excursions and religious devotion.[19] By the time Skae became Medical Superintendent in 1846 there were 466 patients and pressure on accommodation was mounting, particularly from the pauper patients. When Clouston took over in 1873 there was an average of just over 700 patients resident in the asylum.[20] West House was designed by the architect William Burn, with the intention of segregating patients along the lines of social class, gender and the severity of their medical disorder. West House was made up of a cluster of three-storey buildings, with wards (or galleries as they were known), and a separate building (referred to as 'SB' in the case notes) catered for 'refractory patients'.[21] The case books reveal a great deal of movement between the galleries depending on the degree of mental disturbance and the behaviour of the patients. Although the institution as a whole was run with an emphasis on occupying the patients through work, exercise and rational pursuits, an American psychiatrist commented after a visit in 1860 on the discrepancies between the social classes, with the paying patients being 'quite comfortable', the paupers 'crowded too much and the violent, and excited and noisy were together'.[22] The latter were expected to work as soon as they were fit, as an integral part of their therapy.

Under McKinnon and Skae, there was much emphasis on improving the physical status of patients through diet, as well as the benefits of work as therapy, with patients urged to use their trades and skills, something Clouston would hone to a fine art, developing his four 'gospels' of work and play, fatness and self-control when he took over in 1873.[23] For the women patients, work largely meant needlework, though pauper women were also urged to help in the wards and laundry. The emphasis on food was no doubt welcomed by many poor, malnourished patients and was an important component of the treatment of puerperal insanity.[24]

Case notes, the 'patient's view' and the 'doctors' view'

In recent years, the use of case notes, long neglected not least because of the time-consuming nature of this form of research, has enjoyed a growing vogue as part of a broader agenda within medical history to access and understand the patient's view and to explore 'history from below'.[25] It also ties in with a greater emphasis on the use of various kinds of patient record within the history of psychiatry, particularly admission certificates and admission and discharge books, sources that have lent themselves primarily to quantitative analysis.[26] Jonathan Andrews has persuasively highlighted the value of patient histories, referring particularly to the records of the Gartnavel Asylum in Glasgow.[27] As he argues, case notes provide 'the surest' basis for understanding 'the changing nature of the experience of the insane in asylums since 1800'.[28] Akihito Suzuki has employed case records to explore the triangular relationship between patient, family and friends, and doctors in nineteenth-century Bethlem.[29] Other historians have combined an analysis of case notes with admission records to trace the routes of individual patients between Poor Law workhouses and public asylums, providing insight into long-term illness careers, as patients traversed between institutions and at some points back to their families, a process mediated largely by the perceived 'dangerousness' of the patient.[30] The focus of this chapter, the period from the 1840s to the early 1870s, was in many ways the 'golden age' of note-keeping. Before the introduction of the printed form of case book in the early 1870s, case histories could be expansive and give detailed stories, particularly of patients' decline into madness. The case notes offer insights into medical treatment and the practices and details of asylum regimes and patients' responses to these. They also reveal the attitudes of the asylum medical staff towards their patients and their disorders, with this no longer implying 'the automatic identification of the physician's or the asylum's point of view with repression and social control'.[31]

Few studies have looked at case notes to illuminate a specific disorder.[32] It will be demonstrated here that the Edinburgh case histories – complemented by Tuke's articles on puerperal insanity – can greatly deepen our understanding of puerperal insanity and the ways in which it was experienced in the asylum by physicians and patients.[33] The case notes offer a means of exploring how the often diverse general conceptions about the cause and treatment of puerperal insanity, examined in chapter 2, were put into practice in the asylum, the workplace of most alienists. The focus in this chapter is assessments of the character, appearance and physical health of the patients, the link with childbearing, treatment regimes and perceptions of the 'dangerousness' of patients, expressed in delusions, immoral behaviour and violence. Even very difficult patients, who contravened every code of proper behaviour, could be treated with respect, emphasis being placed on their normally good characters, rather than their lapse into lewdness and destructiveness. But, equally, the case notes reveal terrifying spectres of Victorian women transformed by their illness.

Roy Porter opened the discussion on the problems of 'hearing the mad' and initiated research into the patient's view in psychiatry, demonstrating that, as with other forms of illness, 'it takes two to make a medical encounter' and possibly more, 'because medical events have frequently been complex social rituals involving family and community as well as sufferers and physicians'.[34] Porter also forcefully posited that case histories, along with other asylum records, psychiatric textbooks and court hearings, are frequently 'dialogues of the deaf; too often those who have been shut up have indeed been shut up, or at least nobody has attended to what they have said, except to put down their dislocated speech as a proof of derangement'.[35] Harriet Nowell-Smith has also suggested that nineteenth-century case notes could be less than objective, as their recorders tended to emphasise certain symptoms and take on a conventional style and content: 'nineteenth-century case histories transformed individuals' bodies into something statistically regular and understandable.'[36]

Case notes are of course mediated accounts, written by psychiatrists about their patients. They are in part intended to fit individuals into a pre-set and understandable diagnosis. Though we need to be sensitive to these tendencies, given the rarity of patients' accounts, it could be argued that we would poorly serve the majority of nineteenth-century psychiatric patients – for the most part silenced not just by their mental condition but also by their class and limits on their opportunities to communicate – if their case histories were ignored, for we have little else to work with. Moreover, case notes were not necessarily incommensurate with the patient's view. Occasionally, patients are allowed to speak through the case notes, to offer an explanation for their condition and confinement, though more often their family and friends spoke for them. Case histories, often detailed narratives of the onset of the illness, had to be based on something, and that something was the story put together from information in the admissions documents and medical certificates, which were largely reliant on details given by those closest to the patient – family, friends and neighbours.

Akihito Suzuki has argued that in Bethlem Hospital during the 1850s the authority of the family was challenged in identifying the cause of patients' mental collapse, as the doctor sought increasingly to hear directly from his patients.[37] The Edinburgh case notes show that doctors did not always feel 'bound' by what families told them, but they continued to rely a good deal on family and friends speaking for the patients, describing their deteriorating condition and how it affected those around them. After 1857, when more detailed medical certificates signed by two doctors became necessary for admission to Scottish asylums,[38] the evidence was furnished largely through the testimony of relatives, and when Thomas Clouston revised the case book rules in the 1870s he stipulated that information was to be 'got directly from a relative, if possible'.[39]

Though many institutions kept registers and case notes from the early nineteenth century, after 1857 Scottish asylums were compelled to maintain a patients' register, and the case notes also came to function as an administrative record, to display the smooth and careful running of the institution to the Commissioners in Lunacy. The case notes additionally provided some of the raw material for compiling annual reports, with extracts being taken from more interesting cases.

The initial case history, sometimes up to a page in length, taken usually on the day the patient was admitted, has an immediacy and directness about it, as the Medical Assistants responsible for compiling the histories attempted to make sense of the patient's condition, drawing extensively on the information contained in the medical certificates. The case notes are generally multi-authored documents; given the relatively rapid turnover of junior staff in Edinburgh, even patients staying only a few months in the asylum were likely to have had contact with a number of physicians. Skae made himself responsible for checking the entries and for reaching a diagnosis, which was entered at the head of each new case. It is a not unreasonable supposition that the young, recently qualified men writing the case notes may have had limited knowledge and expectations of the mental diseases they were recording, and for the most part their clinical training would take place in the asylum itself. They were there to build up experience, not to apply it, and their medical preconceptions may have been limited. It is also possible that earlier entries on individual patients were not referred to in detail as they were being added to, even by those responsible for regularly monitoring their progress. As the pressure of numbers became more severe, the Medical Assistants may have rapidly read previous notes, and wrote their own entries in haste, only sufficiently in many cases to come up with curt summaries such as 'no change' or 'much the same'. Meanwhile, those having most direct contact with the patients, the ward attendants responsible for much of the day-to-day care, remain largely silent in the case record, except when they report on particularly disruptive behaviour.

Patterns of recording, emphases on particular symptoms and recommendations for treatment are inevitably repeated and build an impression about what could be construed as 'typical' ways in which puerperal insanity was explained, understood and dealt with in the asylum. If we are also to consider the views of the medical staff, this kind of information is useful in developing our knowledge of medical ideas, which were then reflected in responses to patient care. However, because the Edinburgh case notes are based to a large extent on information derived from relatives and incorporate, particularly in the early years, a history which included a wealth of detail on personal circumstances and family life as well as medical evidence, they read too as vivid individual narratives. The words used to describe symptoms and behaviour also vary enormously; there was no fixed language of puerperal insanity. In many ways, manuscript case notes represent the 'coal face' of psychiatric experience, freed from the rhetoric and self-presentation of published works or even annual reports which were intended for public scrutiny. Such records give us an opening to 'explore the person-to-person interaction of the patient and the doctor'[40] at the level of micro-reality, and it is here that case notes differ from other psychiatric records. They represent the day-to-day business of trying to understand and cure mental disorder. They may have been constrained and shaped by preconceptions, but less so than many other records.

An older tradition, which pre-dates interest in the 'patient's view' and is perhaps best represented by the historian-psychiatrists Hunter and Macalpine, called for psychiatry to turn away from hearing the mentally ill, arguing that madness could not be understood by listening to what the mad said, but by 'a reorientation of psychiatry from listening to looking' and observing organic mental disease.[41] Case books offer us the opportunity to do this too, in effect not only to read and hear, but also to look. Though also advocating moral approaches to treatment, the Edinburgh Asylum physicians were somatists to a man, and their rich descriptions of how the patients looked as well as behaved and, in many cases, their appalling physical condition, provide powerful images of the patients' bodily state and appearance prior to and into the era of psychiatric photography. It was argued that puerperal insanity was a particularly visible and recognisable disorder, and this visibility is evocatively brought out in the case notes.[42] During the mid-nineteenth century, as asylums became truly 'medical' institutions – signalled in Edinburgh by the appointment of Dr William McKinnon as the first Physician Superintendent in 1839[43] – a great deal of information was contained in the notes on the patients' physical status, as well as the management of their psychiatric disorder, and the link with bodily health was deemed particularly important in the case of puerperal insanity.

The sheer bulk of manuscript case notes also makes them a valuable resource. All asylums were keeping case notes by the mid-nineteenth century, some more rigorously than others and some have survived better than others. Edinburgh has an almost complete record of case books from 1840 onwards,[44] which have been trawled through to identify around 120 cases connected with childbearing; other cases have been traced via the admissions registers[45] and leads in annual reports.[46] Already by the 1840s, the relatively high incidence of puerperal mania was attracting the attention of David Skae, when he drew up his annual reports. In 1842 just one case was admitted in 'the puerperal state'; in 1843, five puerperal cases were admitted; and in 1849, eight cases were recorded out of a total of 156 female patients (5 per cent), noted in the report as being more than in any other year.[47] The annual reports continued to record between five and ten cases a year, though in 1865 nineteen cases were listed out of 144 female admissions (13 per cent, broken down into six related to childbearing, one to pregnancy and twelve to over-lactation).[48] By the late 1860s insanity linked to prolonged breastfeeding constituted an ever-growing category, and would prompt serious attention from Thomas Clouston when he took over as Superintendent in 1873.

The Royal Edinburgh Asylum case notes have one further feature, which is a boon to the historian ploughing through hefty case books, with their varied legibility. At the top of most entries a diagnosis is clearly recorded: 'puerperal mania' or, less often, 'puerperal melancholia', 'insanity of pregnancy' or 'lactational insanity'. This was due to the rigorous adoption of Skae's system of classification while he was Medical Superintendent, with Skae himself entering the diagnosis.[49] Skae's classification scheme was vital in coding the case histories of the Edinburgh Asylum's patients. It was a simple enough scheme, which correlated mental disorder with any accompanying physical affection – phthisical insanity, syphilitic mania, delirium tremens and sexual mania. Age and life cycle changes also played an important part in the scheme – mania of pubescence, climacteric mania and senile mania.[50] Skae's scheme offered a classification that stressed 'the relationship the different varieties of the disease have to the great physiological periods of life, and to the activities of the body, other than the mental – in other words, it regards the whole *natural history* of the disease'.[51] Skae's classification dovetailed with a broader medical and psychiatric interest in nosology,[52] as well as with a remarkably tenacious and widespread approach to mental disorder, based on the notion that it was to some degree determined by bodily impairment, combined with moral components, stress and predisposition, which was, in turn, reflected in the dual – physical and moral – approach to treatment.[53] Puerperal insanity, with its close link with childbearing and the female life cycle, fitted Skae's schema well.[54] This aspect of his classificatory system also enjoyed a much wider vogue, with many authors of psychiatric textbooks elaborating on distinctions between insanity of pregnancy, puerperal insanity and lactational insanity. Skae's system in its entirety was adopted in only a few asylums,[55] but the classifications of puerperal mania and its associated disorders became fairly standard, largely because these conditions appeared to be relatively easy to identify. However, the highly classificatory approach to mental diseases adopted at Edinburgh during Skae's tenure does not seem to have dampened enthusiasm for developing individual life and medical histories in the case books.

The admissions registers occasionally missed cases of puerperal insanity, listing them as straightforward cases of mania, melancholia or dementia, the link with childbearing being established only in the case notes themselves. The junior medical staff drew up a detailed case history shortly after admission, but only at this point might it have become apparent that the woman had recently given birth or was breastfeeding. Skae's subsequent addition of a diagnosis gave time for information to be gathered, and occasional crossings out and substitutions of new diagnoses were made. A small number of women were admitted in a terrible state as emergency admissions, and few personal or medical details could be discovered. Others were recorded in the admissions registers as cases of puerperal insanity, but when one turns to the case book entry it is impossible to find a reason for this attribution; the women were manic, alcoholic, 'imbeciles' or suffering from general paralysis, but there is no discernible link with childbearing.

Elizabeth Mawn became a patient in May 1864, with varying diagnoses appearing in the case book, 'melancholia', 'mania of pregnancy' and (in another hand) 'phthisical insanity'. Her illness was related to the loss of children born in America, but she was also referred to as 'another Revival case'. She had delusions on religious subjects and had been attending meetings, where she was 'much affected by the insane discussion of the preachers'.[56] Yet it is the very catch-all nature of the attribution, its apparent overuse in some cases, that is so intriguing. Asylum superintendents, and not just in Edinburgh, seem to have been keen to find labels that fitted their patients, and puerperal insanity was one of the few apparently straightforward categories of mental disorder to slot them into.[57] Only those cases have been discarded here where no discernible link could be found with childbirth in the subsequent case history; others were retained in the sample even if the link was tenuous and where, despite Skae's resolute diagnosis, a wide variety of explanations, as well as childbirth, was given for the women's condition.

Puerperal insanity in the asylum

The case records vary greatly in length, depending on how long the patient remained in the asylum and the interest of the medical staff, as well as on the Assistant Physicians' 'conscientiousness, clinical insight and literary power'.[58] The case notes for the late 1840s and 1850s tend to be particularly rich in that they recorded detailed case histories at admission; they were often evocative tales of the women's decline into madness. Though there was some uniformity of content, there was also

space, in terms of a blank page for each new case, as well as the com-
piler's creativity, for a narrative to emerge.[59] Individual entries were
shaped by what the individual Assistant Physician chose to write down
and the information available from the patient, family and friends
and, after 1857, the medical certificates, which were sometimes copied
verbatim into the case book.[60] The Assistant Physicians were expected
to make daily entries for acute and interesting cases, but after this only
quarterly assessments were required, and these could be very brief.
The fullest notes were taken on admission, the case histories referring
to long-term causes of breakdown as well as to the women's recent
deliveries, and during the first days and weeks, when the patients were
most disturbed and treatment most active. Where insanity became
'entrenched', with the women remaining in the asylum for many
years, occasionally until their death, the case notes peter out into terse
entries, relating largely to their bodily health and behaviour. By the
time they had reached this point, such women seem to have been
absorbed into the asylum regime and their disorder blurred with other
forms of long-term insanity. After a few months or years in the asylum,
one gets the impression from the case notes that it mattered little what
the patients were suffering from; their status was measured in bodily
health, in obedience and tractability, the ability to perform simple
tasks, in quietness and cheerfulness or, for some, dementia or outbursts
of frustration and violence.

Each new case was entered on a fresh page and often filled a page or
more with close notes. Standard information was recorded at the head
of the entry: name, age, date of admission, followed by place of
residence, a note on education, occupation, religion, marital status and
family circumstances. The case notes then begin to tell a story, outlining details of the patient's character and general disposition as well as
of the onset of the current attack, describing her childbearing history,
the events of the recent birth, her social and financial position and
family circumstances. An assessment was made as to whether the
patient was dangerous or suicidal, and information given, if possible,
about previous periods of insanity or mental disorder in the family.
The notes thereafter recorded the progress the patient made in the
asylum, as they were moved from ward to ward, depending on their
responsiveness to treatment and compliance with asylum codes and
rules. They served as a record of treatment, noting down drugs and the
use of other therapies, recommendations for extra foodstuffs and
stimulants, and instances of restraint and force-feeding. They noted
the patient's appetite, her pattern of sleep or sleeplessness, ability to

employ herself, attitudes to the staff and other patients, disruptive tendencies and the nature of her delusions. The case notes could embody moral judgements, but also attitudes that we might not anticipate, including respect and a broad understanding and interpretation of the woman's condition.

Admission: character and appearance

The majority of women admitted with puerperal insanity and its related disorders were married to labourers, tradesmen and artisans, farmers and farm workers, hat binders, shoemakers, carters, publicans, gardeners, servants and policemen, and were described as housewives.[61] A small number, however, were employed as needlewomen, dressmakers, shop assistants or mill workers; one woman admitted to West House was a former teacher. Edinburgh was home to the largest number of domestic servants in Scotland, which was reflected in patient admissions;[62] many women, it was noted, had been in service prior to marriage and the majority of the unmarried women were servants. Other employment for single women included mill work, sewing and field labour, and one woman who had recently given birth to her second illegitimate child was a nurse in the asylum.

Though a handful of women were admitted as private patients to East House, and a few were paying patients in West House, the majority suffering from puerperal disorders were admitted as paupers and paid for by their parishes. Lactational insanity seemed to be almost exclusive to West House patients. Tuke pointed to the close link between lactational insanity and poor health, poverty and age, with women aged over 30 with several children being especially susceptible; Clouston would also refer to these factors several decades later.[63]

Religious affiliation was noted but usually without further comment, although religion would play a large part in reported delusions and occasionally religious excitement was said to have contributed to the onset of the disorder. Most of the women were Presbyterian or Church of Scotland, far fewer were Roman Catholic. Levels of education included 'modest', 'ordinary', 'can read and write' and 'of limited education'. Most patients came directly from their homes, but a small number from the poorhouse or maternity hospital, while a few were picked up on the streets by the police after causing a disturbance.

The case notes, provided the information was available – and in a few cases very little was known about the patient – combined the standardised details on marital status, occupation and so on, with a

brief assessment of the patient's 'natural temperament' and character. Families were important in providing this kind of information,
as well as details of the woman's childbearing history, whether this
was her first attack of illness, accounts of any history of mental disorder in the family, of any delusional tendencies or dangers that
she might present to herself or others, and of any treatment that
she had received prior to admission. Early admission to the asylum
was seen as a huge advantage in cases of puerperal insanity, and it
was important for the medical staff to know how long the patient
had been ill.

Elizabeth Mackie had been ill for only a week when she was admitted in April 1864 with 'mania of lactation'. She had delusions of
hearing, with people speaking loudly to her, 'wants the cats taken off
her head', and 'Mrs S. her aunt states that she has been talking much
of Calcraft and of being on the scaffold and that she is always afraid of
some persons going to hunt her, and cries for the Prince of Wales to
take her child'.[64] The illness of Jane Duncanson, an East House patient,
was attributed to her catching cold and developing an abscess on her
breast a month after giving birth to her third child: 'She could not in
consequence nurse her child and became melancholy and restless and
took an idea that she had murdered her child.' Her husband had tried a
change of air and took her on a tour of the Highlands, but she became
worse and stopped eating, and was in consequence admitted to the
Asylum.[65] In the case of Margaret Harper, admitted in 1861 with puerperal mania, it was reported:

> She has a sister who took puerperal mania after birth of her first
> child & she frequently exhibited a dread of the same disease. On the
> 25 April she had her first child and progressed favourably until six
> days ago when she began to be very silly in her manner, by constant
> fondling & talking to it her affection for it [sic], and did not sleep
> well at night. She soon became maniacal being incoherent in her
> conduct. She had got sedatives of all kinds in large doses without
> producing much affect.[66]

Sometimes the family provided very significant evidence, as in the
case of Isabella Hay who, in April 1855, was brought to the Edinburgh
Asylum from the Parish Charity Workhouse where she had been for
two months. 'Three weeks after childbearing she was exposed to cold
and became insane shortly afterwards.'[67] Six days after her admission
Mrs Hay's sister explained 'more fully the history of her case',

attributing her illness to the cruelty of her drunken husband. In her increasing desperation as Mrs Hay felt her mental state deteriorate – something many patients suffering from puerperal insanity seem to have been aware of – she had charged her mother to take care of her children. Her sister also reported that she had been forced to stop feeding the child herself after a week 'for want of milk'. She had been unable to nurse her other three children for longer than six weeks from the same cause. She was indifferent towards the infant, but never tried to injure it.[68] Such reporting also revealed disputes and disagreements about the cause of the patient's condition; families were by no means united in their views, and it is likely that Mrs Hay's husband had a different opinion about the cause of her illness. She was also, the case book noted, 'short-tempered and somewhat intemperate'. Isabella Hay was discharged 'recovered' in August, 'civil and industrious and very desirous of seeing her children'.[69]

The case history also delved, albeit often briefly, into the disposition and character of the patient. The purpose of this appears to have been to position the patient's attack of insanity into the context of her general mental state and outlook, whether she was naturally of 'a nervous disposition', dull, sanguine, melancholic or cheerful. Again, this assessment was based to a large extent on the testimony of family and friends, but also embodied judgements on the ideal features of a woman of the poorer classes: quiet, frank, temperate, active, industrious and able to carry out her household duties. The medical officers did not refrain from making critical judgements in some cases, but they could also be positive and impressed with their patients, praising them for raising themselves above their station or for their efforts to provide for their families.

Jean Main, admitted in December 1852, a former servant, aged 38, married and with three children born within four years was 'Naturally rather dull and melancholy in disposition which is increased by an impediment in her hearing', but was described too as being of average intelligence for a person of her rank, and quiet, sober and industrious in her habits. This contrasted starkly with her noisy and disruptive deportment on admission.[70] Mrs Eliza Paterson or Lumsden, admitted in July 1855, was said to be of good education, cheerful, frank, quiet, sober and of 'a very active and industrious turn of mind'; she had almost single-handedly managed the books and business of a society with about 1,400 members. Even more, although she had never been strong, she had nursed her children and 'her persevering in nursing the last child which is ten months old is supposed to be the cause of her mental affection'.[71] The restoration of positive attributes was also the desired goal of the asylum stay. Mrs Lumsden's case proved frustrating; she attempted suicide repeatedly and escaped from the asylum in August. She was plied with stimulants, fed a nutritious diet and had frequent shower baths, but in November was reported to be 'dull, listless and idle', a sharp contrast to her former industriousness. She was removed by her friends, 'contrary to advice', later that month.[72]

Overexertion in performing household and maternal duties, particularly breastfeeding, was commonly noted in the case books as a cause of breakdown; it was reported that the women were trying too hard to keep their homes in order at the cost of their own health. But neglect of these duties or an inability to perform them also denoted madness. As we have seen in chapter 3, the disruption inflicted on bourgeois households as a result of puerperal insanity was a cause of great concern for both family members and doctors. This anxiety about the breakdown of the household was also mirrored in the records of poorer patients, which referred to women's instability and difficulties in continuing to perform their duties when physically and mentally challenged by childbearing. The women themselves also fretted about their inability to look after their homes and families, and these worries were conflated with guilty obsessions or delusions.

Agnes Hastie, the wife of a Fifeshire farmer, admitted in 1865 with 'mania lactation', was naturally 'frank, steady and industrious' and enjoyed robust health. The death of one of her children was said to have triggered the attack. She refused to eat and 'fancied' that the household was starving and that she was responsible for the utter poverty of the family.[73] Mrs Jessie Eisdale, an East House patient, believed she was guilty of a great crime, had ruined her husband and lost her soul, and was 'much agitated' about her worthlessness and uselessness as a wife.[74] The ability to perform housekeeping functions was seen by the asylum doctors as one of the signs of the restoration of sanity and closely monitored as women were urged to sew or work in the laundry or on the wards.

The appearance of the patient was recorded, in detail in some cases, with attention being paid to hair and eye colour, skin tone, build and facial features, as well as medical symptoms such as furred tongue, staring eyes, faint pulse and general debility.[75] Appearance, character and physical state were assessed together, intermingled in the case notes, and used as a starting point to assess the success of treatment and the improvement or deterioration in the mental health of the patient. This was often linked to the florid behaviour that characterised this disorder, with the women described as excited, loud, disruptive, bad-tempered, dirty, foul-mouthed and overtly sexual, but reference was made to their usual character and deportment too.

Janet Smith was described as sleepless, restless, inclined to use bad language and 'noisy & obstinate & destructive of clothing' on admission in April 1853.[76] After six weeks in the asylum the situation was much worse: 'she was exceedingly mischievous, very destructive, and dirty, she used the most profane and obscene language ... abused everyone who came near her, spat in their faces, exposed herself when possible, passed her motions amongst the bed clothes and did every sort of mischief.' Yet she was naturally 'of a cheerful and frank disposition and of temperate, active and quiet habits'.[77] The normal deportment of Jane Gardener, a 26-year-old admitted in July 1853 suffering from puerperal mania, was described in generally positive terms, even though she was unmarried with three children, and caused chaos in the asylum with her rage and excitement after admission: 'In disposition naturally cheerful and frank, rather intelligent, quick in her temper, sober and active in her habits, of the nervo-sanguine temperament, hair reddish, stout and well-formed.'[78]

Not so for Jane Stirling, admitted in January 1865, the supposed cause of her insanity the '<u>Birth of an illegitimate child</u>'. She was described as dangerous and suicidal; her 'ill-shaped' head hinted at a congenital disorder. She was also dressed in 'flashy clothes', presenting the appearance of a 'professional', was dirty and quarrelsome, very violent and 'the ingenuity of her obscenity was remarkable'. She was removed after two months to Perth District Asylum showing no signs of improvement.[79]

Clearly discernible in the case notes are the influences of physiognomy and phrenology and the theory of temperaments, which claimed that the features of insanity were cast on and could be read on patients' faces.[80] Written descriptions of features and appearance, combined with details of character and bodily state, created vivid impressions of the patients, and, while psychiatric photography would be perhaps more immediately compelling, most asylum doctors, in Edinburgh and elsewhere, continued to rely on case histories to represent their patients.[81] The descriptions taken on admission were also used as a yardstick to measure change in features and expression. Though puerperal insanity was considered to be a temporary condition, certain features were watched for and regarded as typical signs. As seen in chapter 2, Alexander Morison had declared puerperal mania to be one of the most readable of mental disorders, commenting on the victims' exaggerated and changing physiognomy as well as the appearance of exhaustion.[82] Tuke asserted that simply by looking at a woman it was possible for a physician to recognise puerperal mania, citing a case brought by the police to the asylum under a certificate of emergency. There was no information on the woman except that she had been found in a maniacal condition, but the doctors who admitted her 'immediately came to the conclusion that she was suffering from puerperal mania' and offered the following description:

> The bodily symptoms are at direct variance with the mental. She is pale, cold, often clammy, with a quick, small irritable pulse, features pinched, generally weak in the extreme, at times almost collapsed-looking. But withal she is blatantly noisy, incoherent in word and gesture; she seems to have hallucinations of vision, staring wildly at imaginary objects, seizes on any word spoken by those near her which suggests for a moment a new volume of words, catches at anything or anyone about her, picks at the bed-clothes, curses and swears, will not lie in bed, starts up constantly as if vaguely anxious to wander away, and over all there is a characteristic obscenity and lasciviousness.[83]

Tuke argued that such cases were unmistakable, just as a return to the normal set of the features and acceptable behaviour were signs of cure.

The case of Mrs Jane Ferguson, a former teacher, admitted in September 1863, was cited extensively by Tuke:

> Ten weeks ago she was confined naturally of a healthy child which died a fortnight after birth of infantile cholera. She was, according to her mother's account, attacked with acute puerperal mania immediately after the childs [sic] birth, she was very violent & attempted the life of her infant. A week after she attempted to jump out of the window; she continued violent until about 5 weeks since when she settled into melancholy. She suffered from abscess of right mamma ... On admission she was obstinately taciturn, and attempted to swallow a brooch on the first night. She walks about, & when sitting cannot stay quiet for a moment, jumping about & fidgeting.[84]

Mrs Ferguson had delusions concerning her identity, was 'quite sleepless', and an attendant was ordered to be constantly with her. Tonics and a nourishing diet were prescribed and her breast poulticed. After a few days she began to sleep well and the restlessness decreased; she took to singing and playing the piano.[85] The case history also vividly highlighted the patient's changing expressions, bound closely to her behaviour and responsiveness to treatment:

> Her manner was characterized by a considerable degree of amativeness; when moving about she grimaced and conducted herself in a playfully ludicrous fashion. A marked change came over her for the better shortly after the sinus of her breast had been freely laid open, and the wound commenced to heal. The expression of her face, which had previously been rather repulsive, became pleasant and agreeable, she gradually gave up her restless habits altogether, and was at last induced to sew. When the breast had quite healed, the old symptoms disappeared, a little waywardness and intolerance of control excepted, and she soon became the life of the gallery and a great favourite. She subsequently had one or two slight melancholy fits, but within six months of her admission she was discharged perfectly recovered.[86]

Tuke commented that an hereditary tendency was 'proved' in almost a third of patients admitted with puerperal insanity, which could usually be traced to the female side of the family.[87] However, at least until the 1860s, links with heredity are few in the case notes, confined largely to comments such as 'mother was a very nervous woman' or '[a] sister died insane in the Asylum two years ago'.[88] They are interesting notes, but little more.[89] Hereditary disposition and instances of insanity in the family were stressed more strongly by the 1860s; at the same time, ideas on physical appearance hardened, though few patients excited the vehemence of Jane Stirling, with her 'ill-shaped head' reflecting her intrinsically, irredeemably bad character.

Childbearing

Puerperal mania was directly related to the act of giving birth, yet the detail provided on the woman's obstetric history, and even recent delivery, varied greatly. Childbearing was seen in some instances as a precipitating factor only, with the woman's mental state related to other physical challenges and stresses. Puerperal insanity occurred in normal and 'easy' births, as with Mrs Ferguson, as well as tedious, painful or instrumental cases, multiple births and stillbirths. Most admissions linked to childbearing were characterised by mania; Tuke reported on 53 cases of mania and 15 of melancholia occurring between 1846 and 1864. In addition he counted 28 cases of insanity of pregnancy and 39 of lactational insanity.[90] He also concluded that puerperal insanity occurred in nine cases where the labour had been instrumental, four times following a tedious birth, in six where profuse haemorrhage had followed labour, in two cases after the delivery of twins, and two where the child had been stillborn. Almost a third of cases (23 out of 73) followed a complicated labour, indicating 'the tendency of unnatural parturition to produce insanity', either through strain or physical shock, such as blood loss, or through the emotional shock of losing a child.[91]

Mary Sibbald was admitted in 1855 with puerperal mania, nine years after a similar attack. She had borne several children in the meantime without any 'return of malady'. This time, however, she had suffered a massive postpartum haemorrhage; she had no milk and was unable to nurse the child.[92] Mrs Sarah Andrew, married to the asylum's janitor, 'became insane for the first time about a month ago, three weeks after being confined of twins; the labour was complex and instrumental'.[93] Margaret Blackie's attack of puerperal mania was implicitly related to anxiety; she became ill five days after giving birth to a living infant following a history of miscarriages.[94]

The majority of women became disturbed after normal deliveries, itself a disturbing fact for nineteenth-century alienists, but the proportion affected following difficult labours is striking, particularly in the case of forceps deliveries. These seem to have been carried out on a very limited scale in Edinburgh until the late nineteenth century but they accounted for 12 per cent of all cases of puerperal insanity.[95]

Puerperal insanity afflicted first-time mothers as well as women who had given birth several times, but most cases of puerperal insanity supervened following first or second confinements. According to Tuke's figures, 46 per cent of cases of puerperal insanity occurred among primiparae, 21 per cent had borne two children; smaller numbers of women were affected after their third and subsequent deliveries, but one woman developed puerperal mania following her ninth confinement.[96] However, nine out of 28 cases of insanity of pregnancy occurred during first pregnancies, which Tuke attributed to 'the moral exciting causes, anxiety and dread of the coming event, which exist to a greater degree in the inexperienced woman'.[97] For some women their mental disorder was a one-time occurrence; others became mentally disturbed during each pregnancy or after every delivery and were admitted again and again to the Edinburgh Asylum.

The obstetric histories of some of these women read as veritable horror stories of repeated attacks of mental disorder and anguish for themselves and their families. Elizabeth Winks or Love, a mill worker from North Leith, developed mania attributed to over-nursing in 1859 at the age of 37. She was very disturbed, threatening suicide and to choke anyone who came near her and had attempted to strangle her children. She was discharged recovered but fell ill again two years later after the birth of another child. Removed by her husband in June 1861, she had to be readmitted shortly after and was not discharged 'relieved' until November 1862, correct and rational, though still 'queer'.[98]

The link with fright or a sudden shock was recognised as potentially devastating. This was a factor that the earliest commentators, including Gooch, had stressed. Tuke referred to the dangers of cold and chills and the taking of spirits after confinement.[99] In one instance mania was ascribed to 'a great shock when she was 3 months gone in pregnancy [with her fifth child] by an attempt at Burglary being made on her house when her husband was absent'. The woman's confinement was also protracted and she had heavy postpartum bleeding. However, she continued to do well until 'On the 7th day ... when sitting up in bed taking tea, a few friends being present, the discharges suddenly ceased & shortly after she became maniacal'.[100]

Women who were doing well after delivery became deranged following a seemingly innocent occurrence. In the case of Jean Main, 'Parturition had been safely gone through, & she was nursing the child & doing well, when the occurrence of a thunderstorm frightened her so much that she has never been well since'.[101] Another woman admitted with lactational insanity became insane following 'A fright she got by a man in liquor knocking furiously at her door after she had gone to bed'.[102]

In Edinburgh one feature of childbirth had particular resonance, the use of chloroform to ease delivery.[103] Pioneered by Dr James Young Simpson in 1847, his colleagues at the Asylum were eager to rebuff accusations circulating in the medical press that the administration of chloroform during delivery could lead to insanity. Simpson himself had more than a passing interest in puerperal mania, and subsequently published on its aetiology and treatment, citing cases occurring in the Edinburgh Maternity Hospital; he also recommended the use of chloroform as a preventative.[104] David Skae asserted in his Annual Report for 1855 that until that year only one case of puerperal mania out of the 50 admitted to the asylum was associated with the administration of chloroform.[105] Skae concluded from this that there

was no relationship between puerperal mania and the use of chloroform, 'otherwise in Edinburgh, where it is so freely and extensively
used, many cases of Puerperal Mania would have been brought to the
Asylum in which this agent had been given'.[106] Tuke cited two cases
involving chloroform in his statistical analysis of patients, 'a number
so small as to give the strongest denial to any absurd theory regarding
the danger of its exhibition'.[107] One patient treated in the asylum in
1852, however, expressed a different view. Mary Bird was admitted in
August with her first attack of puerperal mania, which came on 14 days
after delivery. She ascribed 'all her sufferings' to chloroform, which
had been administered ten minutes before her tedious delivery was
'terminated with the assistance of forceps'.[108]

Treatment

Once an initial assessment of the patient had been made, treatment
commenced immediately, taking the form more often than not of an
emergency response that addressed the woman's emaciated status and
general debility. The poor bodily health of the patients attracted a
great deal of comment and attention, further challenged as it was by
the strains of childbearing and breastfeeding. The case books described
a range of symptoms: dry skin, poor colour, furred tongue, quick and
rapid pulse, 'feverishness' and 'low health', exacerbated in many
instances by extreme excitability.

Many women's stay in the asylum commenced in the sick room,
where they were given beef tea, custard, eggs, jelly, milk and wine to
build them up, or in Tuke's words, to enlist 'the support of nature to
withstand the wear and tear of the disease'.[109] Mary Sibbald,
admitted in 1855, was very incoherent, violent and thin, having
experienced heavy bleeding after labour. She had no milk and so was
unable to feed her child. Though she was disruptive and turned her
room, including the bedstead, 'upside down', she was also described
as exhausted, pale, weak, her pulse was feeble, mouth parched, teeth
filthy, dull eyed, and showing 'symptoms of sinking from condition'. An abscess on her left breast was poulticed, and she was given
brandy and morphine, force-fed custard and sherry, first by the nose
and then using a stomach pump. After three days she started to take
her food voluntarily.[110]

Many women were refusing to eat when admitted – although often
there may have been little food to be had at home – and it reflected
the sense of chaos about their abandonment of normal functions
around the home and about their own bodies, including eating,

sleeping and evacuating their bowels. Diet, rest, exercise, encouragement to work and regularity in all these activities constituted the cornerstones of treatment. Regular bowel movements were vital to recovery, as was the restoration of regular menses; the latter indicated normality, fertility and improved physical well-being. Force-feeding seems to be a direct contravention of moral therapy, but it was also a response to the semi-starved state of some of the women, though better-off patients were not exempt. In December 1868 Mrs Jane Duncanson, an East House 'gentlewoman', was force-fed with custard following admission. She was said to be 'very delicate looking and somewhat emaciated', thin and anaemic. She remained 'obstinate' about food months later.[111]

If the woman was recently delivered, a great deal of attention was paid to her breasts and lochia. Many women were still breastfeeding until admission when they were separated from the infant, and it was often necessary to draw the milk from their breasts, which were poulticed and bathed. Laxatives were generally prescribed, as regular bowel movements were considered vital to recovery. Margaret Harper was dosed with large quantities of castor oil and given turpentine enemas when she was admitted in May 1861; her friends had caused considerable alarm among the medical staff when they reported that she had not had a motion for five days. Her urine was drawn off by catheter and it was commented that she had taken no food for several days, 'only a little wine to support her'. She was given wine, beef tea and custard and blistered several times.[112]

Despite an overall emphasis on mild therapies, some patients received heroic doses of medicine, or were subjected to counter-irritants, particularly blisters, or bathing of various sorts, though patients were rarely bled. It is not always clear what the rationale behind heavy dosing and intervention was, but it seemed to be used most often in violent or excited patients, or where there was little response to milder therapies. More drastic remedies were attempted when other approaches had failed. Stimulants were used, although concern was expressed that they could aggravate mania; wine and whisky were found beneficial for melancholic cases. Opiates were given sparingly to induce sleep. Sleeplessness was seen not only as a symptom of the condition, but also, more importantly, as a barrier to cure, making the patient more and more exhausted. Their use, however, was limited and opiates quickly withdrawn if they failed to be effective or made the patient's condition worse. Sedatives had a mixed reception, but, particularly after the introduction of chloral hydrate in the late 1860s, were prescribed on a fairly routine basis, in line with many other asylums; it was praised as an excellent calmative in the Annual Report of 1869.[113] Tuke, however, was not a convert, believing that sedatives offered only a short-term solution, a dampening effect, but retarded recovery overall.

Women subjected to a barrage of treatments stand out in the case notes even if they are not necessarily typical. Mary Bird, a former lady's maid, was admitted with puerperal mania in August 1852, following a forceps delivery. As mentioned, she had blamed her sufferings on the use of chloroform at the delivery and these sufferings were manifold. While still at home her right breast had become inflamed and suppurated, and had been incised. She was anxious, sleepless, feverish and refused to eat. The case entry wrote of an improvement after she was dosed with tonics and opiates, but a visit from friends further upset her and it was determined to move her to the asylum. After admission she continued to be noisy and disruptive, and was treated with tonics, stimulants, nutritious food, laxatives and counter-irritation at the back of her neck and head; her breasts were poulticed and she was placed in seclusion.[114]

In some cases a stepping up of treatment seemed to have been a response to disruptive behaviour. Elizabeth Robertson proved to be particularly unmanageable when she was admitted with 'acute mania', ascribed to the birth of a child ten months previously, which she was still nursing. She was said to be dissipated, idle and intemperate, and mania had been preceded by an attack of delirium tremens. After admission she was destructive, mischievous and violent, had broken windows and tables, tore her dress, bedclothes and bed, struck the attendants and escaped from them. She was secluded and purged frequently, bathed, her head was shaved and blistered, she was rubbed with tarter emetic ointment, given shower baths and anodynes. At first she was given extra food and stimulants because she was so thin, but, because of her continued excitement, she was subsequently placed on a low diet.[115]

Agnes Watson, admitted in May 1865 with 'mania lactation', was disruptive, restless, delusional and suicidal. Like Elizabeth Robertson she was said to be intemperate, and like her was a single mother. She was subjected to heavy dosing to subdue her worst symptoms, fed with the stomach pump after admission, and morphine was administered to counteract her restlessness. This failed to help, and resulted in profuse diarrhoea and vomiting. However, she was still being fed with the stomach pump five weeks after admission, when the case entry also related that she was very dirty in her habits. In June cold baths were introduced and her ears were blistered, which she claimed relieved her headaches. The force-feeding stopped and from then on Agnes Watson continued to improve until her discharge in September.[116]

Tuke, as we have seen in chapter 2, objected strongly to some of the treatment methods employed in cases of puerperal insanity, which he must have observed at first hand when he served as Medical Assistant to the Asylum between 1864 and 1865. Puerperal insanity needed little medical intervention in his view, and rather more in the way of rest and food.[117] A case where a patient was dosed for many weeks with cannabis aroused particular criticism. The woman, whom Tuke referred to as B.C., was exhausted, but also very excited and delusional, claiming she had 'brought forth dogs instead of children'. Initially, she improved, but Tuke attributed this not to cannabis, but to exhaustion, combined with the benefits of brandy, custard, beef tea and nursing. He also blamed cannabis for the woman's subsequent relapse into excitement.[118]

A few patients, noted to be 'hypochondriacs' and obsessed with their bodily troubles, demanded heavy dosing. Jean Main was very excited when admitted and 'seemed to have every disease flesh is heir to'. She was said to crave medicine and had been satisfying that craving before she came to the asylum, although the case notes also reported that she was exhausted, pale and suffering from haemorrhoids.[119] After admission she continued to complain, the variety of her diseases was 'immense' and she asked for medicine 'to remove fireballs from her throat, to open her bowels, etc.'.[120]

Many women were treated for physical complaints and complications of childbirth, some severe; phthisis, or tuberculosis, was a major cause of death in the asylum. Tuke commented that women suffering from puerperal mania might not notice their physical diseases, and that the symptoms were often subdued and 'latent', even more so than for other cases of lunacy. He encountered sudden deaths from bronchitis, peritonitis and pelvic cellulitis.[121] Other cases were recorded as dying from 'exhaustion' shortly after admission; Tuke reported seven cases, two of whom died after two days in the asylum, five after three days.[122]

Many of the women spent a great deal of time in the sick room. Ann McDonald was admitted in March 1852 with 'dementia supervening on Puerperal Mania'. The insanity developed after a miscarriage two years previously, and she had spent 18 months in another asylum before being brought to Morningside. She was incoherent, though quiet and easily managed, moderately industrious, tidy and particular about herself. Her bodily condition, however, was appalling; she suffered from diseased teeth – several stumps were extracted – and was susceptible to throat infections and gastric disorders. She was allowed to rest in bed, put on a good diet and given tonics, but her condition worsened. In October she developed a cough and bronchitis and had an attack of severe influenza. By the following February her mental state had deteriorated; she was confused and had developed 'exalted ideas of herself', decorating herself with ribbons and conducting herself as a lady of rank. By May she had a nasty cough and was vomiting, and by July there was little doubt that she was suffering from phthisis. She was emaciated, had diarrhoea and vomited blood. In October she had recovered some strength and was able to move around the sick room, but had declined again by December. In March 1854 she was 'rapidly sinking' and, though rallying unexpectedly several times, Anne passed away on 2 August, two and a half years after admission.[123] Her mental state received little attention as her illness worsened; she was simply a very sick woman who appeared to have been well treated, with efforts continuously made to halt the progress of the disease even as she deteriorated.

Mrs B.B. experienced a natural delivery with only a slight haemorrhage before mania ensued nine days after her confinement. She was described as exhausted and very anaemic, having borne four children in three years, was suffering from acute bronchitis, and her pulse was extremely feeble. She was fed with jelly, milk and Liebig's extract.[124] Dr James Matthews Duncan, one of Edinburgh's foremost obstetric physicians, was consulted, who prescribed wine, brandy and an expectorant mixture, and mustard was applied to her chest. Over the next few days, however, she became weaker and weaker – no further mention was made of her mental state – and she died on 25 April, just ten days after admission.[125] Mary Sibbald, whose initial treatment has been described above, was admitted with puerperal mania in 1855 at the age of 43, and died in the asylum three years later, falling ill only a few days before her death. Her mental state had fluctuated greatly, and when she began to refuse food, she was threatened with the stomach pump. An entry in the case book blamed the medical officers' failure to diagnose her condition on her demented state and refusal to speak, and she died suddenly of what the post-mortem examination showed to be gangrenous peritonitis.[126]

Alongside somatic therapies, moral management was a crucial aspect of treatment, which focused on the encouragement of regularity. First and foremost Skae presented the asylum as a domestic environment; emphasis was placed on work, taking meals properly and good conduct towards the attendants and other patients. Tuke claimed good rates of cure were achieved under moral treatment, due to 'the assurance of protection, the regularity, amusements, and employment alone to be found in an asylum, – above all, the freedom from domestic anxiety and the misapplied sympathy of relatives'.[127] This included, except for the handful of women who gave birth in the asylum, separation from the newborn. Even the few babies delivered in the asylum were usually parted from their mothers, who were considered unfit to care for them or were an actual danger to them. The infant is seldom mentioned in case notes, although older children were allowed to visit their mothers in Morningside; the expression of a desire to see their children was an important sign of recovery.

Employment for women consisted for the most part of sewing and laundry work; the handful of better-off patients were also urged to do fine needlework, play the piano and read. Patients were moved round the asylum in response to their behaviour, in what operated as a system of rewards and punishments. The aim, however, was to keep isolation to a minimum, and the attendants and medical officers seem to have put up with a fair number of breakages, threats and physical attacks before patients were removed to a separate bay.

Efforts were made to keep Elizabeth Robertson, an unmarried field worker, with the other patients despite her enormously disruptive behaviour. She promised to behave herself, but

> broke her faith and everything she could, before she was over-powered ... She cries much, and laments her fate at times, and in a moment perhaps changes into a burst of maniacal fury, exercising for her size almost superhuman strength, and exhibiting the agility and flexibility of a wild beast, then she sinks for a time into a state of dementia, plays with straws and pays no attention to any person or thing ... [128]

After a period of isolation lasting several weeks, which seemed to do her some good, in combination with walks in the airing court, she was brought back to the gallery 'where it is hoped that she may by associating with the convalescents acquire habits of industry & of self-control'. This worked well for a few days, but 'the bustle and routine of the galery [*sic*] has been found too much for her, and as her excitement was threatening to return she was sent back to the Sp By: much against her will'.[129] Agnes Bennett, who was admitted in April 1862 suffering from puerperal mania, was described as moody, variable and violent, and was moved several times from the second gallery to the separate bay after alarming the other patients, on one occasion 'threatening to throw the chamberpot at the other inmates of dormitory'.[130]

Professor Simpson ordered chloroform to be given in order to remove Mrs Margaret Louisa Maitland or Moir, pregnant and suffering from acute mania, to East House in October 1851, as she was so excitable; it was noted that she had already been dosing herself with chloroform to induce sleep. Once at Morningside she remained irritable and excited, stripping off her clothes. 'She has beat herself in the abdomen – pulled her breasts till the milk came[,] her hair is dishevelled[,] her dress thrown off her shoulders.'[131] She improved somewhat and, on 20 February 1852, she delivered a baby after a tedious, but natural labour. The baby was kept with her 'in the hope that maternal feelings might be developed', but soon this was declared a failure and the infant was removed.[132] She remained excited into the summer and after a visit from her husband: 'Her language to the ladies in the same gallery was so abusive that one of them, convalescent, begged to be removed to another'; she 'struck another with a brush'. In July, '[i]t being thought desirable that she should be completely separated from the sphere of domestic ties & the visits of her relatives she was this day transferred to the Chrichton Institution in Dumfries', thus also relieving her family of the burden of her proximity and abandoning the policy which encouraged family visits.[133]

Delusions, morals and danger

The case books were expansive on the topic of patients' delusions, an aspect of the disorder which fascinated the medical staff. By early in the nineteenth century, a distinction was being made between hallucinations of hearing and seeing, and delusions, which were termed 'wrong beliefs'.[134] Though some women heard voices telling them to harm themselves or their infant, many appeared to be suffering from delusions. Suffering is an apt word here as they were deeply distressed by these wrong beliefs.[135] Like Gooch's patients in the 1820s, the women claimed that they were guilty of terrible crimes or had brought ruin or harm to their families; some confused the asylum with a prison. Many had delusions of mistaken identity, confusing other patients with their friends and relatives. Others were convinced that they had dreadful illnesses, like the 'hypochondriacs' referred to above, and that their death was imminent. Single women described how they were about to be married, several claiming to the Prince of Wales, and others, married and single alike, were convinced that the asylum doctors were plotting their seduction. Delusions with a religious theme were also common, again relating to the idea that they had committed a heinous crime and would be punished by God, or that the devil was persecuting them, although many women were noted to be 'overly religious' or 'wished to go to heaven'. The language of religion provided women, often with limited education, with a vivid means of expressing themselves and their struggles.

Isabella James, a domestic servant from Leith, admitted in February 1861, 'imagined God had especially favoured her, was rambling & incoherent' and 'on admission she insisted on standing in the middle of the floor of the room and singing to a monotonous tune verses of scripture etc.'. It was also remarked in the case book that she had lately been in 'poor circumstances' and 'during the confinement was ill-cared for'.[136] Her erratic behaviour was short-lived and following a good night's sleep she woke composed, related the story of her attack and appeared 'quite sane'. Though much better, she was kept in the asylum until April; her husband committed suicide during this time, but this was noted as having little effect on her as far as her mental state was concerned.[137]

Margaret Scott or Steele was admitted to Morningside in a dreadful state in July 1855:

> The present attack which is the first is connected with childbearing and dates from her last confinement which took place about 12 weeks since. She is getting worse & is restless & noisy both by day & night. Says the devil has got her children & that her soul is lost. Is not suicidal or dangerous to others. On admission she was emaciated, pulse feeble, tongue foul. Said she could not take food as she was full up to the neck, & walked up & down the corridor wringing her hands & crying oh dear! oh dear! She will not answer questions.[138]

After a few days refusing to eat Mrs Scott was force-fed beef tea. She remained very miserable, and '[w]hen asked about her children she cried oh you have killed my bonnie bairns and burst into tears'. She was treated with morphine and started to improve, taking food by herself, but was the 'victim of most unhappy delusions she fancies that the meat she eats is composed of the bodies of her murdered children'.[139] When the next entry was made six months later Margaret Scott had improved, employed herself sewing and was in better bodily health. Yet a year later she was still in the asylum, and a visit from her children in March 1857 ended badly with her declaring that they were murdered. Finally, in December of the same year, some two and a half years after her admission, she was discharged recovered, having seen her children without any adverse effects.[140]

Anxieties about the enmity and plotting of family or friends were common; one woman on admission 'said she was now going to be dissected, thought her neighbours were conspiring against her, and that her husband was too intimate with the wife of one of them'.[141] In January 1861, Jessie Jameson was admitted to Morningside, suffering from puerperal melancholia, 'wild and restless, thinking her relations were trying to poison her and one of her children'. She expressed antipathy towards her husband and other relatives, and claimed that she had been given strychnine that had 'made her lose the power of one foot and hand, but this is not really lost, as she can walk a little and squeeze one's hand pretty hard. Professor Simpson gave her medicine wh[ich] counteracted its effects or she would have been dead. Not only did her husband poison the food but air also'.[142] In turn, she claimed that there was poison in her shoes, that her food was being poisoned, that the medicine itself was poison, and 'manifests a great inkling to annoy and irritate the other patients'. She developed a 'paralytic affection' of her right side that made her limp and prevented her doing much. In June she was 'restless and very loquacious and very irritating to fellow patients, who she seems to take a pleasure in annoying. She has some extravagant delusions now that she is Duchess of York and that the medical officer is a bad character of her acquaintance.'[143] In September she admitted her ideas about poisoning were delusions. By March 1862, some 14 months after admission, Mrs Jameson was described as for the most part 'pleasant and industrious', she had regained much of the use of her paralysed side, though she was occasionally querulous, irritating and talkative, silly, vain and slightly demented. It was not until September that she was sent home 'relieved'.[144]

Although it is difficult to gauge the reactions of the medical officers to their patients' delusions, the women were not to be silenced, restraint was rarely used and only if they proved particularly disruptive were they moved to another ward. The delusions ran their course, their interesting features noted down in the case books, and occasionally the patients were questioned about them. The delusions were often tenuously linked to the woman's social and family circumstances, her poverty, a traumatic recent confinement, shock or bad news, or ill treatment. Mrs Jameson, it was noted, 'seems very glad to be taken here out of reach of her husband's machinations' and 'seems to have been treated badly by her husband',[145] while Isabella James's state was related to her poverty and a mismanaged confinement.[146]

The ability to answer questions in a 'sensible' way was taken as a sign of improvement, and the patients, in this respect, were certainly listened to. Silence on the part of patients appears to have troubled the medical men just as much if not more than noisy, outrageous behaviour, just as the self-absorption of melancholia was in many ways more worrying than mania. Tuke suggested that the patients be invited to speak and address their delusions. He likened cure, particularly if it was rapid, to 'waking from a dream', an expression found in other accounts of recovery. Tuke also encouraged patients convalescent from puerperal insanity to talk about their recollections of the events of the preceding months or weeks.

> The account usually given was, that memory was lost for a time, – the period of the maniacal paroxysm, – but they have always given a wonderfully accurate account of proceeding subsequent to its disappearance. The delusions are very vividly impressed on the recollection, and I have been told by convalescents that it was long time before they could argue themselves out of them. It is immensely interesting to watch the gradual re-assertion of reason; to me it is difficult to conceive a greater amount of satisfaction to be derived from the successful treatment of any other disease.[147]

Tuke reported on a woman recently discharged, who during her stay had been violent and noisy and had mistaken him for a doctor who had been attending her, and whom she much disliked. When Tuke visited her, he was 'the object of her most unlimited abuse'. On paying his last visit on the tenth day after admission, he saw a change had taken place; the woman was calm and composed, and, as far as he could judge, sane:

> she apologized for all the bad language and abuse she had given me, and said that she recollected perfectly well the incidents of the last few days, and explained her reasons for her apparent dislike of me.

She never showed another bad symptom, and was discharged in three weeks perfectly recovered.

In Tuke's eyes, recovery with an apology must have been a perfect outcome, and his account is replete with condescension.[148] However, there are indications in both Tuke's anecdote and in the case notes that the women's remarks, even at the height of their delusions, were attended to, their responses actively encouraged, and fewer indications that psychiatry 'covered its ears' or tried merely 'to capture the objective stigmata of mental illness in its gaze'.[149]

As I will argue in chapter 5, few references were made to the moral failings of unmarried women who were brought to the Morningside Asylum. Nor were the patients – married or single – accused of profligacy for having too many children, but this is perhaps because most of the women admitted became ill after the birth of their first or second child. Large families, however, were considered a cause for anxiety in that they exposed women to constant childbearing and breastfeeding, which threatened their physical and mental health. Tuke noted the relatively low rates of insanity of pregnancy and puerperal insanity among unmarried women (14 and 18 per cent respectively of the cases he surveyed). Not surprisingly, married women continued to feed their infants for longer periods and no single women were admitted with lactational insanity.[150] The marital status of the women was noted in the case books, but usually without further comment. Some emphasis was placed, however, on the fondness of some of the women for strong liquor and, less often, laudanum. Tuke commented that in insanity of pregnancy 'moral insanity is by no means infrequent, dipsomania being the most common symptom'.[151] The case notes noted outrageous behaviour or salaciousness with some interest, but also commented that this was a symptom of the disease and 'out of character'.

In some cases, however, the women's insanity was linked resolutely to moral shortcomings. In June 1855 Mary Cameron or Robertson was admitted: 'Of infamous notoriety. After following a course of extravagance and dissipation for a series of years which brought husband[,] family and self to sin and disgrace Mrs R living in Scotland while her husband was in America gave birth to an illegitimate child.' Since her delivery she had been confined in another asylum and had subsequently gone home, but her conduct had made it necessary to place her once again in an asylum. In particular she had 'made violent attempts to embrace gentlemen on the highway who were entire strangers to her'. 'Her mother is mad and has been for many years in confinement and it appears that she was only taken out of an asylum to be delivered of the present subject. Her father[,] Patrick Robertson[,] if not insane was considered by most people eccentric.'[152] During her years in the asylum, Cameron's behaviour was at points described as improved, but she remained destructive and five years after admission in September 1860 was reported as 'now very restless, has a kind of pleasure in destroying everything she can e.g. broke a tree in the airing court – and on one occasion tried to throw the cat out of the window'.[153] She was to become a permanent resident and appeared, in some respects, to have found her place in the asylum, just as her mother had. In 1877 she was still at Morningside and described as 'a nice old lady', and in 1881 she was 'enfeebled in mind – quite silly & childish, but happy and contented, and a great favourite with those around her. Is very stout.' The last entry for Mary Cameron was made in 1888 when she was suffering from repeated attacks of bronchitis, and it can be assumed that she died in Morningside.[154]

A small number of the women were accused of self-abuse, or masturbation; though rare, these women were dreaded, not just because of their prurience and depravity, but also because they needed constant watching. They were put in leather gloves, aprons and tight jackets (known as polkas) to prevent the 'obscenity'. Mrs Wilson, 'a well marked lactation case', with staring eyes, blanched face, nervously excited, delusional, restless, sleepless and anxious, was also described as salacious and 'frequently found flushed in the face & perspiring profusely'. Suspicions were aroused and '[t]he moment the Doctor or any other man enters her presence she makes rush at him, or begs him to come into bed with her. She says the only thing she needs is a Man & if she cd only get that she wd be all right.'[155] Finally, 'on close watching', the attendant found her 'greatly given to masturbation'. She was put in a polka, 'and since this she has never been found perspiring &c.'. This failed, however, to relieve her mental state, and six months after admission, still demented and idle, she was removed against the advice of the superintendent.[156]

Margaret Bell proved to be an especially challenging case, and the asylum doctors clearly despaired of her; her case was also highlighted by Tuke, who treated her privately in between her stays in the Edinburgh Asylum.[157] She was diagnosed in 1863 as suffering from 'moral insanity' as well as 'mania of pregnancy' and described as unsettled, irritable, excitable, intemperate, untruthful, threatening suicide and violent to others, especially her husband, whom she had hit and threatened to choke. She developed mania after each of her four pregnancies and during her fifth she was admitted to Morningside. After admission her tendency to lie, steal, annoy the other patients, feign various ailments and her general attention-seeking were considered very noteworthy, not least her stories that the doctor was trying to seduce her. She gave her husband a letter when he visited her 'telling him that the doctor had attempted her virtue & had placed her in a single room at night that he might accomplish his foul purpose! Happily the husband was very familiar with her falsehoods.'[158] Several months later she was still telling extravagant stories but had ceased to steal, and was attempting to 'do better that she may get home'. Her conduct and demeanour became quite 'natural', and in September she was discharged, kind, pleasant and obliging.[159]

One year later, Mrs Bell was readmitted, following a suicide attempt. Tuke, the case entry reported, had been caring for her since her discharge, but in January her old 'tendency to drinking' had reasserted itself and she had obtained whisky by pawning her own and her husband's clothes. She was discharged recovered after two months,[160] only to be admitted again in May 1865, five months pregnant.

> [Her] conduct has been such as to render her poor husband perfectly miserable. She would pawn every article of her own or her husbands that she could lay her hands on & buy drink with the money. At times she would sit slovenly & half-dressed reading the Bible, at others she was very violent towards her husband if he attempted to control her conduct.

On one occasion she had fled from her husband and hid in a brothel, pretending that she did not know 'the real character of the house'. 'She expresses great suspicion of the husband's fidelity to her; & is intensely jealous of a woman whom her husband had employed to tend his house and children during his wife's incompetency.'[161]

Once in the asylum she formed an 'unfortunate friendship' with another female patient: 'the two of them grumble together to their hearts' content.'[162] In September 1865 Mrs Bell was delivered of a male child after a tedious labour, but this brought no relief. She was described as morally perverted, lying and stealing, telling 'the nastiest stories without a blush', particularly about those who had been most kind to her, showing 'not a grain of gratitude', the 'incarnation of evil'. 'She became quite unbearable in the Sick Room, so the baby was weaned & she was sent to the S.B. for a few weeks – she is now in the 2nd [gallery] – a little less beastly than formerly.'[163]

Few patients stirred comments such as these and Margaret Bell clearly tested the attendants and doctors to their limits, though her efforts to improve her behaviour and exercise self-control were also repeatedly commented on. Mrs Bell finally improved and was discharged in September 1866. She returned to the asylum in March 1867 and remained there until December 1868. Again she was sent home for six months before returning for the last time to Morningside; in November 1869 she was removed to the lunatic ward of the Aberdeen Poorhouse.[164] Dr Turnbull, one of the doctors, certifying her insane in 1865, however, questioned whether Margaret Bell was 'a proper person for an ordinary lunatic asylum'. A clear case of 'puerperal mania' such as hers, he argued, needed restraint, perhaps for several months, but

> I think it might do very serious injury to her mental condition to confine her amongst a mixed class of lunatics for so long a period ... until a house for the reception of such patients alone is established I believe much injury will be done by sending such patients to the ordinary wards of a common Lunatic establishment.[165]

Dr Turnbull's assertion may have been borne out in Mrs Bell's case.

Women suffering puerperal insanity were considered dangerous even if, in most cases, this was only a temporary phenomenon. They were potentially murderous, a danger to husbands, family members, doctors, asylum attendants and particularly their children; the actuality of their destroying their infants will be explored in chapter 6. A great many of the women were also a danger to themselves. Tuke argued that in no other form of mental disorder as insanity of pregnancy was 'the suicidal tendency so well marked'; nearly half of the patients in his study had attempted or meditated suicide. 'In some the attempts were most determined, a loathing of life and intense desire to get rid of it being the actuating motives.'[166] One woman, the mother of several children, took morphia during the third month of her pregnancy with the intention of committing suicide, and bought oxalic acid for the same purpose. She believed that she had caused the death of ten children and that the police were watching her. Mary Oswald became ill during her first pregnancy in 1864. In the fifth month she became very low-spirited and depressed:

> She attempted suicide by drowning, but did not succeed in her intention, from the shallowness of the water, although she persevered for several hours. The attempt was made in the sea, where the

sands were not deeply covered, and extended so far out as to make it difficult or impossible for her to reach deep water, so that as the tide receded she was always left high, if not dry, after each effort to effect her purpose.[167]

She partly recovered from her melancholy state and the baby was born in due course, but 11 weeks after its birth, she deliberately strangled it and then attempted to poison herself with laudanum.[168] Tuke noted, however, that some of the women attempted suicide in a manner which showed that it was not the result of any direct cerebration: 'she may wildly throw herself on the floor, attempt to jump from the window, or draw her cap-strings around her throat, but there is no method about it, it is an impulse, the incentive of which is purely abstract'.[169]

Notwithstanding, great emphasis was placed on careful watching of patients admitted with insanity related to pregnancy, childbearing or breastfeeding, in their manic and melancholic forms. All groups were judged to be at risk, and appear, as comments on the efforts of these women to destroy their children also made plain, to be regarded as crafty and imaginative in their attempts. Eliza Paterson or Lumsden, the woman praised for her bookkeeping skills, had attempted to poison herself with laudanum, had concealed her husband's razors under a pillow with the intention of cutting her throat, and had tried to strangle herself on three occasions. After admission to Morningside the attempts continued, and the case notes remarked that she must be closely watched. She attempted strangulation in the corner of the parlour 'in the presence of the other ladies', and on another occasion she tried to strangle herself with a towel and to choke herself with pieces of cotton rag. She also made a bid to escape, but was found half an hour later in the Jordan Burn drenched from head to foot and exhausted.[170]

Women who had given birth out of wedlock were thought to be at special risk of suicide, driven by guilt and dread of the future, which exacerbated their mental state. Christina Nicol, a former asylum nurse who had one illegitimate child and was pregnant with another, attempted to poison herself with laudanum, and on admission in 1851 said that she 'wished herself dead'. Tellingly, her disorder, which was diagnosed as 'melancholia', was attributed to folly and seduction; she was not mad, the case notes recorded, but shocked and frightened by her pregnancy and in great despair.[171] Agnes Watson was admitted in

May 1865 after several months of nursing an illegitimate child who had subsequently died. She was depressed and violent, suicidal and threatening to others. Her appetite was impaired, 'food refused because "a dead person requires none"'. She imagined that she was being 'devoured by a foul spirit', and that the devil attempted to take her off her bed and throw her out of the window: 'the Devil tells her constantly he had done for her now'. She had attempted to jump out of the window and cut her throat.[172]

Ghastly instances of self-harm were also described. Mrs Jessie Eisdale was admitted to East House in October 1854 with puerperal insanity following the birth of her second child, but also with a longer history of depression and mental disturbance related in the case history to menstruation. She had talked of her own worthlessness and had taken laudanum and acetate of lead in attempts to destroy herself. A few days after admission she suffered febrile symptoms and pain under her breast, and an abscess formed. This was treated and at length she admitted that she had pushed a darning needle into her side by leaning on the bedpost until it had disappeared. The medical officers extracted a three-inch needle.[173] Given that the chief occupation of women in the asylum was sewing, self-harm and suicide attempts frequently involved needles, a direct subversion of the efforts to impose moral management in the sewing room and to establish regular patterns of domestic employment.[174]

Even when violent towards others, patients were held in solitary confinement only as a last resort, great faith being placed on the benefit of contact with others. Actual physical restraint was rarely used. Jean Main had, prior to admission in December 1852, threatened violence to those around her and was heard to say that she would split the skull of anyone who differed from her. She 'fought desperately' in the street to resist being put in the coach to transport her to the asylum. Her violent behaviour continued after admission and she attacked the attendants and other patients, but she was kept on the galleries.[175] Isabella Hogg remained excited months after her admission to Morningside in August 1849, destructive, violent, noisy, abusive and restless, but she could still be persuaded to pull her weight in the laundry. She remained a patient and a threat for several years, and in June 1853 was reported as 'still very irritable & hesitates not to assault anyone offending her: she often gives or rec[eive]s a black eye'. The usual response to these outbursts was to confine her to a room of her own for a few days, which calmed her rage.[176]

Leaving Morningside

The rage of some of the women was little short of stunning, even though this was often combined with bodily feebleness, a combination particularly associated with puerperal insanity: 'On admission she was so weak as to be considered almost moribund, but fearfully excited, causing astonishment as to how so much noise could be produced by one so debilitated.'[177] Though all women suffering from mania or melancholia contravened their social roles and deviated from feminine behaviour, puerperal cases were remarked on as doing so with much more force. Women suffering from puerperal mania unleashed demons, not merely rejecting their domestic functions and their children, but becoming loud and alarming, dirty and unkempt. References to the women's physicality in the case notes – their power and strength as well as their dreadful state of health – form an absolute contrast to maternal images emphasising succour and gentleness. It is here that the case notes are most evocative, presenting not static images, but virtual impressions of these wild women roaming the asylum, expressing themselves in violent actions, overturning furniture, breaking windows, speaking of strange things, swearing, making sexual advances to the doctors, attacking other patients and their attendants. Yet the interpretations of causality and the women's behaviour contained in the case books also went beyond close detailing their strange, violent and outlandish behaviour, to position their illness in the context of their previous good demeanour in many cases, and to link their decline to a broad set of social and familial factors. The women and their friends and families to some extent were listened to and their accounts shaped the case histories. The case notes also described the patients as individuals with an individual assortment of problems, medical and social, though the lengthier narratives of the 1850s start to diminish in the 1860s, becoming terser, pre-dating the more formulaic accounts introduced under Thomas Clouston.

Most of the women admitted to Morningside recovered, and did in fact become what their families and doctors wanted them to be: mothers and wives. Tuke estimated that almost 80 per cent of women suffering from puerperal mania left having recovered within six months.[178] Insanity of lactation, like melancholia, was more intractable, but 72 per cent of these cases recovered eventually, though often after a long asylum stay.[179] Though many women were admitted to Morningside against their will, most seem to have been willing to remain while convalescing, though the recorders may have been reluctant to note pleas to leave.[180] Treatment was not hurried; indeed, some women were retained for long convalescences to ensure they did not relapse and that their physical as well as mental health was good when they were discharged. In the case of B.C., cited by Tuke,

> her progression towards sanity seemed to be effected by fits and starts. She would improve a little, continue in that state for a fortnight, and then make another step in the right direction. About the end of June she was sent out on trial, but returned of her own accord in a few days, saying that she preferred the asylum ... her recovery progressed favourably step by step, and on the 20th of August she was discharged recovered. She has since been confined twice, with no bad results.[181]

Despite Tuke's outrage at some of the treatment methods employed in Edinburgh, most women were left largely to the healing devices that he advocated: food, rest and purging. The aim of physical and moral treatments was to restore the unstable female body and mind to a balanced state, increasing and stabilising weight, waiting for the restoration of the menses, balancing humours through blistering and the use of other counter-irritants, as well as reintroducing women to their roles as housekeepers, workers and mothers. Treatment did, however, hint strongly at the failure of these women as mothers, as milk was drawn from their breasts, as they were fed like children with spoons and feeding tubes, and referred to as being in a childlike state. This collapse of their ability to mother, even those who had struggled so long to breastfeed their children, was perhaps emphasised more than the lapse in their domestic functions. The childlike behaviour and countenances of patients following recovery would also be remarked on, as if women who could not respond robustly to childbearing were not complete women. Few private patients were admitted, but when they were they could expect little difference in approach or treatment; they may have lived in finer surroundings in the asylum but they could still be force-fed and subjected to isolation. The asylum doctors neither anticipated nor achieved rapid recoveries in some cases, and showed patience in dealing with the women's violence and stubbornness. There is less language of authority expressed in the case notes than might be expected, and moderate use of restraint. Though tractability, regularity and compliance were demanded, the medical staff had considerable trouble achieving these.

Restoration was the aim of asylum treatment just as it was in the well-to-do households explored in chapter 3, and the case books recorded successes with satisfaction. Janet Smith, dirty, profane and abusive, following improvement and then a relapse and several stints in the refractory ward, experienced a remarkable change in September 1853, almost seven months after admission. She began to ask after her friends, husband and child, became 'cheerful, kind, tranquil and industrious'. She was also said to have known that she had been very ill and remembered much of what had happened to her. She regained her natural disposition and talked sensibly of her position and prospects. She was removed to 'her husband's protection' in October.[182]

Elizabeth Robertson, referred to at length above, after being hopelessly violent, immense trouble to the attendants and patients, 'pugnacious, destructive and mischievous', of 'superhuman strength', like 'a wild beast', was noted to experience a marked change eight months after entering the asylum. She became collected and tranquil, industrious, clean and tidy, sociable and very contented, a useful assistant in the sick room and 'remarkable for her kindness of disposition and activity of habits'. After being referred to on admission as looking and talking as one demented, her intellectual capacity was praised. In December, Elizabeth was allowed to visit Edinburgh with an attendant, and was discharged, symbolically, on Christmas Eve.[183] At this point the 'strange biographies' which Clouston referred to in his study of the case notes become biographies of normal, hard-working women and mothers, as they return to their families, their homes, their factories and field work, and their domestic lives and labours.

5

Women, Doctors and Mental Disorder: Explaining Puerperal Insanity in the Nineteenth Century

Representing puerperal insanity

> There was ... a depression mingled with her reveries, arising, as it would appear, from real circumstances. She had been an industrious woman, of good character; but she and her husband were poor, and contemplating, probably, the difficulty of providing food, clothes and shelter for a coming family, her husband left her and his home and his country to seek employment in Australia. The sensitive wife, whose mother had been insane, became deranged and melancholic, almost as soon as her poor little child came into the world of want, in which the father was so perplexed how to provide against starvation.

These are the words used in 1858 by the influential alienist Dr John Conolly to describe the route of a poor woman suffering from puerperal mania to the asylum.[1] The discursive account of the woman's illness was accompanied by a series of four engravings copied from photographs attributed to the influential psychiatric photographer Dr Hugh Welch Diamond, Medical Superintendent of the female department of the Surrey Asylum.[2] The photographs showed four stages of puerperal mania. The first shows the woman's quietness and sullenness after admission and the second, a phase of animation and mirth. In the third, improvement was signalled, though 'a tension of the facial muscles ... prevents the experienced Physician from concluding that all the malady has yet passed away' and suggested that a relapse into 'drollery' was imminent. The fourth portrait shows the final stage of recovery prior to discharge, when the woman is presented in bonnet and paisley shawl replacing the coarse asylum clothing of the other three images.[3]

Figure 5.1 Puerperal Mania in Four Stages, H. Diamond (Source: Wellcome Trust Library)

The photographs revealed the mobility of features that Morison had referred to in his descriptions of puerperal insanity in the late 1830s. Morison drew attention to the visually striking nature of the disorder with its powerful and variable facial expressiveness, and also to the sudden change in these expressions as the women began to recover.[4] Morison's claim for the special status of puerperal insanity as a vivid example of how mental disease was stamped on the face of its victims would be shared by those photographing the insane after the mid-nineteenth century, and Conolly claimed that this was the first series by Diamond 'to delineate, by photography, the progressive changes in the countenance in mental disease'.[5]

> They appear to me to be singularly valuable, even in an artistic point of view; and they certainly teach the medical observer more forcibly than words. Repeated contemplation of them reveals several curious particularities ... [6]

Yet Conolly found words necessary to point out what was being revealed in the photographs. In the second image, he described the woman's features as

> not only lively, but mirthful; her mouth is drawn out laterally, the nostrils are expanded, and the lively eyes, the elevated eyebrows, and the merry cheeks and chin are felicitously rendered in the plate. She sits with her hands crossed, and resting on her knee; but she looks as if she might easily be persuaded to get up and dance. She was, indeed, generally singing; and she now took food, not only willingly but voraciously.[7]

It is the photographs in isolation that have been noted by historians of madness, Conolly's commentary limited to his interest in physiognomy and expressions of faith in the power of photography to provide a diagnosis. Elaine Showalter has described the series as 'Victorianizing Hogarthian conventions' of the 'progress'.[8] The progress, however, is Hogarth in reverse, from craziness through to calm recovery. Unlike the Rake, the woman accompanying the story had the potential to recover and return to family life. Once cured, she rejoined her husband in Australia, the passage across the sea representing the start of a new life together. The posed and glib photographs – for indeed that is what they are – reveal, as Showalter points out, a 'set of visual and psychiatric conventions', imposing 'cultural stereotypes of femininity and female insanity on women who defied their gender roles'.[9] This was displayed in the rough and untidy hair and smirking features of the second image showing the woman's silliness and inappropriate behaviour, the useless hands hidden in the folds of her dress, as well as in the fourth portrait where the woman, dressed in clothes inappropriate to her station, is presented dolled up in respectable attire. Her childish expressions in the first two portraits are a stark contrast to the purposeful, and apparently aged, demeanour of the final image, the woman poised to resume her role as wife and mother. Her 'wellness' was, Conolly concluded, 'commemorated by the *fourth* portrait ... with composed features and pleasant honest face, animated still, but no longer excited'.[10] Turning to the accompanying text, however, which explained the 'real circumstances' of the case, a full, sensitive and plausible explanation of the woman's descent into madness is given. And there is a hopeful ending: 'After her voyage to the other side of the world, and at the joyful meeting of this poor family, it is to be hoped, therefore, that the recovered mother will carry back to her husband the face familiar to him in happier days.'[11]

In 1856 Diamond outlined the ways in which photography could aid in the diagnosis and treatment of the insane: photographs were objective and accurate; they enabled doctors to catalogue psychopathologies, and they could reveal to the patient their own pathological state.[12] Mark Jackson has argued that from early in the nineteenth century 'illustrations of the mentally ill began to supplement or replace descriptions of disease in medical literature on insanity', with images being included in medical works on mental disease.[13] Yet around the same time verbal descriptions of insanity were also expanded into biographical sketches 'as scholars became more and more interested in the total structure of a madman's life as a potential explanation for his illness'.[14] For most medical practitioners, however, it was the briefer case history that served to tell the story of the patient's lapse into madness, while most asylums, lacking the mechanism, money and, perhaps, interest, did not begin to use photographs until the late nineteenth century and then largely as an administrative tool. Illustrating the mad generally involved posing and imposing the preconceptions of both photographer and psychiatrist. Biographies too imposed stereotypes, as do the case notes drawn on in this chapter. But it is perhaps the comparative brevity of the case note, and in the case of manuscript notes their immediacy, compared to the lengthy life accounts of the continental biographers, including Esquirol's rich descriptions of puerperal insanity, that limited the imposition of stereotypes.[15] As we have seen in chapter 4, asylum case notes could produce powerful descriptions, which, unlike drawings and photographs, covered the movement, actions, voices and attitudes of the patients as well as their visual appearance. And it is to the text, the case note, once again that this chapter will turn to demonstrate the importance of narrative descriptions in defining and explaining puerperal insanity, and to show how they were set within the broad socio-economic and emotional as well as mental lives of its victims.

Doctors, women and case notes

For many women suffering from puerperal insanity the prognosis was optimistic. However, given the reasons put forward to explain the affliction, their likelihood of falling victim to the disorder in the first place was considerable. A warning was embedded in Conolly's account, which pointed to danger as well as restoration. He talked about the 'severe trial' of childbearing and also of the dangers of the 'joy, hope, anxiety, suffering, and causes of debility' associated with this dramatic change in physical and life circumstances.[16] Other risks, both moral and social, out of the mixed and often conflicting bag of explanations proposed by the early nineteenth century, were excess or poverty, working too little or working too hard, being single or being married, being unprepared for childbirth or being over-anxious about it, being too full of joy and optimism or being miserable and disappointed. There was clearly much more to puerperal insanity than giving birth.

Based largely on case notes, many of which give, if not a lengthy biographical account, then a sense of patients' life stories and the circumstances leading to mental collapse, this chapter delves deeper into the causes put forward to explain the onset and course of puerperal insanity during the nineteenth century, continuing the discussion in chapter 4 based on Edinburgh case histories. A wide range of case notes is drawn on here covering the nineteenth century. The manuscript case books of the Royal Edinburgh Asylum appear once more, together with cases taken from asylums dealing almost exclusively with pauper patients, particularly the Warwick County Lunatic Asylum, and from private institutions for the well-to-do. Case notes based on private obstetric or general practices, and the records of maternity hospitals and asylums, which were published in medical journals or textbooks, have also been drawn on.

The format, presentation and visibility of case notes varied widely. Manuscript case notes were often apparently written at speed by hard-pressed asylum superintendents and their assistants, though, as chapter 4 shows, the initial case history in particular could provide a rich and informative account of the patient's medical history as well as family and social background.[17] Hand-written notes, however, would be read by few – perhaps other local medical practitioners and occasionally Lunacy Commissioners. Published notes were intended for much wider consumption. Selected from manuscript sources, 'doctored' and rewritten, or presented as a follow-up to particularly interesting cases encountered in practice, they were published in journals or textbooks because they were regarded as being of particular interest to a professional readership. Many were presumably substantially redrafted, and recalled and written from memory. Yet, particularly in the early nineteenth century, case notes constituted an important component of textbook material and the contents of medical journals. The building up of practical experience and the presentation of these experiences to colleagues were deemed very important to the emerging fields of obstetrics and psychiatry. Even the published case notes present direct accounts of the day-to-day work of attempting to cure mental disorder and in this way differ from the more rhetorical descriptions of puerperal insanity found in legal, theoretical and literary texts.

Some case notes express the surprise, shock even, of doctors confronted with puerperal insanity for the first time, and their struggle to understand why a contented woman who had just had a trouble-free delivery should become so violently disturbed. Others reflect the accumulation of experience in diagnosing and treating puerperal insanity, and, for those treating the poor, a sense of frustration at the social conditions that provided the ideal opportunity for the disorder to take root and develop. The case notes might go into great detail on the day-to-day progress of the disease and its treatment, particularly under the intense monitoring and therapeutic regimes adopted in private households.[18] Other case notes are brief and rushed, the prognosis reflecting years of experience in treating the disorder and assurance of its outcome – bleak if there had been delay seeking medical attention, if the woman was seen as prone to mental disorder, if there were indications of melancholia or if her family circumstances were particularly bad, but otherwise optimistic. Overall, however, the descriptions of the admission, diagnosis, therapeutic regimes and outcome for patients build into a sad collection, with stories of deep, protracted depression and excitable, disruptive mania, of suicide attempts, self-harm, infanticide, terror, delusions and misery, and a roller-coaster of improvement and deterioration. The case notes provide us with valuable insight into how alienists and midwifery practitioners interpreted the impact of poverty, physical exhaustion and domestic horrors, or, in other cases, excessive luxury and a lack of preparedness for the maternal role, as bearing on puerperal insanity and its outcome. The case notes also open up a world of direct interaction between doctors, female patients and their families.

Showalter has delineated differences in the perception of madness as it manifested itself in men and women in Victorian England, 'associated with the intellectual and economic pressures on highly civilized men, and a female malady, associated with the sexuality and essential nature of women'.[19] It has been suggested that doctors strongly, almost exclusively, imputed biological factors to explain the incarceration of women in the nineteenth century, but in the case of puerperal insanity it was rarely suggested that biology offered a full explanation. From the 1820s onwards, following Gooch's publications on the subject, different weight was placed by different medical practitioners on somatic, social, moral and hereditary causes in addition to the crisis triggered by giving birth.[20] The alliance between 'woman and madness' was seen as complex. It remained partly a product of women's social situation as daughters, wives and mothers, and was spoken of in terms of vulnerability and natural predisposition. But while puerperal insanity and its main features were seen as common to many of them, case notes tended to present women as individuals with individual problems rather than as a homogeneous group of the susceptible, with weak biological profiles. Nor do case notes necessarily reflect a sense of superiority on the part of doctors over their patients; despair at the women's socio-economic condition and general vulnerability is a more apt reflection of collective medical responses to the disorder, though this sense of impotence was also moderated by their confidence in treating puerperal insanity and their efforts to increase, through their publications, professional and public awareness of such cases. Showalter oversimplifies the doctor–patient relationship in nineteenth-century psychiatry by claiming that,

despite their awareness of poverty, dependency, and illness as factors, the prevailing view among Victorian psychiatrists was that the statistics proved what they had suspected all along: that women were more vulnerable to insanity than men because the instability of their reproductive systems interfered with their sexual, emotional, and rational control.[21]

Mary Poovey has declared that the doctor–female patient relationship and the medical profession's superintendence of women in the nineteenth century were governed by a set of assumptions 'that women's reproductive function defines her character, position and value'.[22] Such conclusions have been nuanced and revised for different settings and contexts,[23] and it will be argued here that many doctors working in asylums and private practice not only were sympathetic to women's plight, but also set puerperal insanity within a very broad framework of trouble and misery, partly based on women's intrinsic weakness, but dominated by other challenges associated with maternity and poverty. They were all too aware from the cases they met in their practices of the role of poverty, dependency and illness. They fretted about the prevalence of the disorder – given the reasons for its prevalence, it was not going to go away – but found satisfaction in treating women suffering from puerperal insanity, the majority of whom they could claim as cured within a few months. Doctors acted too in a protective role, particularly seeking to ensure that the women they treated were fully recovered before they left their care. Many women, meanwhile, and not only those in the bourgeois households described in chapter 3, may have seen the diagnosis of puerperal insanity and its treatment as opening up the possibility of respite from household and maternal duties.

Anxiety, poverty and poor health

Of all the mental disorders afflicting women, puerperal insanity was the one tied most closely and obviously to their bodies and physical functions, triggered by the natural event of childbirth, and not necessarily a difficult childbirth, rather than abnormal bodily functions or disease. The close link between the dangerous process of reproduction and puerperal insanity would be emphasised throughout the nineteenth century. Explanations of the disorder, however, as outlined by the doctors recording it, and, on rare occasions, by the women themselves, reveal more complex diagnoses, referring to the impact of poverty on the one hand, and lives of luxury on the other. Jane Ussher has commented that working-class women were too busy to be 'mad', but the vast majority of women diagnosed with puerperal insanity were poor and their illness associated by doctors with their poverty.[24] Also cited in case histories is a weariness which went beyond the physical, which as we have seen in chapter 3 was not confined to the poor, and factors harder to pin down but so prevalent in case notes that it is impossible to overlook them: disappointment, gloom, fear and desolation. While the Edinburgh case notes discussed in chapter 4 are disturbing, those describing cases of puerperal insanity at the Warwick County Lunatic Asylum, an institution opened in 1856 to cater predominantly for paupers and poor labourers, are harrowing. All the women admitted to Warwick Asylum lived in poverty, most were in bad health and many had life experiences that deeply troubled their doctors. The case notes of asylums can be steeped in a rich, expressive language surprising for unpublished accounts and open up a world that went beyond 'anatomy is destiny' to one populated by frustrated, exhausted, sad women, subject to numerous social evils.

As we have seen in chapter 2, puerperal insanity was not only of interest to alienists, but was also the concern of specialists in the closely linked fields of midwifery and the diseases of women. The stakes in claiming competence were high, not least because cure rates were claimed to be around 70 per cent, and explanations for the disorder were worked up from the starting points of obstetrics, psychiatry and general medicine. Puerperal insanity, as chapter 3 has shown, was associated with effete luxury, with well-to-do women depicted as enfeebled by their idle existence and heightened sensitivity, unable to withstand the strains of pregnancy and childbearing. But it was also linked closely to the effects of grinding poverty. Interpretations varied among different groups of practitioners. Midwifery practitioners increasingly depicted childbirth as a dangerous process, with puerperal insanity one of the many disasters that could befall women following delivery. Alienists too recognised the close relationship with childbirth, but, dealing on the whole with a poorer class of patients, were quicker to link the condition to poverty and need, physical exhaustion, malnourishment, the hardships of rearing children born in rapid succession and long periods of breastfeeding. The same association was made between women and mental disorder in general, as Janet Saunders has shown for the Warwick Asylum, where women's slightly raised 'proneness' to insanity was linked to bearing numerous children, prolonged breastfeeding, poor diet and exhaustion, rather than to their biological role *per se*.[25]

A good deal of ground, however, was shared between alienists and obstetricians, both employing broad explanations for the onset of puerperal insanity, attributing it to social factors, family relationships, poor health and stress. Alienists and midwifery practitioners saw maternity itself as rife with problems and not just of a bodily nature. Childbirth marked a time of great change, with new and often unwelcome responsibilities and demands. Both groups expressed optimism about curing their patients, but were equally certain that the condition would not go away and expected it to recur in individual women; many, like James Reid, expressed surprise that the disorder was not more prevalent, given the conditions that caused it.[26] Because of its timing at what should have been a 'moment ordinarily so joyful',[27] doctors felt bound to offer the family some explanation and comfort, particularly when it occurred in upper- or middle-class households in which an appropriate response would be demanded. Midwifery practitioners tending the well-to-do experienced a loss of face when their charges succumbed to the disorder after they had cared for them through childbirth, and hence we are left with lengthy, if not always sincere and often self-justifying, explanations. Blame was attributed with great tact, exonerating the family and the doctor.

Maternity, described by physicians as women's natural role, was also recognised as presenting a severe physical and mental challenge. What should have been a blessing could confront women with hard work and worry about their new responsibilities.[28] William Ellis, Medical Superintendent at Hanwell and formerly of the West Riding Asylum, Wakefield, concluded:

> there is no doubt, but that the various circumstances of hope and fear in which females are necessarily placed at such times, render them more sensitive than usual to the operations of a variety of moral impressions.[29]

The struggle to cope with the new role thrust on them as mothers often manifested itself in problems related to breastfeeding; in especially harrowing instances, some women persisted in attempting to feed their child themselves after their milk was long gone. Samuel Ashwell, an authority on the diseases of women, strongly encouraged women to wean the child if their milk was scanty and their health poor and warned of the dangers of persevering in breastfeeding in such circumstances.[30] In one of his case histories, Gooch reported on a woman 'having nursed her child without feeding it for three or four months, with much unnecessary anxiety and exertion, she grew thin and weak ... and experienced so much confusion of mind, that she could not arrange her domestic accounts'.[31]

Failure to adapt to the maternal role was expressed as 'aversion' to motherhood or 'dislike' of the child, which in extreme cases manifested itself in infanticide; this will be explored in chapter 6. This failure greatly concerned doctors, but they also sympathised with women's difficulties in adjusting to this role. In other cases mothers expressed great fondness for their child, but feared that they would harm it either through an inability to provide for it or from ignorance. Many dreamt or, in a state of delusion, believed that they had murdered their children.[32] Puerperal mania, it was suggested, also occurred frequently among women who had lost children in childbirth or shortly after, or who had suffered miscarriages. Gooch remarked on the case of one woman, aged only 29, who had been pregnant many times but had borne only one living child; the last delivery of a dead child at seven months had resulted in a dreadful case of mania.[33]

For the numerous women treated in large public asylums by the mid-nineteenth century poverty and ill health were common running mates, compounded by seduction, desertion and illegitimate births for some, unhappy domestic circumstances for others. As we have seen in chapter 4, one of the first actions the asylum doctors took when admitting poor patients was to feed them. Most women were referred to as badly nourished, and in some cases as semi-starving. It was apparent that they had often stinted themselves, giving their husbands and children the limited food available, even when they were pregnant or breastfeeding. Conolly talked of the 'half-starved' condition of many women admitted with puerperal insanity to the Hanwell Asylum. One woman, nursing her four-month-old child, explained on admission that 'insufficiency of food was the principal cause of that distress of mind which forms my complaint'. She also said, perhaps expressing the anxiety about being able to continue to feed her child that troubled many poor women, that her milk 'began to fly away on a Sunday, when she was told to repent of her sins'.[34]

The case books of Warwick Asylum contain numerous examples of undernourished women and patients with severe health problems.[35] Emma Wall, admitted in August 1866, described as slight and undersized, 'has been in bad circumstances for the last twelve months – insufficiently fed & clothed'.[36] Maria Alexander, admitted in the same year after being confined with twins, bringing her total number of children to eight, was noted as having suffered from swollen legs and difficulty passing water during pregnancy, and since her confinement had complained of the loss of sight in her left eye and pain in her stomach and bowels. 'Her husband says that he has not been able to afford her sufficient nourishing food since her confinement, and to this cause he attributes her present attack of insanity.'[37] While in the asylum the women's physical ailments were treated, skin conditions, sores and leg ulcers healed, teeth pulled. Such women were literally 'patched up', the doctors realising full well that this was all they were achieving, and that the women were likely to be readmitted with similar problems if they gave birth again.[38]

Specific reasons or events were not always put forward in the case notes to explain a woman's affliction. Rather, it was suggested that a series of disappointments and unpleasant circumstances had resulted in her loss of reason. Childbirth was the trigger for puerperal insanity, but the build-up to this was often set within a much longer time frame. The women had unpleasant and disordered lives; giving birth added yet another burden. The case books of the Warwick County Lunatic Asylum are striking in a number of ways. The patients they describe were, almost without exception, admitted in a shocking physical state and extremely poor, while the language employed to describe them is littered with words such as 'brooding', 'desponding', 'disappointed' and 'desolate'. Stress or anxiety featured frequently in the women's derangement. Harriet Ashmore had 'been peculiar' for two months, 'desponding & threatening suicide', before giving birth to her sixth child. Immediately after being confined she became much worse, even though she had a 'normal & easy' delivery.

> For a considerable period before her accouchement, she was harassed by debt, & had to contend with great poverty, her means were very insufficient for the maintenance of herself & family so that she had to deny herself many of the necessaries of life, & was therefore weak & badly nourished when her confinement came on. Her excitement has partaken of a suspicious character, & her delusions have been chiefly about the persecutions to which she is subjected & the misfortunes that have befallen her family.

Added to this, Harriet's teeth were decayed, her skin covered in scratches, she was anaemic, her breasts flaccid and without milk, and she had an abscess on her right hand. She was dosed with henbane, a narcotic, given brandy and a nourishing diet of beef tea, eggs and milk, had warm baths and was sponged with vinegar. Only a month after admission she was rapidly approaching 'perfect health' and was impatient to return home. 'She remembers her arrival at the Asylum her introduction to the padded room ... as a "misty dream"' and claimed that she 'has not felt so well as she is at present for the last 10 years'.[39]

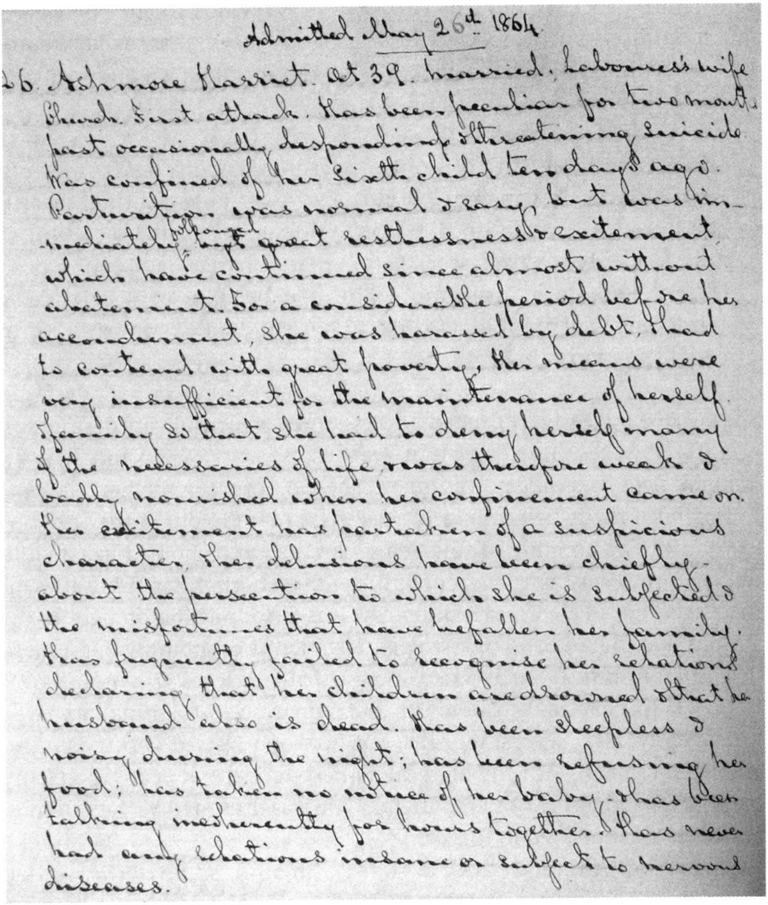

Figure 5.2 Case Note: Harriet Ashmore, Admitted to Warwick County Lunatic Asylum 26 May 1864 (Source: Warwick County Record Office)

Emma Walker's case clearly disturbed even the asylum doctors used to seeing women admitted in an appalling state. Emma was not even of the poorest class of patient, but her mental anxiety was extreme. A 33-year-old dressmaker, she was admitted in November 1864, having been insane for about three weeks. She had been confined with her fifth child about six months earlier, and 'previous to that had been much harassed & distressed by her husbands [*sic*] embarrassed circumstances & threatened bankruptcy'. She did not make a good recovery after the delivery.

> She was still kept in constant anxiety & fear, & took to her old employment, dressmaking, with the view of eking out their means. She nursed her child up till a fortnight ago when the secretion of milk suddenly ceased. A week before that a change had been noticed in her manner & disposition. She had become very irritable & capricious, had exhibited dislike to her children & had partaken of very little food.

She became increasingly excited, restless, incoherent and very destructive, was 'forcibly restrained in strait waistcoats, belts, etc.', and had no 'natural' sleep and little nourishment. She did not recognise her relations, experienced many delusions and attempted to jump out of the window. Her father had committed suicide by hanging, and she had lost a child during a previous marriage, though the five children from her present marriage were all alive and healthy. She was 'on admission brought to the asylum bound hand and foot struggling violently & shouting out. Had to be carried into the ward.' When she was bathed, sores were found over her back and left ankle, the results of severe restraint. Emma improved 'remarkably' after several months, but her sores caused her discomfort; they also rankled with the Medical Superintendent at Warwick, Henry Parsey.[40] An advocate of moral therapy, he was disgusted at the extent of the restraint employed in this case.

Frustration and annoyance at the treatment of some of the women prior to admission was reported in many asylums and the exaggerated nature of the restraint used by relatives and ignorant medical attendants infuriated the asylum physicians. Conolly, who supported a system of non-restraint at Hanwell Asylum, was greatly distressed by the bodily condition and shackling in one case of puerperal insanity, having 'seldom seen so wretched an object brought to the asylum'. The woman had been ill for a year and a half, having become insane when nursing.

She was emaciated to the last degree; she could not walk, or even stand up; her wrists were wounded with the iron handcuffs she had worn; her ankles were ulcerated by the leg-locks; and her toes were in a state of mortification. She had been fastened down in bed and had worn a strait-waistcoat; and I have not the least doubt that she had been half starved. It required many weeks of care to improve her appearance, and to redeem her from wildness and misery; but although often excited she was always perfectly harmless.[41]

Given the mental chaos and physical hardship these women were going through, it was seen as unacceptable, cruel, humiliating and unnecessary to subject them to further torment.

Even women suffering from puerperal mania are described as 'melancholic' or 'gloomy' or 'anxious'. This often appears to refer to overwhelming life circumstances or their general emotional status as much as to their acute mental state. Ellis concluded that most of the inmates at Wakefield and Hanwell, whatever their mental affliction, had been admitted because of their 'distressed circumstances'. 'Parents, in addition to their own personal sufferings from want of the common necessaries of life, are continually enduring the most painful anxiety, from seeing their children, who look up to them for support, undergoing the same privations.'[42]

Doctors might criticise their patients and make derogatory remarks about them, but such comments are more unusual than might be expected, and class-mediated disdain concerning cases of puerperal insanity was recorded rarely in the case books. One exception was the cruel reference to the Irish woman, Helen McCorkle, brought from the City Poorhouse in Edinburgh to the Asylum in 1862, who was 'of coarse and vulgar habits like the rest of her class, very indifferent education & somewhat intemperate habits'.[43] Insanity appeared several days before she was brought to the asylum and in the meantime she ran amok in the workhouse, raving, violent and breaking windows. Her life was also in 'great danger'. Despite bed rest, dosing with stimulants and large quantities of whisky and efforts to nourish her, Helen McCorkle died of bronchopneumonia a few days after admission.[44] In 1866 the Asylum's Medical Superintendent, David Skae, referred in his Annual Report in terms of disgust – perhaps in part to reveal the challenges of his work to the citizens of Edinburgh – to a woman admitted for the third time suffering from puerperal insanity: 'she had given birth to five illegitimate children; all of whom, it may be feared, may partake of the degenerescence of the mother, and contribute to the accumulation of similar burdens upon the rate-payers.'[45] Into the

1860s, though ideas concerning degeneration certainly had an impact, social circumstances and poverty still took precedence when attributing the causes of puerperal insanity. Even in Helen McCorkle's case, her notes stated that she had no hereditary disposition to insanity; she was a drunk and a nuisance and also a very sick woman.

Lactational insanity

The link between good mothering and poor physical condition found particular resonance when doctors wrote about lactational insanity. It was argued that women breastfed their infants for far too long when they were not fit to do so. Cycles of pregnancy, birth and prolonged breastfeeding went on for years, with few or no breaks. Far more than insanity of pregnancy or puerperal insanity, lactational insanity was linked to poverty, physical weakness and poor health and nutrition. Poor women often had little choice than to continue to breastfeed – there was nothing else to give their infants. Well-to-do women, however, were by no means free from the risk of developing lactational insanity if they insisted on breastfeeding for long periods.

Samuel Ashwell presented a detailed case history of one woman's torturous route through pregnancy, childbirth and breastfeeding in his textbook on the diseases of women. Mrs P. was aged twenty-eight when she became insane. A fair, delicate woman, with blue eyes and light hair, at the age of seventeen she had developed chlorosis, a severe form of anaemia. She was cured with tonics and sea air, but relapsed twelve months later. Recovering again under similar treatment, she married at the age of nineteen and before she was twenty had borne her first child. She nursed the child for twelve months, and was again confined just after her twenty-first birthday. She had subsequently borne four living children and had suffered two miscarriages. She had nursed every child herself. By the time Ashwell attended her, she had nursed her last child for eight months, a sure precursor to the disaster that was to follow. Ashwell reported that she was weak, desponding, sharp in manner and had a pale, anxious countenance. Ashwell ordered tonics and a change of air, and the infant was partially weaned, but Mrs P. did not respond; instead, the symptoms worsened and she became exceedingly violent and attempted to destroy her husband and child on several occasions. The child was taken away from Mrs P., her head shaved, a nutritious diet ordered, along with tonics and sedatives, but at length she was removed to an asylum. She was cured there and sent home after four months. Ten months later she gave birth to another child, and a further bout of insanity was prompted when she breastfed the baby for five months. She spent a further five months in an asylum, was discharged and then delivered another child twelve months later. Following the delivery of her last child, her sixth living infant, she was forbidden to nurse the child and her intellect 'remained unimpaired'. Ashwell concluded that weaning was the only effective remedy in such circumstances.[46]

The vulnerability of poor women to lactational insanity generated great concern. John Batty Tuke, Assistant Physician to the Royal Edinburgh Asylum between 1864 and 1865, linked prolonged breast-feeding to efforts to avoid further pregnancy, a factor also mentioned by Thomas Clouston, who became Medical Superintendent at Edinburgh in 1873. Tuke concluded, 'in not a few patients one labour had succeeded another so rapidly as to have weakened the constitution, and in others, the immunity from pregnancy sought for by keeping the child long at the breast, had been dearly paid for by an attack of insanity'.[47] The condition was strongly associated, according to Tuke, with multiparae, women over the age of thirty and with prolonged lactation; his study of Edinburgh patients between 1846 and 1864 concluded that 52 cases out of 54 occurred after six months of nursing. Tuke also emphasised the link with poor health, and in the small number of cases where lactational insanity occurred in first-time mothers, it followed profuse postpartum bleeding or occurred in women who were 'weak, phthisical or anaemic'. Tuke saw asylum treatment as essential in such cases for labouring or lower-middle-class women, and his prognosis where the family had delayed sending patients to the asylum was poor, this being the 'chief cause' of dementia. In any case, a slow rate of recovery was anticipated, although just over 70 per cent did eventually recover; 23 per cent of the patients in Tuke's study, however, lapsed into dementia.[48]

In the final quarter of the century Thomas Clouston became particularly interested in lactational insanity, for him a prime example of the link between poverty and mental illness. It also provided evidence of the value of his therapeutic approach, though this built very much on earlier practices in Edinburgh. While Clouston suggested that rich and poor women were equally liable to succumb to puerperal insanity, lactational insanity, in his view, was a disorder of the very poor. Clouston recorded a total of 52 cases of lactational insanity between 1874 and 1883. He disregarded twelve of these because they were transferred from other asylums or were readmissions. Of the remaining 40, a quarter of the women became disturbed within three months of nursing, a figure, Clouston pointed out, that was quite different from Tuke's results for the mid-nineteenth century. While Tuke attributed lactational insanity chiefly to exhaustion, Clouston linked it to the disturbance of the puerperal period being aggravated by 'the reflex excitation of the brain through the physiological act of suckling the infant'. His diagnosis fitted well with the increased emphasis on brain function in the late nineteenth century, yet he believed that the exhaustion of long-continued breastfeeding also operated as an important cause.[49]

Because Edinburgh was a mixed asylum, Clouston was in a good position to observe differences in class. He noted only two cases of lactational insanity out of the 166 ladies admitted between 1874 and 1882 to the 'higher class departments', while there were among them the usual portion of puerperal cases. By way of comparison, out of the 1,383 pauper and poorer private patients admitted over the same period, there were 38 cases of lactational insanity (3 per cent of all female admissions). Clouston concluded that the better-off had the possibility of treating the disorder at home, and found the low rate unsurprising 'among the well-fed classes, who have servants to work for them, nurses to attend their children, and doctors to tell them when to stop nursing in time'.[50] The rate among his poor West House patients, trapped between wanting and being obliged to feed their infants while their health was being ruined, 'is as might be expected'.

> If the wife of a labourer has had ten children and nursed them all, if she has during all the years those ten pregnancies and childbirths and nursings have been going on had to work hard, if she has had to struggle with poverty and insufficient necessaries of life in addition to this continuous reproductive struggle and family worries, if in addition to all this she has inherited a tendency to mental disease, no physiologist or physician can wonder if she should become insane during the tenth nursing. Indeed, the wonder is that any organism could possibly have survived in body or brain such a terrible strain and output of energy in all directions. Such a woman often enough becomes insane during a nursing long before the tenth. An organic sense of duty and a stern physiological necessity amongst poor women compel them to nurse their offspring. What else can they do? It is well for the offspring, but the mother often enough dies, or is upset in the body or brain in the attempt.[51]

Clouston described a typical scenario where a poor woman with several children became pale and thin while nursing her last child for many months. She became subject to headaches, noises in her ears, giddiness, flashes of light before her eyes, lassitude and nervous irritability due to 'general bloodlessness and brain anaemia'. She became depressed, sleepless, lost self-control and was either lethargic and stupid or suicidal, with delusions that her husband and neighbours were against her. Her health was broken, and nursing, instead of being an 'organic delight', became a burden, marked by exhaustion and irritation. This kind of woman, however, he argued, could be cured if treated in time. And the treatment was simple: wean the infant, give nourishment in abundance together with malt liquor, change the environment for the patient and free her from family cares, give quinine, iron, cod liver oil and tonics, watch for suicide.[52]

Clouston was more optimistic than Tuke about curing such cases (this may have been related to the more rapid admission of pauper women to the Asylum during his superintendence), recording a recovery rate of 77 per cent. His 'gospel of fatness' proved to be particularly valuable in treating lactational insanity, and 'recovery in all the patients was accompanied by a great increase in body-weight, in strength, in appetite, and in fatness'.[53] Clouston argued that to have good mental health, the brain and the chief organs of the body must also be healthy.[54] In terms of treatment, this led to Clouston's holistic approach with its focus on regimen, environment, sleep, regular bowel movements and a diet composed of rich foods, also highly recommended in cases of anaemic insanity and idiopathic insanity, which resulted from want of sleep as well as overwork, exhaustion and starvation.[55] While there might have been more emphasis on physiology under Clouston, treatment remained much as it had been in Edinburgh from the mid-nineteenth century onwards, though with rather more in the way of eggs, milk and beef tea, and he tackled lactational insanity in a direct and appropriate manner.[56]

Cases of lactational insanity which emphasised and re-emphasised too many 'childbirths and nursings', poverty and exhaustion, as well as the alarming delusions which were associated with the condition, feature regularly in the Edinburgh case books, with a particularly large number of admissions occurring in 1865 when twelve women were brought in, worn out and confused.[57] Two cases were admitted in November. Janet Ewing, aged 41, insane for seven months when admitted, had nursed her child for nine months, was anaemic and said to look much older than her years, suicidal, and in dread of poisoning, hanging and being taken to prison.[58] Margaret Reid was a 'well marked case of the mania of lactation'. She had been married for fifteen years, had a family of eight with the last three children having been born within four years. 'Although stoutish, she is far from healthy, her colour is pale and anaemic ... although she has been nursing only three months her constitution has been much tried by her former children being born so quickly one after the other.'[59] According to the medical certificates, she believed everyone to be her enemy, had dashed her hand through a pane of glass to get at her tormentors and tried to escape them by scrambling up the chimney, she was indecent and 'dancing violently', her excitement alternating with fits of depression. After admission, she had to be force-fed and later, when taking food herself, would not touch it until the nurse had eaten some. She refused to go to bed because she believed it was crawling with vermin. However, Mrs Reid recovered quickly, and was sent home after three months, 'cheerful and obliging, her only anxiety being to get home to her family'.[60]

Illegitimacy

Women who breastfed for lengthy periods, medical observers argued, were guilty of foolhardy behaviour and exposed themselves to the risk of lactational insanity, subjecting their bodies and ultimately their minds to intense strain. Yet fundamentally they were excellent mothers who took their succouring role seriously. One might predict that the opinion on unmarried mothers would have been very different, but this was far from the case. It has been argued more generally that 'seduced women became the object of great pity' in the early nineteenth century,[61] and rather than being condemned for their immorality, anxiety was expressed about their vulnerability to mental disorder. Grief, fear and panic after giving birth were recognised as being strongest among unmarried mothers, and even those who destroyed their children elicited a sympathetic response from doctors, judges and juries. Women who committed infanticide were depicted as victims of circumstances, betrayed, alone, in a situation of want, obliged to work yet in terror of their employers discovering their pregnancies, in dread and often ignorant as their confinement approached, abandoned by the real perpetrator of the crime. As we will see in chapter 6, the plea of insanity was used effectively in many such cases.

Like persistent breastfeeders, women who gave birth to illegitimate children risked mental disturbance. In his Annual Report for 1869, Henry Parsey referred to three infants born in Warwick Asylum, one to a single woman; 'in the latter the sense of disgrace attaching to her condition seems to have been the exciting cause of her insanity'.[62] 'Among the moral causes,' John Conolly declared, 'anxiety, the fear of abandonment, and the pressure of poverty, may readily be supposed to be most conspicuous. Some of these causes operate more strongly when the mother is not married ... '[63] Of 415 women afflicted with puerperal insanity and admitted to the pauper Hanwell Asylum up to 1871, 263 (over 63 per cent) were unmarried, 122 (29 per cent) were married and 30 (7 per cent) widowed.[64] The rate of puerperal insanity among unmarried mothers was much lower in Edinburgh: Tuke quoted thirteen cases out of 73 (18 per cent) between 1846 and 1864. Tuke argued that 'the moral causes, shame and disgrace, must have considerable influence over the development of this affection', and also pointed out that none of the single women had 'an hereditary predisposition'.[65]

Women who became pregnant before marriage were also deemed to be vulnerable. Mary Turner's child had been conceived out of wedlock and she had been married for just four months when she gave birth. Her husband had become 'insane' following a head injury just before the child was born. 'Her mother thinks that the knowledge of her condition before marriage was preying upon her mind for she remembers that she was not in such good spirits as usual at the time.' She was described as 'an undersized woman of very childish appearance ... Is continually saying that she wants to go to Jesus – Is pale & feeble & much out of condition.'[66] William Ellis described a case where the woman became insane in her third month of pregnancy, though she was not brought to the asylum until three months later. Two months after admission she delivered a stillborn child and then quickly recovered.

> It appeared from her own statement that she had long been living in a state of concubinage with a man to whom she had borne several children; but so deeply was she now impressed with the sinfulness of her conduct that, though the man repeatedly came to her and urged her to return, no solicitation could prevail; she would not even see her children, unless she was first married. The man was very fond of her ... and he readily consented. The banns were properly proclaimed in the parish church, the parties were married from the Asylum, and she returned with him to her former abode and family, cheerful and happy.[67]

Though lying-in hospitals recorded few cases of puerperal insanity,[68] the small number mentioned in their case books occurred most often among the mothers of illegitimate children, and not necessarily poor mothers. In Queen Charlotte's Hospital,

> [i]t not infrequently happens that young females of a superior class, who are deserted by their friends, are obliged to apply for admission into such an institution, to hide their shame, and for want of means; and of five fatal cases out of the eleven, one only was married. Two were of the class just described, highly educated, of sensitive dispositions, and compelled by the cruel desertion of their seducers, and neglect of their friends, to seek for admission into a public hospital.[69]

Maternity hospital case notes tend to be curt and their physicians clearly interested in dispatching women displaying symptoms of insanity to the asylum or even back to the workhouse from which they had come, but they could also be sympathetic about their plight. James Simpson commented on how unmarried mothers were 'far more under the influence of depressing moral emotions' and more liable to puerperal mania. Their 'terrible predicament' exacerbated the condition.[70] In the case of Christina S., a factory girl of 'a feeble mental constitution', Simpson described how she was in labour for three days, in terrible pain; the child was stillborn. Neither a medical man nor instruments were on hand to relieve her. On admission to the Edinburgh Maternity Hospital she was dirty, had massive tearing and became manic, noisy and troublesome. She died a week after admission.[71] One 'poor girl' was found wandering in the countryside outside Edinburgh after the birth of her baby in the Maternity Hospital. She had 'by some insane impulse' attacked her relations and was brought into custody on a sheriff's order, having made herself 'very troublesome'. She quickly regained her sanity in the Asylum, and was referred to in the case books as 'a remarkly [sic] interesting looking girl, large dark eyes & magnificent hair – with a modest expression & pleasing manner'.[72]

For some women restoration was possible and, after recovery, they took up their former positions. K.C., a nineteen-year-old domestic servant, was described by Thomas Clouston in 1887 as a 'fallen woman' (she had borne an illegitimate child in a maternity hospital), but also 'hardworking' and from a respectable family. Under Clouston's treatment she did well, was put on nourishing food and encouraged to exercise.

> She gained in weight all that time, eating well and spending much time in the open air. Then she began to work, was put to rough

scrubbing and laundry work, so getting rid of her excessive muscular energy. In three months she was fattening, quieting, and working hard. In four months after admission she was stout, sensible, and well in mind and body, menstruation having begun, and she was then sent back to her situation, which had been kept open for her in consideration of her previous good conduct.[73]

Asylum case notes depicted the men who deserted women bearing their children as scoundrels. The women were rarely rebuked for their misfortune. Ann Standbridge was admitted to Warwick Asylum in October 1856 after spending a fortnight in the Union Workhouse. A silk winder, described as 'a sturdy, industrious woman', she 'kept company with a man to whom she expected to be married. Had a child with him about a month ago; suffered much distress and shame at her condition; soon after her confinement began to be strange in her manner, became insane.'[74] Jane Gardener was admitted to the Royal Edinburgh Asylum in July 1853, suffering her third attack of puerperal mania. She was unmarried and her illness was related in the three instances to the death of her father, seeing her sister drown in a well after being unable to save her and her last attack to 'seduction and parturition'. On the first two occasions her attacks of insanity were short-lived, but 'after her seduction her temper and conduct were noticed to be changed, she became melancholy, reflected much on herself and suffered the greatest mental distress'. She was admitted garrulous, restless, incoherent and noisy twelve days after giving birth to twins.

> She is constantly talking about her marriage, and calls for her lover, who she says has not forsaken her. She plans marriages for those around her, and wishes every one in the world married forthwith; occasionally raves about her children and says they are being ill-used for she hears them crying: has wished laudanum to put an end to her existence, & also has talked of leaping over the window.[75]

Jane, according to the case notes, was a troublesome patient, threatening and abusive. She hallucinated about men who were with her in bed at night, kicked the walls and stamped the ground with rage. However, her case is reported in a non-judgemental, matter-of-fact manner, the disorder was allowed to run its course, the case notes remarked on her improvement and industriousness, as well as her bad behaviour. The fact that she had given birth to three illegitimate children is not referred to again, and six months later she was discharged, entrusted to the care of her mother and friends.[76]

The family, in particular husbands

The family was seen as key to the patient's misfortune and disorder in many ways. While the household was defined as the place where women should want to be, for many the household was too dreadful a place to be and its members too dreadful to be with. The family could be instrumental in the admission of patients, providing information to Poor Law officials, magistrates (in the case of Scotland, sheriffs) and certifying doctors and asylum superintendents. As we have seen in chapter 4, they could be important informants in building up the patients' case histories and for some admissions the doctors were heavily dependent on this kind of information. The position of the family, however, was ambiguous. Families were known to misinform asylum doctors about the patient's condition and situation at home, they could be disruptive and they could also be implicated as the cause, or at least partial cause, of the woman's illness. Akihito Suzuki has described how the Bethlem doctors sought to 'liberate' patients from their families, their interference and bad influence.[77] Other institutions shared in this pre-Laingian culture of blaming the family and saw the family in some cases as trying desperately to cover up the ill treatment of the patient.

In other cases family members were deemed to have done their best, though often failing quite abysmally according to the asylum authorities, to treat their wives, mothers and daughters at home, perhaps with the aid of the family doctor. Mixed messages were being sent out by the medical profession about the treatment of puerperal insanity; on the one hand, to keep mild cases out of the asylum, on the other, to ensure that they did not languish too long at home to avoid the risk that their insanity would become entrenched. Tuke made several pleas in his articles for timely removal to the asylum. He argued that more favourable results would be achieved 'if the patients were removed early to the shelter of an asylum, if we could look upon the public to give up old-fashioned prejudices, and look upon such institutions as hospitals for the cure of a disease by no means unamenable to treatment in its earlier stages'.[78] As noted above, in the case of lactational insanity, he blamed lapses into dementia specifically on delay in sending the patient to the asylum.

The reluctance of families to send women to the asylum when their infants were still being breastfed is easy to comprehend. Misguided efforts to cure women outside the asylum included extensive use of blisters and purges – in one case Tuke remarked that 'her wits have been blistered and calomeled away'.[79] The family was applauded for ensuring that women found their way rapidly to the asylum and away from the 'misplaced sympathy of relatives' and their efforts to stand in for physicians.[80] Yet in Edinburgh they were applauded even more if they did not then desert the patients. Unlike many nineteenth-century asylums, the Royal Edinburgh Asylum was close to the city centre and family visits were encouraged; later in the century Clouston boasted that 300 'busy working people' visited their relatives each Wednesday.[81] Patients were usually sent out on a month's trial with their families before being discharged, as they were in many other institutions, including Warwick Asylum.

In private practice too responses to the family were mixed, though medical attendants were constrained to mix tact with authority when they were being paid for their services. As we have seen in earlier chapters, doctors struggled to establish control over their patients in disrupted households as well as over disruptive patients. The family's lack of care and discipline was often described as contributing to the disorder, especially if they bustled, disturbed and interfered with the patient. Thomas More Madden claimed that the

> ordinary exciting cause of puerperal mania is the injudicious kindness of the patient's family and friends, who too often insist on being admitted to visit her. I have seen so many examples of the ill consequences of such visits in causing mental excitement that, as far as possible, I now exclude all visitors from the lying-in room until the patient is able to sit up.[82]

Madden had also encountered many cases linked to poor domestic circumstances in his work at the Rotunda Lying-in Hospital in Dublin, where 'the majority of patients are married women of the poorest class; wives of labourers and artisans, often broken down physically, and depressed mentally, by poverty and hardship', as well as unmarried women. However, one of his cases, a person of a 'very nervous, hysterical temperament', 'had been indulged in every way by her parents as well as by her husband', and this excess of attention and spoiling had, in his view, contributed to her condition.[83] The ideal form of treatment, as chapter 2 has shown, was to separate the patient from disturbing family influences, by either quarantining her within the family home, removing her to private lodgings or banishing the family from the house.[84] The medical attendant's authority replaced that of the husband. In some cases the husband was even cuckolded, and the Edinburgh case notes refer on several occasions to the sexual attention lavished on the asylum doctors, not least in the case of Margaret Bell, who declared to her husband when he visited her in July 1863 that 'the doctor had attempted her virtue'.[85]

Families were the butt of criticism if their doctors identified a hereditary link. While heredity does not appear to have been emphasised as much as social and moral factors even well into the second half of the century, the link between the woman's condition and incidences of insanity in the family was monitored closely. Well before the firming up of theories of degeneration in the late Victorian period, mental disorder in the family was claimed to predispose the patient to mental illness as well as reducing her chances of recovery.[86] Conolly associated puerperal insanity closely to predisposition and hereditary influences, declaring: 'I can scarcely recal [sic] an instance in which the patient's parents were not highly nervous, or some of their relatives insane, or so eccentric as to be on the borders of insanity.'[87] From the mid-nineteenth century onwards asylums, as the principal places of confinement of mental illness in their district, were commenting on the fact that fathers, mothers, sisters, aunts and other relatives had been treated in the same institution.

Annette Skirving was admitted with puerperal mania, very disturbed and suicidal, to the Royal Edinburgh Asylum in December 1852. Her father, it was noted, was a hard drinker and had died insane, her brother was an incurable inmate of the Asylum.[88] Margaret Grossert was listed as an 'HP' (hereditary predisposition) when she was admitted with lactational insanity in 1865; two of her aunts were or had been – the record is vague on this – insane.[89] 'Friends' were also remarked on as potentially bad influences, particularly if the woman had been exposed to 'evil habits', as if the patient could be vulnerable to both hereditary and environment influences, that insanity could be catching through association as well as bad blood.

If families were troublesome, disruptive and predisposed women to puerperal mania, husbands were often described as an outright danger and the root cause of the woman's disappointment and illness. Although marriage should, according to the model of Victorian domesticity, have been an enriching experience for women, the realities of family life, bearing children, often in rapid succession, and the physical, emotional and financial hardships resulting from this, were imputed as causes of the woman's illness just as much, if not more than, the act of giving birth, which is often mentioned as almost incidental. Women who were contented and well balanced when single became worn out and nervous under the strains of married life, when physical toil combined with rearing a family and financial worries.[90] Husbands were the object of implicit and explicit criticism. The well-to-do were certainly treated with more tact, but none the less in case notes were depicted as a source of trouble and kept away from the patient as much as possible. The husbands of poorer women were described as poor providers and hopeless specimens of manhood, indecisive and weak. At their worst they were drunks, womanisers and bullies.

One woman, recorded as having been ill for a year and a half with puerperal mania, was brought to Warwick Asylum in November 1856 when the neighbours who were caring for her could cope no longer; her disorder and the neglect which apparently followed was blamed on her 'so-called husband'.[91] Isabella Hay's sister was able 'to explain more fully the history of her case' when she was brought from the Charity Workhouse where she had been for two months to the Royal Edinburgh Asylum in April 1855. Isabella was described as having always been cheerful, if somewhat forward, industrious and well behaved.

> A predisposing cause is the reckless conduct of her husband who is a confirmed drunkard and treats her very badly. A fortnight after childbirth she felt ill having caught cold, and sent for her mother when she told her that she was sure something was about to happen to her as she had great pain in her head and felt very ill. She charged her mother to take care of her children. She had been four weeks ill at home before being removed to the workhouse. During that time she was generally quiet, except when her husband entered the house, when she cried much becoming quite excited breaking furniture, &c.[92]

Husbands were the object of considerable hatred on the part of their wives. Robert Boyd, who worked for many years at the Somerset County Lunatic Asylum, believed that puerperal insanity was characterised as much by the women's hatred of their husbands as their babies.[93] Husbands also featured extensively in reported delusions. In some cases it is difficult to get a sense of what is real and what is delusion, and the doctors writing the case notes reflected their uncertainty about where the boundaries lay.

Helen McDonald declared on admission to the Royal Edinburgh Asylum in July 1864, suffering from lactational insanity after feeding her baby for sixteen months, that 'her husband is a bad man and that he keeps every woman on the stair where they reside, and that he has many others that he keeps'.[94] On several occasions she threatened to drown herself and refused to clean the house, her children and herself. After over four years in the Edinburgh Asylum, she was no better and was removed to Murray's Royal Asylum in Perth.[95]

Other cases resulted in a happy reunion. Essie McKay was admitted to Edinburgh Asylum in 1865 'full of suspicion' of her husband; she fancied that she was being watched and her food poisoned. However, after a few days she declared herself devotedly attached to her husband and that her fears were groundless, and she 'settled down again into a pleasant, agreeable, witty Irishwoman'. Her husband collected her when she was still convalescing and told the asylum doctor that her attack resulted from drinking. Both childbirth and mania of alcoholism were given as diagnoses in the case book.[96]

Doctors on occasion openly criticised the husbands of their patients. In 1867 the obstetrician James Young reported a case of dereliction of duty by a husband, who called in a midwife rather than a medical man to attend his wife, Rebecca Walker, in her fourth confinement. This, 'accompanied by [his] intemperate habits, aggravated her illness before her confinement; and possibly *that disquietude* of mind, so induced, and irregular meals, may have tended to the production of the disease which so speedily cut her off'. Before her confinement, Rebecca was described as dull and melancholic, and as having 'no spirit for the performance of her domestic duties, and took little or no charge of household affairs'. A week after giving birth she became insane.[97]

Other husbands were well meaning but ineffectual. Ellis related the case of a woman who, ten days after delivery, became insane; she was the wife 'of a kind-hearted labouring man'. As her husband was unwilling to bring her to the Wakefield Asylum, she was kept at home for two months where she deteriorated. However, once in the asylum she was fattened up, purged and blistered, and after four months sent home. The next time she was confined two years later she felt similar symptoms coming on: quick pulse, hot dry skin, confusion of mind and 'cessation of secretions'. Ellis lived nearby and saw her immediately, she was treated as before, the 'violence of the attack prevented' and after a few weeks she was fully recovered.[98]

It was also the doctors' belief that women who had married beneath them were liable to mental disturbance. Madden described a case in his care of a twenty-year-old woman giving birth to her first child, 'who had been married a year previously to a man of very inferior station to her own, and had suddenly passed from a condition of affluence and comfort to one of poverty and privation'.[99] A few days after delivery in the Rotunda she became excited in her manner and 'manifested a strong aversion to the child and to her husband, for whom she expressed the greatest contempt and dislike, although he was a very fond and indulgent husband'.[100] She was removed to Richmond Lunatic Asylum and after six weeks was discharged well. 'About a year afterwards her circumstances became again very comfortable, her husband got a good situation in England, and before going to join him she sent for me to attend her in her second confinement, which ... passed off very favorably [*sic*] without any return of puerperal mania.'[101] In 1839 John Conolly admitted the wife of a poor tailor to Hanwell Asylum, who came to them 'almost a skeleton'. After a few months, she was improving:

> the calmness of recovery began, but the sight of her poorly-clad and half-starved husband who came to see her, and to offer to take her to his house of toil and privation, produced a melancholy impression on her; and even when this gave way to the natural wish to be at home ... destitute as it was, with her husband and little children, we feared for a time to make the experiment; but at length she went out and continued well.[102]

Doctors occasionally challenged women's mistrust of their husbands and antipathy towards them. In other cases they appear to have agreed with the grim view of their spouse. Their antipathy was the feature that persisted, women still expressing their views in no uncertain terms when otherwise 'cured'. Ann Taylor, a labourer's wife, admitted in April 1853 to Warwick Asylum, disliked all her close relations, had 'a great antipathy & suspicion of her family and ... neglected her child'. By July she was talking of her family members in a 'kindly manner', but when her husband and daughter visited her in September she squabbled with her daughter and ignored her husband.[103] Lois Eames, the wife of a farm bailiff admitted to Warwick Asylum in June 1868, had 'always betrayed jealousy of her husband for which there is not the slightest foundation' and almost until her discharge in March the following year she continued to express her antipathy towards him.[104]

Women who were doing well lapsed following ill-timed visits from their husbands. Many women blamed their husbands for their incarceration and no more so than when the patient was sent to a private asylum.[105] Isabella Campbell Foster of Hyde Park, London, the wife of a barrister, was brought to the Ticehurst House Asylum in April 1858. The case book reported that Isabella had been a violent child. Her first attack of insanity was due to 'mortification' when her elderly father married a young girl and threw Isabella out of the house. A few months later she married and thus regained a good position, but she could never forget her father's conduct. Until the birth of her first child Isabella remained ill tempered, worse around the time of menstruation, though she showed no signs of mental disorder, 'but as soon as childbirth had taken [place] these attacks came on, and they have been worse after each of her four confinements'.[106] She was furious with her children and servants, but saved her greatest anger for her husband, brother and the medical men who had treated her. As the months passed, her condition greatly improved, but she was not interested in her children and, though the case book reported that 'no one can express more kindly feelings towards a wife than Mr Foster, or shew a stronger disposition to do anything for her comfort and benefit', her agitation towards him continued. In November arrangements were made for Isabella to leave, but she wished to remain at Ticehurst and she stayed for three more months. She finally left in March 1859 to live in London; it is not clear whether she rejoined her husband and children.[107]

In dramatic scenarios husbands turned up at the Royal Edinburgh Asylum demanding the return of their wives. They claimed that they were unable to manage at home and that the children were missing their mothers, but the Medical Superintendent often refused to relinquish them. On many occasions husbands – and sometimes other family members – did remove patients. In November 1862, 'against the strongest advice', the brother and husband of Agnes Bennett removed her from the Edinburgh Asylum. Just one week later Agnes was returned 'as headstrong and unreasoning and impulsive as ever they have been glad to send her back'.[108] In October 1862 the husband of Agnes Jenkinson resolved to take her home after she had been in the Asylum for over three months 'in spite of warnings that she was quite unfit for household duties and cares'. After nine days she was returned unsettled, obstinate, idle, untidy, noisy and destructive. However, in February the following year she was moved again by her husband despite warnings that she was not fit to leave.[109] In July 1864 Ellen

Vernon was visited by her hat binder husband and a friend, who 'most obstinately & pig-headedly insist in taking her out, against all advice – She herself, although most anxious to go, agrees with the physician in thinking it inadvisable'. Though Mrs Vernon was working steadily, was very agreeable '& expresses herself very grateful for the attention she received', her general appearance of exaltation showed that she was not completely well. She was finally removed by her husband one month later.[110]

* * *

On occasion doctors became exasperated with the women they were treating; some had apparently brought the condition on themselves through their lax lifestyles and liability to expose themselves to risk of mental disturbance, or, alternatively, because they tried too hard as mothers. There was sympathy for the patients, but it was bounded. While social and moral causes of insanity were still seen as being so significant – as they seem to have been at least until the third quarter of the nineteenth century – there was little that could be done in the long term, at least for poor asylum patients.

It is possible that some women became fonder of the domestic space of the asylum with its decent food and relative comfort than their grim homes.[111] Though most were discharged within a year of admission, some left their households for ever and remained in the asylum until their deaths. Women leaving private asylums occasionally departed for a refuge other than their own homes, wealthier patients frequently accompanied by one of the attendants to whom they had grown attached.[112] Asylum regimes and domestic management emphasised bed rest, nutritious food and limited use of drugs. The curing regime was marked by the assertion of authority but also patience on the part of doctors, allowing women time to heal, regain their strength and come to themselves, and it also granted doctors the time to carry out a complete cure. Peter Bartlett has commented that the asylum prided itself on offering 'food and respite'[113] and this sense of respite is yet stronger in the case of puerperal insanity, an illness so directly related to home and family. Rather than the social protest and subversion suggested by Showalter in the case of hysteria, women suffering from puerperal insanity could be seen as reacting through their illness to their difficult lives and roles as mothers. Doctors responded by offering a setting and time to recover, though the aim of this was to place women back into their prescribed roles.[114]

Some convalescing women 'relapsed', like Conolly's subject, illustrated in the third engraving at the start of this chapter, who experienced a temporary setback before once again becoming 'industrious and tranquil'.[115] This seems to have been an expected feature of the curing process, a pause, a sense of becoming ready to face life outside the asylum again. The insistence on a full convalescence, meanwhile, was emphasised by the reluctance of the asylum doctors to relinquish their patients before they were completely cured.

Within the closed world of case notes written in the large public asylums, where few would view them, there is much less of a sense of doctors playing to an audience, and even in published case histories doctors could be candid about the problems of treating puerperal insanity. Though puerperal insanity was presented as a great horror and doctors claimed success in being able to cure their patients, these claims were muted. Part of the reason for this was that doctors saw themselves as being up against something much larger than women's biological vulnerability, facing challenges of motherhood itself, problems of family life and, for many of their patients, the unbearable strains of poverty and miserable lives. While optimistic about curing their patients, they expected the disorder to recur in individual patients and expressed surprise that it was not more prevalent. Puerperal insanity, as broadly defined and explained by medical men, could be seen as a catch-all, encompassing all women – the nervous and sensitive, dependable and reliable, simple and clever, rich and poor. But it also allowed doctors, and in rarer instances women, to explain how the miseries and frustrations of their lives had led to a dire form of mental alienation, and it recognised that motherhood, in association with other factors, could be disruptive, over-demanding, overwhelming and disappointing.

6

Dangerous Mothers: Puerperal Insanity and Infanticide

In May 1867 the *Warwick Advertiser* reported to its readers on the 'Sad Death of a Child'.[1] The newspaper account, which included a summary of the inquest proceedings, described how a local woman, Elizabeth Barnwell, had drowned her infant boy in the Warwick and Napton Canal while out walking. The story of the rescue of mother and infant from the water and the unsuccessful attempt to revive the child was recounted with dramatic effect. It was agreed that Elizabeth was attempting to destroy the infant, but the surgeon who had attended at the scene after the drowning explained to the coroner how the mother was at the time of the event suffering 'great mental excitement'. After various witnesses presented their evidence, it was agreed, by coroner, surgeon, newspaper reporter and witnesses to the crime, that, at the moment when Elizabeth plunged into the water, she was suffering from mental derangement resulting from her recent childbirth. This was confirmed by the fact that there was no attempt at subterfuge or concealment on her part. Elizabeth neither confessed nor hid her crime or tried to disguise it as an accident. Her story was related as one where her state of mind was responsible for an action which was totally out of character; prior to this sad occurrence Elizabeth had been an orderly, respectable woman and loving mother. The plea went undisputed. The coroner expressed particular sympathy for the accused, who, he declared, had suffered a good deal of distress before and after the incident. The defence of temporary insanity appears to have been clearly understood by medical, legal and lay participants. Elizabeth's defence, which led to her committal to the Warwick County Lunatic Asylum, was her state of mind, taken by all to be sufficient explanation for the murder of her child.[2]

The drowning of Elizabeth Barnwell's infant occurred during a period of considerable concern locally and nationally about the ease with which women were concealing their pregnancies and births and then doing away with their newborn infants.[3] Together with maternal ignorance on infant feeding, the poor supply of cow's milk and the problems of wet-nursing, infanticide contributed to a mortality rate among children that, according to the *Lancet* in 1858, 'out Herods Herod'.[4] During the mid-1860s the public outcry about the high incidence of infanticide reached a peak, fanned by the expanding and often sensationalist press and the appointment of populist coroners[5] and contrasting starkly with the 'boast to have extirpated infanticide in India'.[6] Infanticide was declared as prevalent in 1864 as at any time in history:

> no one who sees the newspapers can have failed to observe almost daily, instances of new-born children found, it may be at a railway station, it may be on a door step, but always under circumstances leading to the belief that the child so found had been destroyed during birth, or immediately afterwards.[7]

'It has been said of the police, with too much truth,' reported Mrs Baines in the *Journal of Social Science* in 1866, 'that they think no more of finding the dead body of a child in the street than of picking up a dead cat or dog.'[8] Such horrors were summarised in evidence presented to the Royal Commission on Capital Punishment in 1866, and coincided with the furore over baby-farming which also erupted in the late 1860s.[9] Illegitimate children were seen as being most at risk, particularly when born to young servant girls. Moreover, it was recognised that only a very small proportion of murders reached a coroner's inquest or the courtroom.[10] 'But who shall declare the number of infants murdered in secret?' questioned the *Social Science Review*. 'Who shall tell how many cases of infanticide are never brought into the light of day? We fear their name is legion.'[11] There were very few expressions of concern about women being wrongly accused, but frustration and anger were provoked by acquittals on what were claimed to be the most slender of grounds. Women were getting away with the murder of their infants, it was argued, because of a failure to detect the crime in the first place and also because of events taking place in the courtroom itself, including the indiscriminate use of the insanity plea.

In 1803 the Offences Against the Person Act overturned the harsh Stuart law of 1624, which had decreed that the mother of a bastard child was guilty of murder, and liable to the death penalty, if she tried to conceal the birth by hiding the body of the infant. The onus fell on the mother to prove that the child had been stillborn or had died from a natural cause. After 1803 infanticide was dealt with like any other murder, with the mother assumed innocent until proven otherwise. If a murder charge failed, as it often did, the jury could return a verdict of 'concealment of birth', which carried a penalty of up to two years' imprisonment. To prove murder, the prosecution had to establish that the infant was fully born and existing independently of the mother's body at the moment that the crime took place. A further Act of 1828 stated that a charge of concealment could be brought without reference to the 'separate existence' rule; even if the child was stillborn, the mother could be liable for concealing the birth and hiding the body. The 1861 Offences Against the Person Act allowed the prosecution to bring an independent charge of concealment of birth, while leaving it as an alternative verdict to a murder charge. In fact, no woman was hanged for the murder of her newborn after 1849, and few were convicted of murder by mid-century, most being acquitted or found guilty of the lesser offence of concealment.[12]

The legal situation was deemed highly unsatisfactory as public revulsion at the number of infanticides mounted during the 1850s and 1860s. The severity of the penalty for infanticide, death by hanging, compared with the mildness of the punishment for concealment, was identified as a particular failing of the law. Meanwhile, the difficulties of establishing separate existence and of proving that murder had taken place, which even the growing body of forensic science failed to address adequately, made juries less likely to find mothers guilty of infanticide. The punishment was also deemed 'unequal': 'it falls on the wretched mother, who may have been more sinned against than sinning, while the equally or more guilty father cannot even be brought under the power of the law.'[13]

The position of medical men in the debate on infanticide was an ambiguous one. They were often at the forefront of the intense public reaction against infanticide, exposing the problem and searching for remedies, with the *Lancet* being particularly prominent in this campaign. Doctors were also closely associated after 1864 with the take-up of the infanticide issue by the National Association for the Promotion of Social Science. In 1865 the Harveian Society undertook a survey of the social aspects of newborn child murder, led by Charles Drysdale, of the Malthusian League, Ernest Hart, later editor of the *British Medical Journal*, and the former military surgeon and supporter of the Contagious

Diseases Acts, J.B. Curgenven. George Greaves, a Manchester obstetrician, William Burke Ryan, author of a socio-medical study of infanticide,[14] and Edwin Lankester, coroner to Central Middlesex, were closely connected to a campaign which was to enhance the image of the medical profession, dealing as it did with such a critical social issue.[15] Ryan presented his case, condemning the prevalence of the crime of infanticide evocatively and bluntly, in 1862.

> In the quiet of the bedroom we raise the boxlid, and the skeletons are there. In the calm evening walk we see in the distance the suspicious-looking bundle, and the mangled infant is within. By the canal side or in the water, we find the dead child. In the solitude of the wood we are horrified by the ghastly sight, if we betake ourselves to the rapid rail in order to escape the pollution, we find at our journey's end that the moldering remains of a murdered innocent have been our travelling companion; and that the odour from that unsuspected parcel too truly indicates what may be found within.[16]

Many doctors, however, had long doubted women's ability to do away with their offspring knowingly. A lecture delivered by the renowned surgeon and man-midwife William Hunter in 1783 and published posthumously the following year was often used as a reference point. Hunter considered himself a good judge of women's characters, and, as Mark Jackson has argued, he legitimated his authority in cases of infanticide by referring to cases taken from his clinical practice, linking social commentary with forensic medicine to elicit sympathy for accused women.[17] According to Hunter's interpretation, pregnant women were deserving of compassion, the absent father of the child the real criminal, the mother weak, credulous and deluded. The destruction of bastard children, 'when committed under a phrensy from despair', when the women were 'violently agitated', should, Hunter concluded, 'as it must raise our horror, raise our pity too'.[18] In 1831 the forensic expert Charles Servern argued that if proof was unequivocal, mercy should be shown,

> Scarcely deeming it possible that any individual of the female sex, distinguished usually, and deservedly for the strength and ardour of the natural affections, could stifle the loud voice of nature and imbrue her hands in the blood of the helpless being to which she has just given birth; a crime which when committed can only be

accounted for by the temporary frenzy of despair, or a degree of moral turpitude, of which few even of the most degraded of women could think without horror.[19]

By the nineteenth century, doctors were accused of offering support to judges and juries already tending towards leniency and of failing to present their medical evidence conscientiously.[20] The role of doctors in the witness stand was to become more significant, particularly as the forensic evidence they were required to offer their opinions on became increasingly complex. As Roger Smith has demonstrated, the insanity defence was used more often during the course of the century, with medical evidence being required on the state of mind of the accused at the time the crime was committed.[21] Individual and social problems, Smith has argued, were described in terms of medical concepts, while questions of 'irresistible impulse' and 'responsibility' played a greater part in legal debates.[22] However, as the examples in this chapter show, the medical evidence tended to be more than medical, often explaining the women's actions as part of a succession of personal disasters and problems, some health-related but some linked, for example, to family worries and financial difficulties. Doctors were keen also to offer their views as character witnesses. The vested interests of the emerging group of alienists, engaged in establishing asylums and urging lunacy reform, had a part to play as well, as the courtroom became a place where they could demonstrate their status and expertise, sometimes claiming, by means of their medical knowledge, an expertise superior to that of the legal profession.

It was as if the plea of insanity following childbirth dropped into the waiting laps of the medical profession, already predisposed to be lenient and dispute women's ability to harm their offspring unless they were suffering from a severe form of mental disturbance. Puerperal insanity, which was 'not a rare disease, and it may take the form of homicidal mania, threatening the life of the child', offered a more refined way of presenting the insanity plea.[23] Women suffering from this form of insanity were deemed either to be unaware of what they were doing or to have temporarily lost control of their actions. Infanticide represented the antithesis of female nature, a total rejection of maternal ties, duties and feelings. Puerperal insanity could explain this, with the mother becoming, as a result of her mental disorder, confused, despondent or driven to a murderous fury. As we will see later, the plea of puerperal insanity could fit different infanticide scenarios, including that of the unmarried mother who quickly destroyed her newborn infant after delivery, and the married woman, often exhausted by breastfeeding, who murdered an older child. Judges and juries appear to have been convinced by this explanation, while other witnesses eagerly took it up and fed into it.

This chapter will explore how puerperal insanity was used to defend women accused of infanticide and concealment, as well as the treatment of women convicted of infanticide in the asylum, drawing particularly on evidence from Warwick where it has been possible in some cases to link court appearances and inquests to asylum admissions. It will be argued that alienists as well as other medical practitioners saw infanticide as an anticipated accompaniment, almost a symptom, of puerperal insanity, while the Medical Superintendents at the Warwick and Edinburgh Asylums were keen to explain that the temporary nature of the insanity could cast doubt on the defence plea in such cases. The case of Elizabeth Barnwell, for which the records are particularly rich, will be discussed in detail. Her story unfolds from childbirth through to the crime, inquest and coroner's report, committal to the asylum, her treatment there, and her subsequent recovery and release. The case illustrates the important role of medical and lay witnesses, and demonstrates why it was so difficult to pursue mothers who murdered their children through the courts when the insanity plea was applied. In Elizabeth's case no attempt at pursuit was made. It will be suggested that the use of puerperal insanity as a defence plea satisfied many of the ambiguities surrounding infanticide. It mediated between the wrath provoked by high levels of child murder and the sympathetic approach towards the mothers who committed the crime.

Linking puerperal insanity and infanticide

The propensity for harmful behaviour on the part of mothers following childbirth had long been recognised by doctors and courts.[24] Following the challenge of giving birth, involving intense physical effort, pain and disruption of the delicate reproductive organs and nervous system, and in many cases an increased strain on family resources, mothers were seen as liable to become deranged, peculiar, neglectful or violent. One seventeenth-century woman described how Satan had urged her to drown herself in a pond near Leeds, 'but by the providence of God, having a great love to a young infant I had then ... I looked upon the child, and considered with myself, what shall I destroy myself and my poor child?'[25]

Mark Jackson has cited a number of cases occurring in the eighteenth century where the woman's state of mind was considered grounds for acquittal. These included a woman tried on the Isle of Man in 1774. There was no evidence that the child had been born alive and 'in the course of the trial, the woman was found insane'.[26] Dana Rabin has claimed that by the eighteenth century 'infanticide by married women was considered so shocking and so unlikely that the only motive assigned to it was insanity'. Mary Dixon, tried and acquitted in 1735, told the court that she was 'weak and faint' on the morning that the baby was found, while one witness, Mrs Bousset, 'could find no sense in her', and together with the midwife 'thought she was dead'.[27] Puerperal insanity gave a rationale and explanation for behaviour which had previously been remarked on simply as being 'senseless' or 'phrenzied', although it too was described in elastic terms to fit many different scenarios and symptoms. As puerperal insanity began to be used with increasing authority as a defence plea in cases of infanticide or concealment, this, in turn, bolstered the position of doctors presenting themselves as experts on the disorder.

In 1820 Robert Gooch produced the first detailed account in English of puerperal insanity.[28] Just two years later, in 1822, a trial taking place at the Old Bailey is cited by Joel Peter Eigen as the first instance of a 'gender-specific psycho-physiological debility: puerperal insanity' being used in a defence plea in the trial of a married woman for infanticide.[29] Rather than 'announce' that the woman was simply deranged, insane or delirious, the medical witness, surgeon Dalton, sought to provide a context and explanation for the crime: hereditary predisposition to insanity, the birth itself and irritation caused by 'the breast extremely full of milk' at weaning. Combining the role of medical and character witness which many doctors were to take in such trials, Dalton testified that 'she was an affectionate wife, and had correct parental feelings'.[30] As descriptions and discussion of puerperal insanity passed during the mid-nineteenth century from obstetric and psychological literature into forensic texts, it was used as a common and frequently successful defence strategy in infanticide and concealment trials.[31] Not only experts on insanity and midwifery, but also general practitioners and surgeons appear to have been well aware of the liability of women to develop mental disorders after giving birth and were confident in pronouncing on the defendant's state of mind. It was men like Dalton and the local surgeon in Elizabeth Barnwell's case who were frequently called to the witness stand as well as eminent forensic experts.

Accounts of puerperal insanity abounded, with stories of unseemly, disruptive behaviour and violence to self, husband, attendants and particularly the newborn infant, which manifested itself in normally peaceful and caring women. In this sense, violence towards the infant was part of a much broader spectrum of alarming and threatening behaviour. All authors on puerperal mania described the potential to harm on the part of mothers, though some, like Fleetwood Churchill, who remarked that the mother became 'forgetful of her child', played this down.[32] They warned those dealing with such cases to remove all dangerous items, even spoons, bed sheets and handkerchiefs, and certainly all sharp objects: 'and the patient should never be left for a moment alone, or especially with her infant.'[33] Conolly included in his list of premonitory symptoms of puerperal mania 'an observable carelessness as regards the infant, as though without a want of affection for it: sometimes the indifference to it is complete; and sometimes it becomes an object of aversion'.[34]

The language of misrule and danger changed little over the decades. In 1859 Forbes Winslow described how the woman's temper 'changes completely, and family affection is apparently changed into the bitterest hatred; and this is particularly observed as regards the child, which the mother often attempts to destroy'.[35] In the 1880s the eminent psychiatrist Thomas Clouston was describing how childbirth,

> One of the most joyous times of life is made full of fearful anxiety, and the strongest affection on earth is then often suddenly converted by disease into an antipathy: for the mother not only 'forgets her suckling child', but often becomes dangerous to its life.[36]

Both melancholia and mania could lead to infanticide, the former through hopelessness, apathy and neglect, the latter through a marked, though often short-lived, loss of reason and violent, explosive behaviour.[37] It was mania that appeared most frequently in infanticide cases with its 'temporary' nature and sudden onset, typified by the struggle mothers felt between not wanting to harm their infants and an inability to prevent themselves from doing so.[38] In 1848 James Reid, Physician to the General Lying-in Hospital in London, an author on puerperal insanity who was much cited in forensic texts, described how 'the mother is urged on by some unaccountable impulse to commit violence on herself or on her offspring', while at the same time being 'impressed with horror and aversion at the crime'.[39]

The infant is usually the object ... in puerperal insanity; an impulse to destroy, haunts the mind continually, and struggles with maternal tenderness, which as strongly checks the act. The sufferer, in some cases, implores that the infant may be removed from her, lest she should altogether lose her self-control, and is heard praying to Heaven to prevent her from yielding to the temptation.[40]

The alienist and author on medical jurisprudence, James Cowles Prichard, cited just such a case. In 1824 the mother of two children in humble circumstances applied to the Hitchen Dispensary 'in consequence of the most miserable feelings of gloom and despondency, accompanied by a strong, and, according to her account, an almost irresistible propensity, or temptation, as she termed it, to destroy her infant'. The feeling came on a month after the birth and she begged to be 'constantly watched'. She recovered, but, when she had another child a few years later, again 'was assailed by the propensity to destroy it'. During the worst phase of the illness, 'she is perfectly aware of the atrocity of the act to which she is so powerfully impelled, and prays fervently to be enabled to withstand so great a temptation'. She retained a great affection for the child, at the same time even identifying the instrument that she would use to destroy it, fearing to handle a knife even at meal times.[41]

Infanticide was seen as the worst possible scenario in an episode of disruptive and violent behaviour, an episode that could be brief and vicious, but could also extend over many months of miserable depression. It could be argued that doctors saw infanticide as an actual symptom of puerperal insanity, so prevalent was the extraordinarily deranged and dangerous behaviour of the women, which was also, as seen in earlier chapters, expressed in sexual obscenity, filthy language, self-neglect, attempted suicide and violence to others.[42]

Even if their infants had not been badly injured or killed, the women often ignored them or handled them roughly. Mary Turner, admitted to Warwick Asylum in 1865, having become 'sullen & sleepless, destroyed her clothing & stripped herself, denied her husband and child & on one occasion threw the child from her'.[43] A year later a single woman, Elizabeth Tandy, noted to be 'a congenital imbecile' as well as suffering mental disturbance following childbirth, was moved from the workhouse to the Warwick Asylum after becoming excited and unmanageable, noisy and at times violent. She had 'beaten several of the inmates & has ill-treated her child so that it had to be taken from her'.[44] In 1871 Thomas Madden related a case where the woman became disturbed a few days after a tedious forceps delivery at the Rotunda Lying-in Hospital. The child was removed from her, and at night she was given soothing medicine; the woman feared that the child was being removed in order to kill it and that she would also be poisoned. At 4.00 am Madden was called and found the woman 'standing erect in the centre of the bed, almost perfectly nude, holding her infant tightly clutched to her heart ... she was shaking the child about with great violence, at the same time protesting vehemently that she would not kill it'.[45]

So disordered in their senses were the women suffering from puerperal insanity that, even when they had not injured the infant, they believed that the child had perished, and in many cases that they had destroyed it and were awaiting punishment for this dreadful crime. Their delusions as well as their nightmares often focused on the theme of the death of the infant or even the entire family. In 1766 the eminent physician Dr John Munro recorded in his case book the mental struggles of Mrs Wilson, who 'thinks herself very wicked, & is afraid she shall murder her husband, or herself or her child, for whom she says she does not feel the affection she formerly did thou she loves her beyond any thing'.[46] James Reid reported a case 'where the patient having fallen asleep in good health, awoke suddenly, crying out that her child was dead, and became maniacal from that moment'.[47] A patient suffering from puerperal mania referred to by Gooch was placed in a private madhouse near London after trying to commit suicide by drinking acid, swallowing a sponge and pricking her arteries and veins with pins. She was convinced that she had committed crimes that had brought ruin and disgrace on herself and family, which had occasioned the death of her husband and children.[48] Gooch reported on other instances where the mother was convinced that she had brought about the ruin of the family, the death or suffering of her children, and would be punished on earth and also by God for her sins.

The onset of the disorder was in some cases associated with anxiety about the child's failure to flourish and fear for its death, which, given the high infant mortality rate, must have figured prominently in many women's minds. It could be argued that these fears were exacerbated by the dread associated with their forthcoming confinements and their potentially dreadful aftermaths, for themselves and their infants, a dread fanned, as we have seen in chapter 1, by advice literature which, while encouraging families to organise proper medical attendance at the birth, also placed responsibility for disasters squarely on the shoulders of the mother. Thomas Bull in his *Hints to Mothers* did much more than hint at the harm mothers could cause through neglecting their responsibilities, the health and constitution of the offspring being dependent upon the conduct of the mother:

> Should she, however, be careless and negligent upon this head, and fail in attention to the measures which her new condition demands ... her child will inevitably be variously and injuriously affected, these causes operating through her system upon that of the child.[49]

While denouncing the notion of maternal imagination affecting the child in the womb, Bull urged calm and tranquillity. Women unable to maintain this calm throughout pregnancy risked injuring their child. He cited the example of a woman who was greatly depressed during pregnancy due to the 'worry' of her husband about supporting another child. 'In consequence of this mental harass[ment] and disturbance, she was confined shortly after the completion of the seventh month. The child was puny and fretful, and continues so.' At eight months it was a 'wasted miserable-looking object', who rarely smiled.[50] Though described by many physicians as a superstitious practice, even the dreadful burden of maternal imagination lingered well into the nineteenth century in the minds of many women.

In 1855 a case was admitted to Worcester Lunatic Asylum of a woman subject to severe nervous attacks during pregnancy who was said to have 'overexerted herself feeding the pigs ... after the birth of her child some of the neighbours had observed some marks upon the child said to be the likeness of one of those animals that preyed much upon her mind & gave her great distress'. She became insane two weeks after the birth and wanted to destroy herself and the child, but subsequently recovered.[51] Robert Boyd referred to a woman admitted to the Somerset County Asylum suffering from puerperal insanity who believed herself 'bewitched'; as late as 1870 Boyd believed that the 'subject of mental emotions during pregnancy causing malformations or disease in the foetus is still an unsettled question'.[52] The infanticide scenario was more complex than mothers simply being suddenly driven to murder their infants by their insanity, and was rooted in dread and anxiety about hurting their children, a dread which for some women had persisted throughout pregnancy. Even in convalescence from puerperal insanity, months after recovery had commenced, doctors had their work cut out to persuade mothers that their infant was alive and well.

Aside from agreement on the fact that it was a condition closely linked to childbirth, for the rest, as we have seen in chapter 2, puerperal insanity was an untidy, elusive disorder. Despite being pursued by many 'experts' in the fields of psychiatry and obstetrics, no firm conclusions could be reached regarding its onset, preconditions, causes, prevalence, precise timing and duration, where it should be treated, how it should be treated, if it was more likely to affect first-time mothers or women who had borne many children, the chances of recurrence, and whether it would prevail more among undernourished, mistreated and deserted poor women than among well-to-do, but feebly constituted ladies for whom childbirth was considered a massive physical and mental shock.[53] The latter were considered less likely to kill their children as they would not be left unattended, particularly if the illness was detected, when domestic servants and nurses would protect the child from harm. For those lacking such buffers, it was reported time and again how women suffering from puerperal mania had taken the opportunity to murder their infants during moments of carelessness. The newborn, sometimes against the advice of a doctor who had warned of such an outcome, had been left alone with the mother, in some instances for just a few minutes.

It was this very vagueness that gave the insanity plea its flexibility in infanticide cases, particularly in terms of timing, but also in defining the kinds of mothers that could be affected. It also led to grave concerns that the plea would be used indiscriminately, for example, on behalf of women murdering older children long after they had given birth, or even that the disorder would be related to a previous confinement. In 1854 John Charles Bucknill, Superintendent of the Devon County Lunatic Asylum and author on the relationship between crime and insanity, pushed the connection to the extreme when he referred to a woman, then under his care after making a recent attempt on her child's life, who had had an attack of puerperal mania 22 years ago. Though he attributed the exciting cause of the act to an argument with her aunt and to ill treatment by her husband, her much earlier episode of mania was considered to have a bearing on her crime.[54] If the woman suffered attacks of insanity even years previously, it was argued that she was also likely to be predisposed to harm her offspring, driven by her vulnerability to mental disorder as well as an intrinsic instability in her character.[55]

Puerperal insanity, responsibility and medical evidence

Confidence in puerperal insanity and its use as a form of defence plea would not be crushed by anxieties about its elasticity. By the mid-nineteenth century Alfred Swaine Taylor, in his *Principles and Practice of Medical Jurisprudence*, was describing cases distinguished from 'deliberate child-murder by there being no motive, no attempt at concealment, nor any denial of the crime on detection. Several trials involving a question of puerperal mania have been decided, generally in favour of the insanity, within the last few years.'[56] Citing as his authorities Samuel Ashwell, the well-known author on the diseases of women,[57] and the alienist, George Man Burrows,[58] Taylor described the 'sudden impulse' prompting women to commit the act of murder, 'so that the legal test of responsibility cannot be applied to such cases'.[59] He went on to cite the case of Mrs Ryder, tried in 1856:

> There was an entire absence of motive in this as in most other cases of a similar kind. The mother was much attached to the child, and had been playing with it on the morning of its death. She destroyed the child by placing it in a pan of water in her bedroom. The medical evidence proved that she had been delivered about a fort-night previously – that she *had had an attack of fever*, and that she had *probably* committed this act while in a state of *delirium*. She was acquitted on the ground of insanity ... it was *evidently* a case in which the insanity was only *temporary*, and the prisoner might be restored to her friends on a representation being made in the proper quarter.[60]

This case combined a very loose description of the disorder with a certainty that the crime resulted from it. This was to typify the use of the insanity plea in infanticide trials. Emphasis on the lack of sub-terfuge or denial also became a defining characteristic of the disorder, as did, for married women at least, comments on the mother's attach-ment to the child and her lack of motive. They were consistently referred to as 'good' and 'fond' mothers. In the case of Mrs Ryder, it is possible that the delirium associated with puerperal fever was being confused with puerperal mania. It is not at all clear which condition she was suffering from.[61]

This elasticity of definition contrasted sharply with the increasing precision employed in seeking to establish criteria to prove when and how the deaths of newborn infants had occurred and whether they had been born alive, which occupied, for example, some 40 pages in Alfred Swaine Taylor's textbook. Proof of whether the child had breathed independently of the mother, or whether strangulation was deliberate or accidental (caused by the umbilical cord being wrapped around the child's neck), whether bruising had been purposefully inflicted or had resulted from the birth, or whether the child had died while a part of its body was still undelivered, was difficult, if not impossible, to ascertain. 'Proof' of the existence of insanity, meanwhile, was readily accepted, often on the flimsiest of evidence. Reading accounts of trials, it is possible to detect a palpable sigh of relief from judges, juries and witnesses if the woman could be declared insane. This also relieved the medical witness from the burden of working out exactly how the infant had died, as the woman was likely to be acquitted.

In some cases, however, witnesses, juries and judges could be left in little doubt about whether a murder had taken place, but were still prepared to consider the role of insanity. One trial that received much publicity in both the lay and medical press took place at the Essex Lent assizes in 1848, when a 37-year-old married woman, Martha Prior, was tried for the murder of her infant by almost decapitating it with a razor.[62] The *Lancet* reported that all the evidence went to show the existence, at the time of the murder, of puerperal mania, which came on a few days after delivery. One of Prior's neighbours, Mary Portway, when giving evidence, described the prisoner's restlessness, pains, discomfort and distracted behaviour. Prior told Portway that 'she was going to die, and wished me to tell her sister to be a friend to her children; she said that her breasts were very bad ... She seemed to me not quite right in her mind; she was flushed in her face.'[63] Prior persisted in telling Portway 'that she was going to die; she said I might shoot her, or get some one else; that she was going to hell, and might as well go at once'. When she saw her again later the same week, Portway described, in remarkably muted tones which typified many trial testimonies, how 'the child lay by her with its throat cut. I said, "Mrs. Prior, what have you been doing?" She said, "what I meant to do, and you may serve me the same". I removed the child to the cradle; I thought her mind was not as it should be.' Mary Portway also stated that the prisoner was a peaceable, well-conducted woman. No professional man, claimed the *Lancet*, summing up the case, would have had a moment's doubt concerning the existence of insanity.[64]

Mr Thomas Bell, the surgeon who delivered her and then prescribed medicines to treat her pains and disorders, declared Martha Prior to be labouring under puerperal mania. Bell complained that he had not been sent for earlier when Prior became prostrated following heavy bleeding. Yet he had cautioned her friends and relatives against trusting her with her child and wanted a nurse to be procured. 'I ordered that the child should not be taken to her, and she should not be left alone; her countenance was haggard and vacant; there was a great tendency to irritation of the body.'[65] Bell's instructions were not observed, and Prior obtained the murder weapon under the pretext of wanting to cut her nails. Prior was declared conscious of what she was doing, but unable to control her actions.[66] She was reportedly calm and collected after she had destroyed the child, claiming that it was what she had intended to do all along.[67] In this case credence was given to the testimony of a neighbour, which was not unusual. Nor was Portway's confidence in declaring that Prior's 'mind was not as it should be'. Signs of insanity included 'irritation of the body' and 'prostration'. Prior had bled heavily, she was vacant and had made no attempt to hide or deny what she had done. Bell's evidence, meanwhile, perhaps showed signs of an eagerness to escape any accusation of negligence or lack of care on his part.

Many infanticide trials depended on the evidence, not of a forensic expert or medical man experienced in treating insanity, but on the opinion of a surgeon, general practitioner or midwife, and a collection of witnesses, neighbours, friends and passers-by, all of whom found it appropriate to comment on the woman's state of mind and felt equipped to testify to the presence of insanity. Medical witnesses were often giving evidence on an ad hoc basis.[68] They may have come across a couple of similar cases during their practices, but could not claim expertise in treating mental disorders, although they may have had broad experience in obstetrics.[69] Yet they had no qualms about diagnosing and firmly ascribing child murder to puerperal insanity.

Other trials involved well-known experts, specialists in mental disorder and asylum superintendents, who were brought in to examine the accused and pronounce on the case, without prior knowledge of the defendant's medical history or circumstances. This led to further complications. Puerperal insanity was deemed a temporary condition, which could mean that the accused woman had recovered by the time she reached the courtroom, or even by the time she was examined by a physician. The suggestion of the existence of 'temporary insanity' meant that it was impossible to be sure about the mental state of the accused at the moment of the crime and whether she was 'responsible' for her actions. This could lead to unpredictable judgements. Doctors as well as judges were not always convinced, but erred on the side of accepting the insanity plea. Trial proceedings, judges' addresses to the jury and verdicts suggest the willingness of judges and jurors alike to support the idea that the effects of giving birth left women in a state of mental excitement, which was sufficient to explain their crime.

In summing up the Prior case, however, Lord Chief Justice Denman expressed some of the frustrations felt by judges about the authority claimed in court by medical practitioners and the enormous assumptions they asked the court to make. In the words of the *Lancet*, Denman threw unjust imputations on the reputation and professional standing of the surgeon Mr Bell and doubt on the insanity plea. Denman declared Bell's opinion 'rashly formed. How could one human being speculate upon the mind of another ... I don't understand how any man, with reasoning powers and scientific acquirements, can say that she acts upon a sudden impulse, when there is nothing that shows alienation of mind, and a great deal that shows a deliberate purpose.'[70] Yet the jury, responding as Denman predicted and acting 'upon the testimony of the medical gentleman', returned a verdict of 'not guilty' on the grounds of insanity. Denman concluded by remarking that 'such opinions were too often given by scientific men upon too slight foundation for the safety of the public'.[71]

However, medical men too could be left in considerable doubt in such cases, even those as experienced as John Charles Bucknill in treating insanity and acting as a medical witness. In the same volume where he had presented the link between infanticide and a case of mania occurring 22 years previously, he expressed concern about the Prior trial, finding Denman's comments not unreasonable, and objecting to the loose use of the term 'uncontrollable impulse', which was, he argued, the root cause of every crime.[72]

The Warwickshire trials

Many cases involving puerperal insanity as the defence plea featured the epitome of the infanticidal woman, the young domestic servant, accused of concealing her pregnancy, giving birth unattended and destroying the infant soon after. Between 1860 and 1865, 21 women were tried for infanticide or concealment in Warwickshire, 16 of whom were domestic servants.[73] The idea of the act of abandonment or infanticide taking place while the woman was in a frenzy of pain and shock was given credence by descriptions of a transient form of puerperal mania that could take place at the moment of birth itself.[74] This phenomenon proved particularly useful in explaining the actions of frightened young women giving birth alone, as it was ostensibly linked to first confinements. Such women frequently declared ignorance of the fact that they were pregnant at all, or claimed they did not realise that they had given birth, that the baby had come too soon, or that they were sent into a frenzy by the pain of childbirth. The single mother was also more likely to conceal the birth, but often in a very flimsy, distracted way, hiding the child under the bed, in a box or in piles of clothes, or abandoning it in the privy, the scene of many unwanted births.[75]

The insanity plea was also used in cases that tended to feature rather less in the national outcry about the prevalence of child murder, where the accused was a married woman, often with a number of children. In these cases the crime would usually take place when the child was several weeks or even several months old. The stakes were higher as these were clear cases of infanticide. It was generally agreed that the onset of puerperal insanity was within the six-week 'puerperal' period, peaking around a few days to two weeks after delivery, but it could also manifest itself much later, when it blurred into lactational insanity. The two periods of intense danger for women, when they were most liable to become mentally disturbed, had been defined by Robert Gooch and others as shortly after birth and then again several months later when their health had been broken by continued efforts to breast-feed when they themselves were in a weakened state. The risk of developing lactational insanity was deemed greater for married women, as they were more likely than single mothers to continue nursing their infants. These women were usually described as being devoted mothers who were fond of their children.

A plea of mental derangement was more often than not also tied in with a 'poverty defence'. The women were penniless, malnourished, had borne too many children in quick succession, were sometimes physically disabled or very sickly, or were victims of husbands who mistreated them or did not provide properly for their families.[76] In other cases, like that of Elizabeth Barnwell and Selina Cranmore (to be discussed below), the circumstances were not regarded as particularly difficult, and the women were described as simply acting out of character and as being afflicted with a temporary loss of reason. The mother in such cases tended to do away with the child quickly and cleverly, while she had been carelessly left unattended. Typically, she had begged not to be left unguarded, having felt a compelling urge to destroy the infant. Again, hardly any attempt was made to hide the crime, which was soon discovered by the neighbours or family.

While the view that most infanticides never reached court is probably correct, the perception of the high rate of acquittals has not been borne out by evidence taken from Warwickshire. Out of eight cases of infanticide recorded in the Calendars of Prisoners for Warwickshire between 1839 and 1857, only three, appearing together in court in the summer of 1856, were acquitted.[77] The five other infanticides reaching court between 1839 and 1857 were found guilty of murder, manslaughter or concealment and sent to prison for terms of up to two years' hard labour.[78] Of the eleven women charged with the lesser crime of concealment between 1850 and 1855, four were acquitted. The rest received custodial sentences, ranging from three days' imprisonment to six months' hard labour.[79]

The cluster of acquitted cases in 1856 sparked a burst of interest in the local press. In one case a verdict of accidental poisoning was reached. Regarding the other two cases, in his opening remarks Mr Justice Cresswell advised that the prisoners should be acquitted, on the ground that the acts resulting in the death of their infants were not 'wilful'.[80] The case that attracted most attention was that of an 18-year-old domestic servant, Sarah Harris, described as being deranged at the time that the crime took place. Sarah Harris had given birth at her place of work in Birmingham while her mistress was in the house, apparently unaware of what was going on. After a detailed account of the movements of the accused around the house and her complaints of feeling 'unwell', and of her even being tended by her employer, she gave birth alone. Harris then suffocated the baby in her skirt before hiding him under the bed. Her employer's suspicions were aroused by strange noises 'like cats', and she raised the alarm and called a midwife, Mary Durnell, to the house. Durnell testified that Harris was 'in a wild state like a mad person', and she had great difficulty to keep her in bed, 'she was so frantic'. For a long time, she was unable to utter a sound but when she did speak it was to declare 'Oh dear, I have murdered the baby'. The midwife found the baby under the bed wrapped in a black skirt, and an examination confirmed that Harris had recently been delivered. The defence lawyer declared that the prisoner could not be held responsible for what had occurred during her 'furious delirium', and the forensic examination could not draw any firm conclusions. The child had suffocated and had lacerations around the mouth, but these, it was pointed out, could have occurred during delivery. Justice Cresswell in his pre-trial remarks had expressed the same view and Sarah Harris was duly acquitted.[81]

Absence of forensic clarity often coincided with cases where the woman's state of mind was in doubt, making it extremely difficult to obtain a conviction. In May 1860 Catherine Malins was brought before the Warwickshire coroner so weak and exhausted from the birth of her child that she was considered unfit to participate in the proceedings.[82] Malins, a young woman who lodged with Mr and Mrs Blackwell in Warwick, claimed that she had unwittingly delivered the child in a privy. The surgeon who was summoned to the scene believed that the infant had not been stillborn but had died shortly after being delivered. It was unclear whether the child had been born in the privy or had been put there subsequently. Malins denied knowing that she was pregnant and described how the child had 'cried and cried while it was falling'. The case, it was concluded, could not be brought to trial, there being insufficient evidence to show how the infant had died.[83]

As the momentum of the campaign to tackle infanticide grew both locally and nationally during the 1860s, it attracted more press coverage. 1865 saw another cluster of cases in Warwickshire, and the interest these caused was fuelled by accounts of more distant infanticides in London and even the continent. The case of Esther Lack, a London woman who had murdered her three children, sparked a cynical response in the *Warwick Advertiser* in September 1865, under the caption 'insanity and crime'.[84] Lack had cut her children's throats at night while her husband was working. Lack claimed that she feared that they would die of starvation. Her crime was attributed 'to debility of constitution, caused by the delivery of three infants at a birth some seven or eight years ago'. Since then her eyesight and strength had failed her. The *Warwick Advertiser* concluded its report by stating that the case 'would probably be resolved by psychologists into an abnormal action of the brain'.[85]

Despite such outcries, acquittals continued. Caroline Russell, one of the three prisoners appearing in court in the Warwick Summer Assizes of 1865, was accused of attempting to murder her illegitimate female child by strangulation. Seventeen-year-old Russell gave birth in secret at the home of her employer. Like Sarah Harris, there was a long story of goings-on in the house, odd noises and obvious excitement on the defendant's part. Mrs Brown, another servant, found blood on the floor and, persuaded by Caroline Russell that she was unwell, produced a mop and cleaned up. Mrs Brown went up later to Russell's room and was asked to get a bath for her. Mrs Brown asked, 'Caroline! what have you been doing?' and she replied 'Nothing at all'. Long after suspicions had been aroused, the cries and blood were correctly interpreted and a doctor called. Dr Tibbets found the child with an apron string around its neck, tied tightly, in a box in Russell's room. The child was successfully revived. It was concluded that Russell had tied up the child but had probably not intended to take its life. She was discharged, her youthfulness and erratic behaviour apparently influencing the jury in reaching their verdict.[86]

Infanticide and the asylum

The press and public could be cynical about the role of doctors in infanticide and concealment trials, believing that a plea involving puerperal insanity could be shaped into any form to suit the occasion and evidence. Some doctors took the witness stand only rarely in such cases, particularly those who simply happened to be called in to the scene of the crime, and the ease with which they linked the crime to mental disorder may have seemed gratuitous. Asylum superintendents, however, had considerable experience of puerperal insanity and saw infanticide as a tragic, but not unanticipated accompaniment and 'symptom' of the condition.[87] Skae (Medical Superintendent to the Royal Edinburgh Asylum) described how 'in almost every case of puerperal insanity there is some morbid change in the maternal feeling towards the child'. Whatever 'mysterious sympathy' existed between a mother and her baby vanished in the insane mother, and was replaced by hatred and homicidal urges.[88] In his Annual Report of 1848, Skae emphasised the need to move patients suffering from puerperal mania rapidly to the asylum, to protect their infants from harm as well as themselves from deterioration. He also commented on the rapid recovery of one woman admitted after murdering her child, which would, he argued, leave her open to suspicion about whether she had ever been insane.

> In the female referred to, well-marked symptoms of mental derangement were manifested soon after the birth of her child, and continued to shew themselves in occasional outbursts of passion, threatenings, sullenness, and irritability. Yet she was left alone with her children, and but imperfectly watched, until at last, under a sudden impulse, she destroyed her infant, and made an attempt upon her own life. This case suggests other very serious reflections. After her admission she remained calm and rational, exhibited no delusions or paroxysms of excitement, – she conducted herself with perfect propriety, and it would have been extremely difficult for any one, not knowing her previous history, to have detected a trace of mental derangement. I cannot help asking the question, had the child been illegitimate, or had its death occurred in circumstances of any kind which were calculated to excite suspicion, how would it have fared with this female had she been placed at the bar of our public tribunals?[89]

Skae went on to relate another infanticide case where the circumstances had been carefully reported by the woman's relatives. The
woman, J.T., was married with three children, well educated, sober and
industrious, but nervous. The disease was of three months' standing.
She was tormented by hallucinations of hearing.

> [V]oices occasionally prompted her to destroy her children. She at
> the same time felt an impulse to obey these internal suggestions;
> and she would spring suddenly out of bed in a phrenzy of anguish,
> and earnestly petition her friends to tie her hand to prevent the per-
> petration of a deed from which her reason and conscience recoiled
> with horror. Her friends state that during the time she was subject
> to these homicidal impulses, her intellectual faculties were more
> powerful and acute than ordinary, and she reasoned upon and
> deplored her miserable condition in a most affecting manner.[90]

Skae was outlining yet another instance of the struggle reported by
many women intent on destroying their infants, and pleading with
those around them to prevent them from doing so. Yet he argued that,
in this instance, the woman was under no delusion regarding her
judgement of right and wrong and displayed intellectual acuteness.
How, then, could she escape the conviction of infanticide and the last
penalty of law?

In this case the woman forgot her homicidal impulse rapidly and on
admission her thoughts turned solely to suicide. She was described in
the case notes as terribly fearful, believing that her soul was doomed to
perdition and that she was about to be buried alive. She remained
unresponsive to treatment for some time, taciturn, dull, inactive and
melancholy, and had to be fed with a stomach pump. However, she
made no attempt at suicide and after three weeks began to improve
and speak of her home and family 'with strong affection', expressing
concern about their welfare and desperation to see them again. Since
then she had thought and spoken of nothing else, and 'night to day
she cries upon her "Jamie", and her "bairns" to come and take her
home'.[91] The woman in this sense was recovered, her mothering
instincts restored. Skae agreed in conclusion that in some cases the
plea of insanity had been applied laxly in criminal cases, but stressed
the need for legal definitions, medical opinion and popular sentiment
to combine in recognising the true features of irresponsible mania.

Dr Henry Parsey, Medical Superintendent of Warwick County
Lunatic Asylum, was careful in presenting evidence to an inquest into

child murder to spell out in detail how puerperal insanity could attack and retreat relatively quickly. He had seen many cases of puerperal mania in his asylum work and believed its link with infanticide to be a close one. In April 1858 he was called to give evidence in a case of child murder which had taken place in the village of Berkswell, near Coventry. The woman accused, Selina Cranmore, would subsequently become his patient. The case was reported in the *Warwick Advertiser* on 10 April: 'The peaceful quietude of the retired village of Berkeswell ... was disturbed on Thursday last by a report that a respectable married woman named Selina Cranmore, had murdered her infant son aged five weeks, by strangling it with a piece of tape.'[92]

The circumstances of the case were unusual in that her husband, a bricklayer, was in 'pretty good circumstances'; he had a small business and they 'lived together comfortably' with their three children, to whom she was 'a kind and affectionate mother'. The inquest had been held on 8 April, when Parsey presented his evidence based on an examination of the accused. He concluded that Selina Cranmore was of unsound mind. She had been entrusted with managing the family finances, and formed the impression that her husband was in difficulties and about to be ruined. Her neighbour had hanged himself on the same day that her child had been delivered and she believed that she had hanged him. Parsey was of the opinion that her hallucinations pre-dated the crime: 'I should think she was at that time labouring under an uncontrollable impulse, which is not at all uncommon in women some weeks after childbirth.' She also told Parsey that she felt the baby to be a burden she could not maintain, though she had not felt this about her other children. The coroner pointed out to the jury that it was important to listen to Dr Parsey's evidence, because, though now insane, by the time Selina Cranmore was brought to trial she might be of sound mind.[93]

Selina Cranmore was brought to Warwick Gaol to await trial. After a week she was visited in prison by Parsey and removed to the Warwick County Lunatic Asylum. On her admission, Parsey was keen to acquire further insight into the case and her mental distress. Her husband had been ill during the previous autumn, which had made her very anxious. Before the child was born she was depressed and worried, wishing she were dead. She often had the feeling she would not last the day, or that the meal she was eating would be her last, or that her children would starve. A few days after her confinement she was found by her husband banging her head on the floor. She had not been attended by a doctor at her delivery, but when one was sent for four days later he advised that she be closely watched. When Parsey visited her in prison she had spoken 'feelingly' of the dead child, who appeared in her dreams. On admission to the asylum, she was still deeply worried about her husband's finances and believed her children would be utterly neglected.[94] When she appeared in court in August, she was acquitted on the grounds of insanity and sent back to the asylum.[95] Selina Cranmore remained in the Asylum for an entire year from her first admission before she was discharged in April 1859, though the case notes already pointed to her recovery prior to her trial in August the previous year and she was declared to be of sound mind by Parsey and a local surgeon in September 1858.[96]

Experiences in the asylum varied. Some women recovered rapidly, though they may have been detained longer if they relapsed or in order to satisfy a feeling that justice had indeed been done. The asylum doctors appeared eager to understand why the crime had been committed and how it related to the woman's mental state, but little mention of the crime itself was made after admission. In October 1848 Jane Anderson, who had murdered her seven-week-old child, was admitted to the Royal Edinburgh Asylum. Of her eight children, six had died (the implication is from natural causes), and after the birth of her last child she became seriously disturbed and maniacal.

> She started suddenly out of sleep in a state of terror, fancying that she was about to be torn to pieces by wild beasts. Sometimes she fancied that she was about to be carried away and put to some violent death ... She was allowed to remain in her own house carefully watched, until this morning, where she murdered her Infant by cutting its throat with a Razor. She afterwards made a few scratches with the Instrument upon her own throat.[97]

A penknife was found in her pocket on admission. She had been insane eight years earlier, which was attributed to grief at the death of her daughter, and the case book noted that she had taken opiates to excess since her first attack. She also had a sister who was insane. On admission, she was put to bed and fell into a sound sleep: 'Either she has forgotten that she has murdered her infant, or she wishes to conceal the fact, for, she said, in answer to a question regarding it that "her youngest child died this morning: it was always a weakly child".'[98] The sheriff examined Mrs Anderson four months later. It was remarked that she had behaved very well while in the asylum, and occupied herself sewing, reading and taking exercise. She did not allude to the murdered infant and never indicated that she had any recollection of having caused its death. 'She has never spoken spontaneously about her Infant; but when her attention has been directed to it during conversation, she has shown some little exertion, but has expressed ignorance of the manner of its death.'[99]

The good character of women who had murdered their children was referred to during their asylum stay as it had been during their trials. So too was the nightmarish process in some cases of reflecting on their crime, for while some women did not acknowledge what they had done, others knew full well and were greatly distressed. In 1876 Mrs Smith or Hunter was admitted to the Stirling District Asylum. Aged 32, she was a 'decent, sober, hard-working woman, fond of her husband and of her children'.[100] She had been married eleven years and had in all seven children, three of whom had died in infancy. This fond and industrious woman had killed her seven-month-old child by striking her on the head.[101] She openly confessed her crime. Soon after her husband left for work on the morning of 8 August she ran into her neighbour's house crying that Lydia was dying. Her neighbour rushed to the house and found the child was dead. Mrs Smith was said to be in great distress, weeping and calling out, 'Oh, Annie! I have killed my Lydia.'[102] It was reported at the trial that Mrs Smith had become disturbed two months before she attacked her child. She was restless, sleepless, intensely nervous, refused to eat, was thin and haggard, wished to leave the house in the middle of the night, and that 'nervousness came on her like a fit'. She had been ill after the birth of her second child eight years previously, which was then attributed to puerperal fever, but it was now hinted that it might have been a previous attack of mental disturbance. The report continued that she was subject to an 'uncontrollable impulse', but was free from delusions. The neighbour's evidence was an important factor in the trial in describing her frenzied state. Mrs Smith pleaded guilty, but she was not sentenced in consideration of her state of body and mind, evidence showing that since the birth of the child she had been 'subject to nervous fits, sleeplessness, and weakness'.[103] She was placed in the charge of her mother, but a few days after the trial was admitted to the Stirling District Asylum in a terrible state of regret and remorse.

Entire horror of her crime seemed to possess her mind. In no patient I ever saw was a look of utter despair so stamped on the face. Among all the varied types that fill an asylum ward, hers was the face that invariably attracted attention, and provoked inquiry. A mental state beyond all rousing, an entire listlessness of body ...

In fact, her state was such as might be imagined in one who had a full appreciation of the terrible act she had been guilty of.[104]

She stayed for weeks in this state, twice having fits. And 'in spite of much care to prevent it, her story oozed out, and it was a singular study to notice the consequent aversion from her of her insane companions'. At first her state provoked sympathy, but then she became a 'sad, solitary figure'. The asylum doctors feared Mrs Smith might attempt suicide and she was carefully watched, but she too watched for her opportunity and was found dangling from a window sill on the second storey by another patient who attempted to hold on to her, before Mrs Smith lost her grip and fell to the ground. She sustained dreadful head injuries and remained insensible for six hours, but against expectations revived. After this her 'horror-stricken look' left her and she gradually recovered her strength and spirits, gained weight and looked well and cheerful, and appeared to be cured in body and mind.[105]

Other women fared less well, including Hannah Harris, a married Coventry woman, who attempted the life of her child and then tried to cut her own throat with a razor. While awaiting trial she became gloomy and taciturn and needed constant watching to prevent her removing the dressings on her throat. Little information was given to the doctors at Warwick Asylum when she was brought there from prison in February 1866, but she was said to be very ill and needed constant watching.[106] It was recorded that her 'suicidal propensity [was] still very strong; delusions of a very distressing nature connected with a vague clouded recollection of the attempted destruction of her child; sometimes piteously proclaiming that she did not do it, it is a mistake'. She mistook the identity of her fellow patients and had immense trouble sleeping; on one occasion she fell asleep standing and fell and cut herself. She was also said to be 'mischievous & spiteful – breaking windows & crockery & striking the patients & nurses'. When visited by her husband, it was reported 'that she had been in bad circumstances previous to her confinement & that other members of her family were insane'. She did not do well in the asylum, 'passing into a state of dementia'; she was described as being vain and at times abusive. By 1874 she had no recollection of being brought to the asylum or of the crime with which she was charged. After ten years in the asylum she was more tractable, capable of working in the laundry, 'cheerful but childish in manner'. In May 1877 the case book noted that she was weak-minded but capable of being discharged, yet the asylum doctors felt unable to do so while she was still menstruating; 'had the catemania ceased she might be discharged & sent home, while however there is a risk of her bearing children her detention is, in view of her past history, necessary'.[107] In July she was removed to Warwick Gaol, after eleven years in the asylum, on the order of the Secretary of State, and no more is heard of Hannah's case.[108]

While working as Assistant Physician to the Royal Edinburgh Asylum between 1864 and 1865 John Batty Tuke became particularly intrigued by the case of Mary Oswald, a 21-year-old married woman, who, following efforts to drown herself when pregnant with her first child, strangled the child eleven weeks after delivery and then attempted to poison herself.[109] Oswald survived and was placed in prison, where Drs Maclagen and Skae visited her. They saw her on various occasions, when she seemed happy and contented, 'exhibiting symptoms of morbid exaltation, talking of the prison being a palace to her'.[110] Certificates of committal were granted and she was taken to the Royal Edinburgh Asylum in March 1864. On admission it was verified that her sister and aunt had been insane; her aunt had been confined in an asylum on three occasions.[111] When admitted Mrs Oswald made herself at home and settled quickly to work, though her mind was weak and she was facile and reserved.

> A few weeks after admission, I got her to converse about her child, and her motives for destroying it. She was not in the least confused, nor did she seem to appreciate her position. She said that her impression at the time was that it would be happier if it was dead, and that she attempted suicide so that her husband might not be taunted with having a murderess for a wife. She expressed no remorse or regret. She continued in this state for nearly two months, when she again became depressed and melancholy, crying bitterly at times as she said, about her child, but in no way alluding to her own guilt.[112]

She relapsed once or twice during menstruation, but soon improved, saw her husband often and joined in all the amusements, becoming a 'great favourite both with attendants and patients'.[113] Six months after admission she was discharged on the authority of the Procurator-fiscal as recovered. Two years after her discharge, she was confined again and Tuke reported that she 'passed through her pregnancy without a bad symptom and made a good and perfect recovery'.[114]

In the long term, Mary Oswald, according to Tuke, did well: she recovered and did not relapse when her second child was born. However, Tuke's survey of cases ended in 1864 and Mary Oswald became disturbed again on several occasions and a regular inmate of the asylum. She was admitted for a second time in July 1871, when her mental collapse was attributed to over-lactation. She was not suicidal on this occasion but was 'dangerous to her children' and had attempted to strangle them. Reflecting on the first instance of insanity in 1864, the case history reported that the first child was born only a few months after marriage, which had seemed to affect her mind. There was a rich and varied history of insanity in the family, with many more members than her sister and aunt apparently being insane, eccentric or drunkards. She seems this time to have been brought to the asylum as a precaution, due to the fact she had killed one infant, was again nursing and had recently attempted to strangle her children.[115] In August she was removed by her friends against the advice of the asylum doctors,[116] but was readmitted a year later, in August 1872, again nursing, in poor health, melancholy and suicidal. Within two months she was fit for release and discharged.[117] In 1873, then aged 30, she was admitted for the fourth time. She was committed by her husband who feared for their children's safety. This time no link with childbirth or nursing was made; she was melancholic and suicidal, and she was discharged cured in March 1874.[118] Mary Oswald's case illustrates forcibly why doctors saw puerperal insanity and nursing as being so closely bound up with efforts to harm the infant and other children. In this case caution was urged, and she was admitted not only so that she could be treated, but to separate her from her children.

The asylum was, it could be argued in these cases, just that. It was not only a refuge, a better place to be than prison, but also the asylum superintendents, as shown by Skae and Parsey's views on the subject, were willing to argue the women's corner, to explain how the disorder drove them to commit horrible crimes and then left them in some cases so rapidly that it exposed them to accusations of guilt and made those who did not understand the condition strongly suspect that they had never been insane.[119] The asylum also served the purpose of protecting the child from harm by separating the mother from it when she was regarded as a danger to its welfare and safety. As we have seen in earlier chapters, the asylum offered care and often better conditions than the homes of poor women, and was certainly far superior to the alternative for infanticides, the prison, a fact that Bridget Butler may have been alert to when she attempted to switch institutions in 1860.

Butler had been sentenced to three years' imprisonment in Warwick Gaol for attempting to murder her illegitimate child (though she was married, her husband was serving a gaol sentence and the child was clearly not his). The matron and governor of the prison reported that she appeared to be subject to epilepsy, and Butler claimed that she had spasms in her right arm and leg and was unable to move them.[120] On admission to the Warwick Asylum, however, any hints of mental disorder were completely rejected; she appeared to have had a fit after admission, and 'since this time she has attempted to have fits but they have merely been attempts'. The asylum doctors believed she was shamming and, ignoring the evidence of the prison officials and surgeon who had certified her insane, packed Butler back to prison.[121] It is impossible to judge in this case whether Bridget Butler was insane, suffering from epilepsy or had nothing wrong with her, but it opens up the possibility that she was aware of the relatively comfortable conditions in the asylum.

The case of Elizabeth Barnwell

Of the infanticide trials linked to mental derangement in Warwickshire, it is the case of the young frenzied servant Sarah Harris in 1856 and that of Elizabeth Barnwell in 1867, married and respectable but suffering from temporary insanity, which stand out. In the case of Sarah Harris the forensic evidence provided no certainty about whether the child had died during or after delivery, whereas for Elizabeth Barnwell there was little doubt about how the infant had met its end. In both cases the insanity plea was eagerly embraced. The fact that the two women were deemed insane seems also to have satisfied the need for retribution, the women having been seen to have suffered greatly, particularly Elizabeth Barnwell, who was to reflect on her dreadful act during her four-month incarceration in the local asylum.

In June 1867 Elizabeth Barnwell was admitted to the Warwick County Lunatic Asylum, recorded as suffering from 'mania, puerperal'. On her order for admission Thomas William Bullock, surgeon of Warwick, stated:

> I have attended her for a month past when her child was drowned in the canal. She has since that time told me she would commit suicide … she maintains at this time that she shall do away with herself if someone does not prevent her.

Her friends and neighbours testified that she had been seen with a rope trying to hang herself; she had a deep red mark around her neck from the attempt. Elizabeth (or Eliza as she seems to have been known) had also asked for poison. She was said to walk about the house wringing her hands and crying.[122]

Eliza Barnwell's route to the asylum had begun some six weeks previously with the opening of the inquest into the death of her child.[123] The first witness to be called, John Berry, a boatman, reported that while working on the canal he saw her out walking. Shortly after he heard a splash and, on running to the spot, saw her struggling in the water. He managed to pull her out by putting a boat hook through her clothes. Once out of the canal Mrs Barnwell said, 'Oh my baby is in the water. I was going for a walk up to Mrs. Seywell [a friend], for I have been very bad.' John Berry stated that after she had said this Eliza Barnwell appeared not to know what she was doing. The witness rushed to the water, saw blood on the surface and fished the baby out with a net. The infant was bleeding at the nose, but was still alive. Frantic efforts were made to revive him; he was rubbed vigorously, given brandy, wrapped in a blanket and quickly brought to the nearest doctor, but was dead on arrival. The local surgeon, Mr Bullock, was called to see the child, and concluded that the body revealed no signs of violence except for a small bruise to the head. He then turned to the mother:

> She was very ill, with cold extremities and suffering under great mental excitement. I have attended her ever since. She is now better but still in a low nervous state. She is quite unable to attend here to-day. In my opinion the child died from suffocation by drowning. The child was little over a month old. I had not attended her [EB] previously. She is better to-day than she was yesterday. She sits up in bed and talks very wildly about the child and asks for it. She was attended by a midwife, and a nurse was with her for a fortnight.[124]

Eliza Barnwell's husband, a carpenter, added in his evidence that she had seemed to be 'going on well since her confinement'. Yet in the days leading up to the incident she had said she felt 'very low spirited'. 'She was particularly fond of the child,' her husband added, 'it being a son, and the other child being a girl.' The coroner explained to the jury that he thought it unnecessary to adjourn the inquiry so that the mother could attend. He wished to avoid causing her the pain and upset that this would invariably entail. He also strongly advised the jury to return a verdict of accidental death, as the 'mother did not have any bad feeling against the child, but was very fond of it'. The jury duly returned this verdict.[125]

On being admitted to Warwick Asylum six weeks later, Eliza's condition had deteriorated. The case book reported how, 'On May 3rd while doubtless under the influence of some morbid impulse she threw herself and her infant which she was carrying into the canal. The infant was drowned.'[126] The word 'infanticide' does not appear in the record. It was concluded that the attack of insanity had commenced after Eliza's confinement in April. Since then she had been in a melancholy condition and had threatened suicide. On admission it was remarked that she had a peculiarly sad expression. She was disruptive and fancied that she could see her child. Her appetite was poor, her bowels constipated and she complained of pains in the head. She cried a good deal, but was willing to employ herself.

Eliza was dosed with morphine and laxatives. By August she was starting to show a slow improvement and had lost much of her depression. Her demeanour was cheerful, but 'not natural'. She employed herself by sewing, slept well, but was still constipated. By October she was making better progress, was in good spirits and her expression was considered 'normal'. She was very industrious at sewing and also assisted on the ward. Eliza was released on a month's trial, which was standard practice at Warwick, and on 1 November was discharged 'recovered'.[127]

I have dwelt on Eliza Barnwell's story partly because it is a complete account, which we can follow through from childbirth, to the crime, inquest, diagnosis and verdict, committal and recovery. Her case typifies many of the features of the puerperal mania-infanticide link, in particular showing how readily this association was accepted, not only by medical men, but also by the family and neighbours whose reports and opinions build so strongly into the evidence presented at inquests and trials. What was more unusual is that the three key witnesses in Eliza's case were men: the surgeon, the boatman and her husband. More often than not it was the female neighbours, fellow servants, the lady of the house or the local midwife who told the story as they saw it in court. Such women would be far more likely to be involved with the accused at the time of birth or shortly afterwards. They may also have been seen as more appropriate witnesses and as being better able to pronounce on the woman's state of mind and on her generally good behaviour before the crime was committed, and to attract the sympathy of judges and juries.

Eliza was described as being low spirited, and the incident was deemed unmeditated, a spur of the moment action. The mother had no motive – indeed, she was said to be more fond of the child than of his sister – and she did not attempt to hide what she had done, or even to pretend that she had fallen into the water by accident. The surgeon who was called in was unlikely to have had much expertise in treating mental illness, but he had no hesitation in attributing the event to the 'great mental excitement' that the mother was experiencing. Surgeon Bullock could not resist the aside that a midwife had attended the mother, rather than a medical man. Her husband pointed to Eliza's fondness for the child. The coroner was sympathetic and did not want to subject Eliza to further suffering. The case was deemed 'sad' in the newspaper account, but neither tragic nor shocking. On entering the asylum, Eliza was described as suicidal and melancholic, repentant too and under the sad delusion that her child was still alive. Her pathway to good health was typical of puerperal mania: slow improvement, an eventual restoration of a natural and cheerful demeanour, the desire to work, and finally recovery and discharge back to her husband and surviving child.

* * *

Even at a time when there was so much concern about the problem of infanticide and mothers getting away with destroying their infants, this account shows above all the ease with which the label of insanity could be accepted by medical men and laymen, judges, witnesses, jurors, neighbours and the public. As Zedner has argued, during the nineteenth century psychiatric discourse built on to traditional, exculpatory discourses to provide a formidable case for the defence in infanticide cases.[128] Even without the pleas of poverty, hunger, ill-treatment or broken hearts, the crime was described as being the result of temporary derangement coupled with a lack of wilful intent or subterfuge, with witnesses reflecting on the mother's fundamentally good character and affectionate nature. This may have linked to a broader tendency towards leniency for women in other areas of law around the middle of the nineteenth century, including divorce, credit disputes and even murder. The new Divorce Court set up in 1857 with avowedly patriarchal goals to protect the rights of propertied men, often served in practice to protect women from their spouses; 'faced with overwhelming evidence of what they came to agree was the "unreasonable" behaviour of husbands'.[129] In the county courts too it was shown that women not only participated in economic activities denied to them by law, but also 'deployed a series of received stereotypes about the inherent frailties of female "nature" to their own – and their husbands' advantage'.[130] During the mid-nineteenth century there was also rising sympathy for women who murdered their husbands, based on the notion that it was women who were often in need of protection, and fewer women ended up on the scaffold for this offence.[131]

During the 1860s heightened concern about the perceived epidemic of infanticide dovetailed with sympathy for infanticidal women.[132] This sometimes divided opinion. When the subject of infanticide came up at the annual meeting of the Social Science Association in 1866, for example, there was a rift between those advocating compassion and sympathy for infanticidal women, and those who felt that they 'represented the very antithesis of womanhood'.[133] This conflict was also evident in medical approaches to infanticide. The appointment of Thomas Wakley, editor of the *Lancet*, as coroner to West Middlesex in 1839 led to more convictions in cases of infanticide, and Wakley with other London physicians were prominent leaders of the anti-infanticide campaign.[134] Yet the same journal, for example in the case of Mrs Prior, supported medical witnesses giving evidence in infanticide trials. The *Lancet* on more than one occasion refuted the views of judges who questioned the idea of temporary insanity and challenged the validity of medical testimony, pointing to the fickleness of legal opinion with respect to the insanity plea, which allowed it 'a certain run', and then rejected it altogether.[135] Judicial views were equally complex and ambiguous. Judges abhorred the crime of infanticide and expressed concern about its unchecked increase, but also, as in the Warwickshire Assizes, advised juries to acquit and take on board evidence explaining the woman's state of mind. It could be argued that there was more convergence of views, medical and judicial, than divergence, while recourse to the insanity plea could satisfy remaining doubts. It explained how the terrified servant girl was pushed in her frenzy to destroy her newborn child, or why naturally good mothers for a short time became murdering demons.

For medical men, puerperal insanity was depicted as an almost 'normal' side-effect of giving birth, with pregnancy and childbirth fraught with many forms of danger to self and others. Infanticide was an end station on the track of puerperal mania, which was linked to harmful behaviour of many kinds, which might or might not end in a threat to the life of the infant. The story of Eliza Barnwell ended penultimately in the asylum. But many of the women acquitted were seen as being in need of neither punishment nor a cure, since doctors testified that, though insane when the crime had been committed, the women were fully recovered by the time the case came to trial. In cases where women were taken to asylums the story was played out in full: they were physically restored to health, encouraged to work and to prepare themselves once again for their roles as wives and mothers.

Figure 6.1 M.B. Melancholia. Infanticide (Source: Bethlem Royal Hospital Archives and Museum)

Occasionally, such restoration is depicted in photographs, including several taken by Henry Hering of Bethlem Hospital patients at the time of Dr Charles Hood's superintendence in the late 1850s.[136] The photograph of M.B. is described as 'Melancholia. Infanticide'. It shows the woman, apparently in a state of convalescence, sitting quietly in a Windsor chair, the furniture of the domestic hearth, engaged in the task of sewing, respectably dressed and restored to proper Victorian femininity.[137]

Verdicts involving puerperal insanity seemed to agree that such women represented the very antithesis of womanhood, yet also allowed them to be treated with compassion.[138] However, it could be posited that madness and infanticide were seen as very much part of femininity and maternity, its very embodiment. Infanticide could not be dissociated from the domestic environment; many of the infanticidal acts discussed here took place in the family home or place of employment, some even between domestic chores, and many married women who destroyed their infants would ultimately return to their homes. Physical and mental fragility, widely agreed to be latent in all women, was greatly strained by childbirth and often resulted in collapse and madness; indeed it was this very combination that typified puerperal insanity.[139] Women's dangerousness was mediated by setting it within a wider context associated with challenging maternal, economic and family circumstances, but it was still a very real danger, sometimes wild and raging, but more often in the case of infanticide a fumbling, absent-minded danger or a struggle between good and evil. For many women, infanticide, ultimately redeemable, was not seen as the antithesis of but as an intrinsic part of motherhood.

7
From Redemption to the Dark Age: The Demise of Puerperal Insanity

Motherhood was deemed dangerous in numerous ways in Victorian Britain, able to leave the highest and lowest born, Queen, commoner and pauper, weakened and mentally disturbed. Pregnancy, birth and lying-in were fraught with physical and mental hazards, hazards that persisted for women who breastfed their infants for long periods. Puerperal insanity made its victims dangerous in all manner of ways, to the household, to themselves, to their family and particularly to the newborn. Their physical state could bring them to the point of collapse. Their delusions were dreadful and alarming. While melancholia crept stealthily into the household, mania tore into it, destructive, noisy and violent. Women falling victim to puerperal mania revealed their dreadful power and, unfettered in their actions, wildly abandoned their domestic and maternal functions. However, puerperal insanity remained a largely domestic disorder, treated at home, or if not there, then in the increasingly domesticated space of the asylum. Though attempts were made by families to maintain privacy when their mothers, wives and daughters were afflicted with insanity, highly publicised courtroom appearances of women who had murdered their infants while disordered thrust them into the public arena. Though well-to-do sufferers from puerperal insanity clearly fared better in terms of the environment in which they were treated, little else divided the social classes; rich and poor were seen to be susceptible, treatment regimes were similar, and, though for many women these marked a period of respite, the aim was to restore all women to their homes and proper domestic functions.

Puerperal insanity was a real condition with real sufferers, but it was also very much a product of the Victorian era. The disorder fascinated the medical profession and captured the public imagination. The nineteenth century represented an optimal environment for puerperal insanity to gain recognition and to flourish, and not only because of the increased apprehension surrounding childbirth. The evolution of a domestic ideology and ideal of motherhood alerted families and doctors alike to any deviation from this norm. Some women were seen as more vulnerable by reason of their poverty and want on the one hand, or excessive luxury and heightened sensitivity on the other, but the disorder could afflict all women regardless of rank, wealth, geography, marital status, age or childbearing history. Medical men alert to the condition watched from the lying-in room and the asylum for its occurrence, keen to claim it as their area of expertise and competence.

The increased isolation of women in the lying-in chamber, following the exclusion of female helpers, may have exacerbated their anxiety about motherhood, while fears of harm to themselves or infants lingered throughout the century linked to the very real and persistently high death rates of both mothers and babies. The actual prevalence of puerperal insanity in the nineteenth century is hard to estimate, but is likely to have been augmented by the great receptiveness to the disorder and also no doubt by confusion about its aetiology. One of the few doctors bold enough to make an estimate on the frequency of occurrence, and that based on a handful of disparate statistics, came up with a rate of one case of puerperal insanity for every 469 deliveries,[1] comparable to current estimates of one case of postpartum psychosis for every 500 deliveries.[2]

Women falling prey to puerperal insanity and its associated disorders of insanity of pregnancy and lactation were treated sympathetically by the medical profession and courts, the treatment meted out by both was mild and leant towards respite, allowing time for nature to restore them to their nurturing role. The women themselves were seen as highly vulnerable. The sombre philosophy which emerged during the early nineteenth century, which claimed that childbirth put women's bodies and minds at severe risk, was, however, in one way beneficial because it offered an explanation for mental breakdown and, in the worst-case scenario, for that most distressing of crimes, infanticide. Mental collapse in childbearing women was seen as likely, and it was also probable that this would include acts of despair and violence. This led to anticipation, alertness to the possibility of mental illness, which built on and refined older notions of women's susceptibility to mental disorder around childbirth. In case histories and court records women are presented as victims not only of their unstable bodies, but of social and environmental influences, crippling poverty, hunger, poor home circumstances, ineffectual or violent husbands, interfering relatives, and overwhelmed by the role of being a mother. Maternity was held up not only as a high ideal but also as deeply problematic and challenging. In cases of infanticide, the emphasis fell not on the voiceless infant, but on the forces that drove women to commit terrible acts of violence. Yet doctors and courts kept their eye on the bigger picture; the breakdown was deemed temporary, the women were described as 'good mothers', they were redeemable and capable of finding their place again in the home, caring for their families. What appeared to be the total collapse of motherhood was in fact not irreparable, and doctors put themselves forward as agents not only of cure but of balance; they would put the household to rights with the woman at its centre.

While there was much debate during the nineteenth century about the causality and preconditions of puerperal insanity, doctors agreed that it existed and was prevalent, and that it was not likely to go away. However, during the late nineteenth century that is exactly what started to happen, and it began to be written out of psychiatric textbooks and out of asylum records. Today psychiatry remains ambivalent, although much clearer on the existence of a milder form of disorder, 'maternity blues', which is said to be experienced by between half and two-thirds of women shortly after childbirth, and milder postnatal depressive disorders, which affect around 10 per cent of women.[3] Such disorders were given little consideration in the nineteenth century, with the focus very much on severe mania and melancholia. Yet now, doubt is cast on the separate existence of psychosis connected to childbearing. Ian Brockington and his colleagues have argued that, while puerperal insanity was 'one of the clearly recognized psychiatric entities during the nineteenth century', in the twentieth it became 'a casualty of the Kraepelinian diagnostic system'.[4] The eminent psychiatrist Emil Kraepelin played down the link between mental disorder and physical states, urging colleagues to focus on prognosis, and, in 1899, making diagnosis quite simple, divided insanity into two major groups: manic-depressive illness and dementia praecox. Eugen Bleuler later proposed the term schizophrenia for the latter condition.[5] By the turn of the twentieth century Kraepelin had suggested, along with many other labels, abandoning the term 'puerperal mania': 'Where mania really appears in the puerperal state, it is, like every other kind of mania, only a link in the chain of attacks of maniacal-depressive insanity.'[6] According to Kraepelin's rigorous interpretation, the mere coincidence of insanity and childbearing was insufficient to account for the separate disorder of puerperal insanity. This view is still current. A recent influential psychiatric textbook stated:

> In the nineteenth century, puerperal and lactational psychoses were thought to be specific entities distinct from other mental illnesses. Later psychiatrists such as Bleuler and Kraepelin regarded the puerperal psychoses as no different from other mental illnesses. This latter view is widely held today on the grounds that puerperal psychoses generally resemble other psychoses in their clinical picture.[7]

It also states that there is 'no clear relationship between psychosis and obstetric factors', though the incidence of one case per 500 births is recognised as significantly higher than the rate of psychosis for non-puerperal women of the same age.[8]

This ambiguity can have grim consequences with respect to provision of care for women who do, textbook or no textbook, become disturbed after giving birth. The symptoms described by nineteenth-century doctors are certainly recognisable today, even though the material settings, place of childbirth, treatment of the condition and the language used to describe it have changed dramatically. There are still urgent debates on causality, on whether postnatal mental illness is related to previous mental health crises and depression, single motherhood or social isolation, poverty, poor levels of maternal care or hormonal factors, on regimes of treatment, with electro-convulsive therapy advocated by some psychiatrists, and the location of care in mother and baby units or psychiatric wards. The debate about mental disorder and motherhood continues in the public arena, bound particularly to concerns about its link with harm of the infant and attempts to destroy it, a topic which continues in the twenty-first century, as it did in the nineteenth, to attract great public and media interest. Postnatal mental illness has not lost its power to shock, whether it is the previously contented mother thrown into deep depression going far beyond the realm of baby blues, or the unmarried mother still, incredibly, giving birth in secret and then murdering or abandoning her child.

Eerie echoes of the nineteenth century and earlier linger on. In a recent shocking case in the United States, Andrea Yates described the voices telling her to destroy her children:

> My children weren't righteous. They stumbled because I was evil. The way I was raising them, they could never be saved ... Better for

someone else to tie a millstone around their neck and cast them into a river than to stumble.[9]

From early twenty-first-century Texas, with five children drowned in their bathtub, back through nineteenth-century courtrooms, servants' bedrooms, canal banks and outhouses, back even to Margery Kempe's struggle with the devil in medieval England, the dialogue of internal struggle is remarkably similar. Charged with first-degree murder, even though the prosecution conceded that she was mentally ill at the time she drowned her children, Andrea Yates was sentenced to life imprisonment. Although she had a long history of mental illness, including four psychiatric hospitalisations, suicide attempts and visions of violence, her family distanced themselves from her problems; her husband 'lacked empathy' and ignored warnings about the risks of having more children.[10]

By the third quarter of the nineteenth century a number of asylum doctors questioned the separate status of puerperal insanity. In 1866 W.H.O. Sankey, Lecturer on Mental Diseases at University College, London and late Medical Superintendent of the Female Department at Hanwell Asylum, argued that puerperal insanity, along with hysterical mania, phthisical mania, nostalgia and maine à potû,[11] was indistinguishable from other kinds of madness, possessing 'no distinctive character, according to my experience, sufficient to constitute them a distinct kind of disease'.[12] J. Thompson Dickson, Physician to St Luke's Asylum, London, echoed this view, arguing that 'there is nothing peculiar in the insanity of child-bed, rendering it a disease peculiar to women ... and that the so-called *puerperal* insanity is ordinary insanity, appearing at, and only slightly modified by the child-bearing circumstance'.[13] He went on to relate several cases admitted to St Luke's which he reclassified as dementia, hereditary insanity or, in one case, an attack prompted by epilepsy and evidence of insanity before the confinement; 'it appears much more correct to speak of the cases as insanity appearing at the puerperal season, than to use the term "puerperal" in an adjectival sense, as though the insanity was a special form peculiar to child-bearing'.[14] It was opinions like these, rather than the new theories of nosology coming from the continent, that would mark the demise of puerperal insanity.[15]

Even as it denied puerperal insanity's separate existence, however, psychiatry began to dominate publications and treatment. Obstetricians made fewer claims to the condition, and, as more patients were treated in asylums, and as explanations of mental disorder shifted from stressing moral and physical causes to emphasising brain function, so did the close, physiological and social links between motherhood, the process of childbearing and madness begin to unravel. Accumulating asylum numbers represented a shift away from treatment by individual practitioners, the family doctor or obstetrician who had attended the delivery, to a situation where puerperal insanity became absorbed into general asylum regimes and therapeutics. Within the asylum, regimes of force-feeding, the administration of morphine, the use of powerful sedatives and vaginal injections with carbolic began to dominate, treatments that would have alarmed physicians like John Batty Tuke with his advocacy of beef tea and rest.[16] More and more women of 'good social condition' were also admitted to the asylum as it came to provide a more acceptable locus of care. Though some physicians still argued that home care was better for many patients, middle-class families, finding it difficult to continue to pay for private medical care, allowed their female relatives to enter the large public asylums. For many psychiatrists, their doubt about the asylum as an appropriate place of treatment was replaced by certainty that it was.

Yet, there were continuities. Sympathetic responses and therapies highlighting a patient, expectant approach did not vanish, although treatments shifted in general towards heavier dosing and sedation. Nor were social and environmental causes lost from view; Nakamura has shown that these played a role in explaining puerperal insanity well into the late nineteenth century.[17] As we have seen in chapter 5, in the 1880s Thomas Clouston continued to place emphasis on nutrition and environmental factors in his approach to treating lactational insanity. Many patients were kept at home for lengthy periods before removal to an asylum, and some psychiatrists still questioned the validity of the institutional approach, advocated cooperation with obstetricians and acknowledged the overlap in disciplines. In 1886 D.M. Macleod, Medical Superintendent of the East Riding Asylum, claimed that most women – around three-quarters of all cases – were treated at home.[18] The psychiatrist George Savage, writing in 1884, explained that the question 'as to which class of patients should be sent to asylums, and when' remained pertinent. There was strong feeling, for example, against sending a young married woman to an asylum when insanity developed after the birth of her first child, because the child might suffer socially and the mother would learn to dread future confinements: 'If the friends have ample means, if their home is in a healthy district, and if the doctor can see the patient twice daily at least for the first few weeks, it is possible to treat almost the most serious case at home.'[19] Even if it was found necessary to send women to asylums, 'they must not be kept under control too long, but should be sent home as soon as symptoms of danger have passed'.[20] Clearly a class dimension was involved, for ample funds were needed to pay a physician for his diligent attention.

Paradoxically, and despite the continuing role of home treatment, at the very time that its existence was being denied, more and more women were entering asylums diagnosed with puerperal insanity. This rise in cases of puerperal insanity, or purported cases, dovetailed with broader anxieties about the impact of degeneration, overcrowded asylums and the alleged increase in insanity in all its forms.[21] Between 1855 and 1860 admissions to Bethlem under the category of puerperal insanity accounted for one-eighth of the total female intake.[22] By 1886–88 some 18 per cent of female admissions to the Warwick County Asylum of childbearing age were assigned a puerperal cause (over 11 per cent of all female admissions).[23] Between 1889 and 1891 over 14 per cent of the women admitted to the Rainhill Asylum, Liverpool were said to be suffering from puerperal insanity.[24]

Explanations for the prevalence of puerperal insanity began to be framed increasingly around the rhetoric of heredity and degeneration, referring to a form of failing linked less and less to maternity and environmental factors. Rather than all women being vulnerable, a particular kind of woman was liable to the disorder, one with a hereditary disposition. While domestic influences, worry associated with pregnancy and moral causes were still referred to in the dictionaries of psychiatry which appeared towards the end of the century, a new list of causal factors emerged: 'hereditary tendencies to neurosis, advanced age at first pregnancy, frequent pregnancy, especially in those who are nervously degenerate, previous nervous illnesses.'[25] *Quain's Dictionary of Medicine* claimed in 'many cases puerperal insanity is the expression of the presence of neuropathic diathesis, and its subjects are "degenerates" whose nervous system gives way first at the puerperal epoch'.[26] Rainhill Asylum, with its huge pauper population, embodied the pessimistic stance on hereditary influences, claiming a hereditary link in 26.6 per cent of pregnancy cases, 18.7 per cent in puerperal cases and 30.4 per cent in those related to lactation.[27] The hereditary taint was linked to immorality, adding to the gloomy diagnosis and burden of responsibility of poor women. It was reported in 1886 that Rainhill Asylum had fewer than usual cases of puerperal insanity amongst women giving birth out of wedlock. However, Joseph Wiglesworth, Assistant Medical Officer to the institution, nothing daunted, referred to the 'old offenders', women who had attacks of mania after their first confinements out of wedlock, and then became insane again after each subsequent delivery of an illegitimate child. He dismissed the idea of shame and grief causing their condition, preferring to attribute their mania to the same 'defective inhibitory power' that caused single women to become pregnant in the first place.[28] This was a far cry from the accusations referring to the abandonment and neglect of the seducer who had left his victim to give birth alone earlier in the century.

Medical science came forward to offer enlightenment on causality, focusing increasingly on preconditions such as blood poisoning and renal disorders, as well as complications of childbirth.[29] Clinical approaches and changes in recording protocols, meanwhile, led to more rigid case book narratives. When Thomas Clouston took over the superintendence of the Royal Edinburgh Asylum in 1873 he introduced uniform printed case books with recording schedules, including a brief section for the patients' 'history', with blank spaces for information on habits, disposition, previous episodes, heredity, duration of the present attack and its causes.[30] There was a consequent diminution of the patient's story of her decline into madness; less space to record it and perhaps the sort of creativity involved in weaving a story would have been frowned upon under the new recording ethos. Clouston urged the recording of clinical facts, temperature and weight, and also more precision, 'the facts to be entered in as definite and scientific way as possible, e.g. not "does not sleep well" but "only slept four hours last night"; not "is gaining in weight", but "has gained three pounds last week"'.[31]

By the 1880s A. Campbell Clark, Medical Superintendent of the Glasgow District Asylum, Bothwell had adopted the language of late nineteenth-century psychiatry, emphasising clinical details and the physical status of the women, including temperature charts with his case histories and stressing the importance of brain pathology, uterine disease and renal complications, as well as being significantly ruder about his patients. He tabulated heredity links carefully. Though he argued that puerperal insanity – a category he, like Menzies, Wiglesworth and Clouston, did not deny – was not 'beautifully simple', 'the lines of causative conduction are so innumerable, reflective and interminable, that finality of research is not to be looked for', a powerful link was made with heredity 'from many points of view'. Some 26 out of the 38 cases he had treated at the Glasgow District Asylum had, he argued, a heredity basis and Clark believed this to be an underestimate.[32] Two of the patients who recovered 'could never at their best, be very much exalted above the type of educable imbeciles, although their mental and physical development were sufficient to allow them a "bare pass" in the world at large'.[33] The women are described as 'sluggish and stupid', of 'low type of intelligence' or 'has stupid obtuse look as if she did not know the language'.[34] However, the case histories also mention such factors as the neglect of house and family, a lack of affection for the husband, a propensity to dance and talk incoherently, and details of delusions, and echoes of older case note narratives sit awkwardly with detailed clinical components. One woman admitted in 1881 related how 'at night the devil came and put his "clutches" on her face, and that she heard his chains rattling. Smelt sulphur on her bedclothes ... said she heard the death warning – unearthly voices telling her she was to die and go to hell.'[35]

The ambiguity surrounding puerperal insanity and the emergence of different paradigms of causality and treatment at the end of the nineteenth century still have ramifications today. To explore the fortunes of puerperal insanity in the twentieth century, the implications of the rise of psychoanalytical approaches to therapy against the unremitting march of biological psychiatry, the abandonment of the asylum in the last quarter of the century and the move to care in the community, as well as major changes in obstetric services, including the shift to hospital births, all against a background of vast changes in women's maternal, social and economic lives, is beyond the scope of this book, but offers tantalising possibilities. By the end of the nineteenth century, however, the more expectant approach, which saw puerperal insanity as a condition to be watched for and guarded against as it invaded the family home, was to a large extent lost. The bridge that puerperal insanity represented, that had connected, but also separated, obstetrics and psychiatry appears to have been taken by psychiatry. The subsuming of cases of puerperal insanity into asylum regimes along with other forms of mental illness meant that it was treated less and less as something apart, requiring specific treatment in a specific environment. Women were no longer deemed susceptible or sensitive but defective, not individuals with individual stories but part of a larger problem of mental decline. The rich emotional landscape of fear, despair and misery, which marked much of the discourse on puerperal insanity throughout the century, was obscured by the gloom of hereditary insanity, the separate existence of the disorder denied by many. The gentle, painstaking regimes of care were to a large extent lost, and puerperal insanity took on a darker and more menacing aspect.

Appendix

TABLES ILLUSTRATIVE OF PUERPERAL INSANITY IN THE
ROYAL EDINBURGH ASYLUM.

Total of so-called Puerperal Cases, . 155

SUBDIVISION.

Insanity of Pregnancy,	.	.	.	28
True Puerperal Insanity,	.	.	.	73
Insanity of Lactation,	.	.	.	54

155

TABLE I.—*Insanity of Pregnancy.*

1. Ages at which Attack occurred.	2. Pregnancy during which Insanity supervened.	3. Month of Pregnancy during which Insanity supervened.
15 years, 1	In 9 cases during 1st pregnancy	In 3 cases during 3d month
19 „ 2	4 „ „ 2d „	5 „ „ 5th „
21 „ 2	3 „ „ 3d „	1 „ „ 6th „
22 „ 3	2 „ „ 4th „	9 „ „ 7th „
23 „ 1	2 „ „ 5th „	1 „ „ 8th „
25 „ 1	1 „ „ 6th „	9 not recorded
26 „ 3	1 „ „ 8th „	—
29 „ 4	6 multiparæ the number of	28
31 „ 2	— pregnancy not recorded	
32 „ 2	28	
34 „ 1		
35 „ 1		
36 „ 1		
39 „ 1		
42 „ 1		
43 „ 1		
44 „ 1		
—		
28		

4. Mental Symptoms.	5. Result.	6. Length of Time under Treatment.
Mania, with Exaltation, . 2	Recovered, . 21	Under 3 weeks, 1
Melancholia, . . . 15	Died (3 years after	„ 4 „ 1
Dementia, with Melancholia, 5	admission), . 1	„ 5 „ 2
Dipsomania, . . . 4	Under treatment, 1	„ 2 months, 5
Moral Perversion, . . 2	Became demented, 5	„ 10 weeks, 1
—	—	„ 3 months, 2
28	28	„ 5 „ 2
		„ 6 „ 3
		„ 2 years, 2
		„ 3 „ 2
		„ 4 „ 1
		Remain demented, 5
		Died, . . . 1
		—
		28

In 13 cases suicidal tendencies were exhibited.
„ 12 cases hereditary predisposition was ascertained to exist.
„ 4 cases the patients were unmarried.

TABLE II.—*Puerperal Insanity.*

1. Age at which Attack occurred.	2. Number of Confinement on which Insanity supervened.	4. Period after Confinement at which Insanity developed itself.
Cases.	Cases.	Cases.
At 20 in 5	On 1st in 34	Under 1 day in 9
21 „ 2	2d „ 16	„ 2 „ 6
22 „ 2	3d „ 5	„ 3 „ 1
24 „ 6	4th „ 3	„ 4 „ 4
25 „ 11	5th „ 2	„ 5 „ 4
26 „ 5	6th „ 2	„ 6 „ 3
27 „ 3	8th „ 1	„ 7 „ 10
28 „ 5	9th „ 1	„ 8 „ 3
29 „ 2	Multiparæ, but exact	„ 9 „ 2
30 „ 3	number of confine-	„ 10 „ 4
31 „ 1	ment not recorded, „ 9	„ 11 „ 1
32 „ 3	—	„ 12 „ 2
33 „ 3	73	„ 13 „ 2
35 „ 5		„ 14 „ 4
36 „ 3		„ 15 „ 1
37 „ 2	3. Showing tendency to Recurrence.	„ 16 „ 1
38 „ 3		„ 18 „ 2
40 „ 2		„ 21 „ 2
41 „ 1	In 58 cases the present was 1st attack	„ 24 „ 2
43 „ 2	10 „ „ 2d „	„ 26 „ 1
Not known 4	4 „ „ 3d „	„ 6 weeks in 1
—	1 „ „ 4th „	„ 10 „ 1
73	—	In 7
	73	the exact period is not recorded more particularly than " a few days after the birth," &c.
		—
		73

5. Symptoms (Mental).	7. Period of Residence of those Discharged Recovered.	8. Causes of Death, and length of Residence of Deceased.
Acute Mania, . 53	Under 1 month, 2	1 Pelvic Cellu-
Melancholia, . 15	„ 2 „ 7	litis, . 2 days
Acute Dementia, . 4	„ 3 „ 10	2 Exhaustion, 7 „
Epileptic Mania, . 1	„ 4 „ 5	3 Peritonitis, 2 „
—	„ 5 „ 6	4 Phthisis, 9 months
73	„ 6 „ 12	5 Exhaustion, 3 days
	„ 9 „ 5	6 Phthisis, 2 years
6. Results.	„ 12 „ 3	7 Bronchitis, 8 days
	„ 18 „ 1	8 Phthisis, 2 years
Died, . . . 6	„ 1 year, 2	
Became Demented 7	„ 2 „ 1	
Discharged Relieved, 2	—	
Recovered, . . 56	54	
—		
73		

Puerperal Insanity—Continued.

Showing the Duration of Insanity previous to Admission.

9.

No. of Cases.	Insane previous to Admission.	Recovered.	Relieved.	Became Demented.	Died.
2	For 2 days	0	0	0	2
2	" 3 "	1	0	0	1
3	" 4 "	3	0	0	0
1	" 5 "	1	0	0	0
5	" 6 "	4	0	1	0
4	" 7 "	3	0	0	1
2	" 8 "	1	0	0	1
2	" 9 "	2	0	0	0
1	" 10 "	1	0	0	0
2	" 14 "	2	0	0	0
6	" 21 "	5	0	1	0
4	" 28 "	4	0	0	0
5	" 6 weeks	5	0	0	0
4	" 2 months	3	0	1	0
3	" 3 "	3	0	0	0
4	" 4 "	3	1	0	0
5	" 6 "	3	1	1	0
6	" 9 "	4	0	2	0
2	" 1 year	0	0	1	1
2	" 2 "	0	0	0	2
8 not ascertained, of whom		8	0	0	0
73		56	2	7	8
					73

In 13 cases the patients were unmarried.
" 22 cases hereditary predisposition was ascertained.
" 25 cases suicidal tendencies were evinced.
" 9 cases the labour had been instrumental.
" 4 " " " tedious.
" 2 cases twins had been born.
" 2 cases chloroform had been administered.
" 6 cases profuse hæmorrhage had succeeded labour.
" 2 cases the child was still-born.

TABLE III.—*Insanity of Lactation.*

1. Age at Time of Attack.		2. Nursing during which Insanity appeared.	3. Month of Nursing during which Insanity appeared.
Age.	No. of Patients.	In 8 during 1st nursing.	In 2 during 3d month.
19 .	1	4 „ 2d „	6 „ 6th „
20 .	1	9 „ 3d „	4 „ 7th „
22 .	2	5 „ 4th „	2 „ 8th „
25 .	4	6 „ 5th „	6 „ 9th „
26 .	4	2 „ 6th „	6 „ 10th „
27 .	5	1 „ 7th „	5 „ 11th „
28 .	2	1 „ 8th „	6 „ 12th „
29 .	5	2 „ 9th „	2 „ 13th „
30 .	6	1 „ 10th „	2 „ 16th „
31 .	3	1 „ 12th „	13 cases month not recorded.
34 .	4	14 cases the women were multiparæ, but exact confinement not recorded.	—
35 .	3		54
36 .	3		
37 .	2	—	
38 .	2	54	
39 .	2		
40 .	2		
42 .	1		
Not known,	2		
	—		
	54		

4. Results.	5. Symptoms.	6. Length of Time under Treatment of those Recovered.
Died, . . 1	Acute Mania, . . 10	3 weeks in 2 cases.
Became Demented, 12	Melancholia, . . 39	1 month „ 3 „
Under Treatment, 2	Dementia, . . . 5	2 „ „ 4 „
Recovered, . 39	—	3 „ „ 6 „
—	54	5 „ „ 4 „
54		6 „ „ 5 „
		7 „ „ 4 „
		8 „ „ 4 „
		Over 9 „ „ 7 „
		—
		39

In 17 cases suicidal tendencies were evinced.
   „ 14 cases hereditary predisposition was ascertained.
   „ 2 cases profuse hæmorrhage had occurred after labour.

Notes

Introduction

1 John Raymond (ed.), *Queen Victoria's Early Letters*, revised edition (London: B.T. Batsford, 1963), p. 74.

2 Gordon N. Ray (ed.), *The Letters and Private Papers of William Makepeace Thackeray*, vol. 1 (London: Oxford University Press, 1945), 'To Mrs. Carmichael-Smyth', 4–5 October 1840, p. 483.

3 Ibid., vol. 2, 'To Edward Fitzgerald', 10 January 1841, p. 3.

4 Howell Evans, 'Puerperal Mania Occurring Suddenly Three Days after Delivery Treated by Opiates and Purgatives', *Medical Times*, 15 (14 November 1846), 145.

5 Edinburgh University Library: Lothian Health Board Archive, Royal Edinburgh Hospital, LHB7/51, Case Books (hereafter CB): LHB7/51/9, 1851–55, Janet Smith or Curle, admitted 1 April 1853, p. 411.

6 Prior to Victoria's first confinement, her Physician-Accoucheur, Dr Robert Ferguson, had prepared for publication a highly influential text on the diseases of lying-in women. Robert Gooch, *On Some of the Most Important Diseases Peculiar to Women; with Other Papers*, with a prefatory essay by Robert Ferguson, M.D. (London: The New Sydenham Society, 1831). Ferguson attended all of Victoria's deliveries together with Sir Charles Locock.

7 Evans, 'Puerperal Mania', p. 145.

8 LHB7/51/9, CB, 1851–55, Janet Smith, pp. 411–12, 448.

9 LHB7/51/11, CB, 1855–58, Janet Smith or Curle, admitted 8 October 1855, p. 155. Chapter 4 is based on Edinburgh Asylum cases.

10 Isabella Thackeray's illness is described in greater detail in chapter 3.

11 For an excellent overview of recent studies of postpartum depression, see Patrizia Romito, 'Postpartum Depression and the Experience of Motherhood', *Acta Obstetricia et Gynecologica Scandinavica*, 69, Supplement 154 (1990), 7–37.

12 This shift in definition could be compared with the history of chlorosis and 'mystery' of its disappearance. See I.S.L. Loudon, 'Chlorosis, Anaemia, and Anorexia Nervosa', *British Medical Journal*, 281 (20–27 December 1980), 1669–75; Karl Figlio, 'Chlorosis and Chronic Disease in Nineteenth-Century Britain: The Social Constitution of Somatic Illness in a Capitalist Society', *Social History*, 3 (1978), 167–97; Helen King, *The Disease of Virgins: Green Sickness, Chlorosis and the Problems of Puberty* (London and New York: Routledge, 2004).

13 Robert Gooch, *A Practical Compendium of Midwifery; Being the Course of Lectures on Midwifery, and on Diseases of Women and Infants, Delivered at St. Bartholomew's Hospital, by the late Robert Gooch, M.D.* (London: Longman, Rees, Orme, Brown, and Green, 1831), p. 290.

14 George Man Burrows, *Commentaries on the Causes, Forms, Symptoms, and Treatment, Moral and Medical, of Insanity* (London: Thomas and George Underwood, 1828), p. 368.

15 Thomas Denman, *Observations on the Rupture of the Uterus, on the Snuffles in Infants, and on Mania Lactea* (London: J. Johnson, 1810), p. 63.

16 By Robert Gooch, in his short treatise *Observations on Puerperal Insanity* (London, 1820) (extracted from the sixth volume of *Medical Transactions*, Royal College of Physicians, read at the College, 16 December 1819).

17 See e.g. Lucy Bland, *Banishing the Beast: English Feminism and Sexual Morality 1885–1914* (London: Penguin, 1995); and Jane Lewis, *Women in England 1870–1950: Sexual Divisions and Social Change* (Brighton: Wheatsheaf, 1984).

18 Anne Digby, 'Women's Biological Straitjacket', in Susan Mendus and Jane Rendall (eds), *Sexuality and Subordination: Interdisciplinary Studies of Gender in the Nineteenth Century* (London and New York: Routledge, 1989), pp. 192–220, quote on p. 196.

19 Important exceptions to this are two unpublished PhD theses: Shelley Day, 'Puerperal Insanity: The Historical Sociology of a Disease', University of Cambridge, 1985; and Lisa Ellen Nakamura, 'Puerperal Insanity: Women, Psychiatry, and the Asylum in Victorian England, 1820–1895', University of Washington, 1999. See also Catherine Quinn, 'Representations of Puerperal Insanity in England and Scotland, 1850–1900', unpublished MA dissertation, University of Manchester, 1998; and Jennifer Ryan, 'Confinement after Confinement: The Status of "Puerperal Insanity" in the Nineteenth and Twentieth Centuries', unpublished Intercalated BSc dissertation, Wellcome Centre London, 1999. Catherine Quinn submitted her PhD thesis on puerperal insanity between 1860 and 1922 as this book was going to press; Quinn's thesis has a particularly valuable survey of the Devon asylums and explores patients' perspectives on the disorder from the late nineteenth century onwards: see 'Include the Mother and Exclude the Lunatic. A Social History of Puerperal Insanity c.1860–1920' (University of Exeter, 2003). Nancy Theriot's article on puerperal insanity focuses on late nineteenth-century America and the relationship between male practitioners and women patients: Nancy Theriot, 'Diagnosing Unnatural Motherhood: Nineteenth-Century Physicians and "Puerperal Insanity"', *American Studies*, 26 (1990), 69–88. A useful introduction is provided in Irvine Loudon's article 'Puerperal Insanity in the 19th Century', *Journal of the Royal Society of Medicine*, 81 (February 1988), 76–9. See also my previous publications, '"Destined to a Perfect Recovery": The Confinement of Puerperal Insanity in the Nineteenth Century', in Joseph Melling and Bill Forsythe (eds), *Insanity, Institutions and Society, 1800–1914* (London and New York: Routledge, 1999), pp. 137–56; and 'At Home with Puerperal Mania: The Domestic Treatment of Insanity of Childbirth in the Nineteenth Century', in Peter Bartlett and David Wright (eds), *Outside the Walls of the Asylum: The History of Care in the Community 1750–2000* (London: Athlone Press, 1999), pp. 45–65.

20 Elaine Showalter, *The Female Malady: Women, Madness and English Culture, 1830–1980* (London: Virago, 1987; first published New York: Pantheon, 1985), pp. 57–9, 71–2.

21 These include Elizabeth Lunbeck, *The Psychiatric Persuasion: Knowledge, Gender, and Power in Modern America* (Princeton, NJ: Princeton University Press, 1994); Jane Ussher, *Women's Madness: Misogyny or Mental Illness?* (New York, London, etc.: Harvester Wheatsheaf, 1991); Charlotte

MacKenzie, 'Women and Psychiatric Professionalization 1780–1914', in London Feminist History Collective (eds), *The Sexual Dynamics of History* (London: Pluto Press, 1983), pp. 107–19; and Joan Busfield, *Men, Women and Madness: Understanding Gender and Mental Disorder* (London: Macmillan, 1996); see also the excellent overview essay of Nancy Tomes, 'Historical Perspectives on Women and Mental Illness', in Rima D. Apple (ed.), *Women, Health, and Medicine in America* (New Brunswick, NJ: Rutgers University Press, 1990), pp. 143–71. Yannick Ripa, in her book on women and madness in nineteenth-century France, mistakenly stated that puerperal insanity was caused by the unhygienic conditions in which women gave birth and aligned it closely with puerperal fever, dismissing insanity related to childbirth as too difficult to study: *Women and Madness: The Incarceration of Women in Nineteenth-Century France* (Cambridge: Polity Press, 1990), pp. 53–4. Several recent studies have focused on literary sources and female illness, particularly the 'invalid culture', but not on the relationship between childbirth and mental disorder, including Miriam Bailin, *The Sickroom and Victorian Fiction: The Art of Being Ill* (Cambridge: Cambridge University Press, 1994); Helen Small, *Love's Madness: Medicine, the Novel, and Female Insanity, 1800–1865* (Oxford: Clarendon, 1996); Pamela K. Gilbert, *Disease, Desire, and the Body in Victorian Women's Popular Novels* (Cambridge: Cambridge University Press, 1997); and Jane Wood, *Passion and Pathology in Victorian Fiction* (Oxford: Oxford University Press, 2001).

1 The Birth of Puerperal Insanity

1 *The Book of Margery Kempe*, translated and introduced by B.A. Windeatt (London: Penguin, 1985). Kempe lived between c.1373 and c.1440.
2 Ibid., pp. 41–2.
3 The suggestion that Kempe experienced some form of 'postpartum disorder' has been challenged by Freeman et al., with an alternative explanation that she was suffering from episodes of mania and melancholia, culminating in mystical visions. Though the authors strive to put Kempe in her 'proper medieval context' and are correct in arguing that attaching the modern label of postpartum psychosis to her case is misleading, none the less parallels in Kempe's description of her disorder with accounts from other centuries from the eighteenth to the twenty-first are striking. See Phyllis R. Freeman, Carley Rees Bogarad and Diane E. Sholomskas, 'Margery Kempe, a New Theory: The Inadequacy of Hysteria and Postpartum Psychosis as Diagnostic Categories', *History of Psychiatry*, 1 (1990), 169–90. For an account of postpartum psychosis which seeks to locate the disorder in different historical contexts and compare its aetiology, see I.F. Brockington, G. Winokur and Christine Dean, 'Puerperal Psychosis', in I.F. Brockington and R. Kumar (eds), *Motherhood and Mental Illness* (London and New York: Academic Press/Grune and Stratton, 1982), pp. 37–69.
4 *The Holy Life of Mrs Elizabeth Walker* (London, 1690), pp. 25–6. Cited Anne Laurence, 'Women's Psychological Disorders in Seventeenth-Century

Britain', in Arina Angerman et al., *Current Issues in Women's History* (London and New York: Routledge, 1989), pp. 203–19, quote on p. 209.

5 Michael MacDonald, *Mystical Bedlam: Madness, Anxiety, and Healing in Seventeenth-Century England* (Cambridge: Cambridge University Press, 1981), p. 108.

6 Anne Laurence, *Women in England 1500–1760: A Social History* (London: Phoenix, 1996), p. 80.

7 John Pechey, *A General Treatise of the Diseases of Maids, Bigbellied Women, Child-bed Women, and Widows* (London, 1696), p. 170. Cited Laurence, 'Women's Psychological Disorders', p. 210.

8 Mark Jackson, *New-Born Child Murder* (Manchester and New York: Manchester University Press, 1996), p. 20.

9 *York Courant*, 8 June 1742. Cited ibid., p. 131, n. 55.

10 Jackson, *New-Born Child Murder*, especially pp. 120–3. Depositions in infanticide trials provide some of the best evidence of seventeenth- and eighteenth-century understandings of insanity related to childbirth. See also Peter C. Hoffer and N.E.H. Hull, *Murdering Mothers: Infanticide in England and New England 1558–1803* (New York: New York University Press, 1981). See chapter 6 for the relationship between puerperal insanity and infanticide.

11 John Woodward, *Select Cases, and Consultations, in Physick. By the Late Eminent John Woodward ... Now First Published by Dr. Peter Templeman, 1757* (London: Davis & Reymers, 1757), pp. 259–65. The full case, from which this summary is drawn, is published in Richard Hunter and Ida Macalpine, *Three Hundred Years of Psychiatry 1535–1860* (London: Oxford University Press, 1963), pp. 338–41. See, for maternal imagination, Dennis Todd, *Imagining Monsters: Miscreations of the Self in Eighteenth-Century England* (Chicago and London: University of Chicago Press, 1995); and Herman W. Roodenburg, 'The Maternal Imagination: The Fears of Pregnant Women in Seventeenth-Century Holland', *Journal of Social History*, 21 (1988), 701–16.

12 Woodward, *Select Cases*. Cited Hunter and Macalpine, *Three Hundred Years of Psychiatry*, pp. 339–40.

13 Ibid., p. 340.

14 Ibid., p. 341.

15 John Leake, *A Lecture Introductory to the Theory and Practice of Midwifery* (London: R. Baldwin, 1773), p. 25.

16 Jane Sharp, *The Midwives Book. Or the Whole Art of Midwifry Discovered* (London: Simon Miller, 1671), ed. Elaine Hobby (New York and Oxford: Oxford University Press, 1999), p. 191. Elaine Hobby notes that 'fits', 'strangling' and 'Rage' refer to hysteria.

17 Ibid., p.193, 'swoonding' = swooning; 'watching' = insomnia; 'doting' is not explained, but presumably refers to longings.

18 Martha Mears, *The Midwife's Candid Advice to the Fair Sex; or the Pupil of Nature* (London: Crosby and Co. and R. Faudler, c.1805), pp. 15, 21–2, 87, 28, 33.

19 Martha Ballard, practising in Hallowell, Maine between 1785 and 1812, was dismissive of weakness and mental disorder following birth. She was impatient with her old friend and neighbour Elizabeth Weston, who became intensely anxious during her last pregnancy at the age of 45. After delivery she was 'of the mind Shee Cannot take care of hir infant at home' and a

month later remained weak. Mrs Williams fell 'in a Deliriam by reason of a mistep of her Husband'. The 'Deliriam' seems to have been emotional, but one cannot be sure, and Ballard remarked only on her swift recovery: Laurel Thatcher Ulrich, *A Midwife's Tale: The Life of Martha Ballard, Based on Her Diary, 1785–1812* (New York: Alfred Knopf, 1990), pp. 195–6, 191–2.

20 Shelley Day cites a handful of mainly uninfluential continental works published from early in the eighteenth century, including a cluster of German dissertations: Shelley Day, 'Puerperal Insanity: The Historical Sociology of a Disease', unpublished PhD thesis, University of Cambridge, 1985, p. 153.

21 William Smellie, *A Treatise on the Theory and Practice of Midwifery*, vol. I, third edition (London: D. Wilson and T. Durham, 1756), pp. 395–6.

22 Ibid., vol. III, *A Collection of Preternatural Cases and Observations in Midwifery* (1764), pp. 469–70. 'The college' presumably refers to the College of Physicians.

23 William Hunter, 'On the Uncertainty of the Signs of Murder, in the Case of Bastard Children', *Medical Observations and Inquiries*, 6 (1784), reprinted in William Cummin, *The Proof of Infanticide Considered: Including Dr. Hunter's Tract on Child Murder, with Illustrative Notes; and a Summary of the Present State of Medico-Legal Knowledge on that Subject* (London: Longman, Rees, Orme, Brown, Green, and Longman, 1836). See chapter 6, p. 170 and Jackson, *New-Born Child Murder*, especially pp. 115–23, for Hunter's contribution to the debate on infanticide.

24 John Clarke, *Practical Essays on the Management of Pregnancy and Labour; and on The Inflammatory and Febrile Diseases of Lying-in Women* (1793), second edition (London: J. Johnson, 1806), p. 1.

25 Ibid., pp. 15, 21.

26 Quoted in James Cowles Prichard, *A Treatise on Insanity and Other Disorders Affecting the Mind* (London: Sherwood, Gilbert, and Piper, 1835), pp. 311–12 and taken from John Ferriar, *Medical Histories and Reflections*, published between 1792 and 1798. James Cowles Prichard claimed that Ferriar made the only useful attempt to explain the disorder in this period.

27 William Pargeter, *Observations on Maniacal Disorders* (1792); reprint edited by Stanley W. Jackson (London and New York: Routledge, 1988), pp. 58–61.

28 Thomas Trotter, *A View of the Nervous Temperament; being a Practical Inquiry into the Increasing Prevalence, Prevention, Treatment of those Diseases commonly called Nervous, Bilious, Stomach & Liver Complaints; Indigestion; Low Spirits, Gout, &c.* (Boston: Wright, Goodenow and Stockwell, 1808), pp. 95, 92.

29 For a full account of John Haslam's career, see Andrew Scull, Charlotte MacKenzie and Nicholas Hervey, *Masters of Bedlam: The Transformation of the Mad-Doctoring Trade* (Princeton, NJ: Princeton University Press, 1996), chapter 2. Haslam's name has been strongly associated with the scandals at Bethlem, particularly the case of James Norris, but his opinions and writings while resident apothecary to Bethlem were highly influential, also among lunacy reformers such as Pinel and Tuke.

30 John Haslam, *Observations on Insanity: with Practical Remarks on the Disease, and An Account of the Morbid Appearances on Dissection* (London: F. and C. Rivington, 1798), p. 108.

31 Ibid., p. 110.

32 Ibid.

33 Michael J. O'Dowd and Elliot E. Philipp, *The History of Obstetrics and Gynaecology* (New York and London: Parthenon, 1994), p. 624. See Judith Schneid Lewis, *In the Family Way: Childbearing in the British Aristocracy, 1760–1860* (New Brunswick, NJ: Rutgers, 1986), for Denman's practice amongst the upper class.

34 Thomas Denman, *An Introduction to the Practice of Midwifery*, second edition (London: J. Johnson, 1801), vol. 2, chapter XIX 'On Mania', pp. 494–503. The first edition, published in 1794–95, had no details of mental disorders in childbearing women, though it included long sections on diseases and disorders following childbirth and the management of women in childbed.

35 Thomas Denman, *Observations on the Rupture of the Uterus, on the Snuffles in Infants, and on Mania Lactea* (London: J. Johnson, 1810), pp. 37–70.

36 Ibid., p. 37.

37 Denman, *Practice of Midwifery*, second edition, vol. 2, p. 494; Denman, *Observations*, pp. 37–8.

38 Denman, *Practice of Midwifery*, second edition, vol. 2, pp. 498–9. For the potential confusion of puerperal insanity with childbed fever, see chapter 2, pp. 41–2.

39 Denman, *Observations*, p. 50.

40 Denman, *Practice of Midwifery*, second edition, vol. 2, pp. 434–5.

41 Ibid., vol. 1, p. 163. Cited Anne Digby, 'Women's Biological Straitjacket', in Susan Mendus and Jane Rendall (eds), *Sexuality and Subordination: Interdisciplinary Studies of Gender in the Nineteenth Century* (London and New York: Routledge, 1989), pp. 192–220, on p. 197.

42 Digby, 'Women's Biological Straitjacket', p. 197.

43 Eardley Holland, 'The Princess Charlotte of Wales: A Triple Obstetric Tragedy', *Journal of Obstetrics & Gynaecology of the British Empire*, 58 (1951), 905–19, quote on p. 905. The triple obstetric tragedy refers to the death of Princess Charlotte, her baby and the subsequent suicide of Richard Croft.

44 For a full description of Princess Charlotte's death and the debate following this event, see ibid.; and Lewis, *In the Family Way*, pp. 182–7.

45 Ellen Ross, *Love and Toil: Motherhood in Outcast London, 1870–1918* (New York and Oxford: Oxford University Press, 1993), p. 92.

46 For the persistence of maternal deaths, see Irvine Loudon, *Death in Childbirth: An International Study of Maternal Care and Maternal Mortality 1800–1950* (Oxford: Clarendon, 1992).

47 For the takeover of childbirth by male practitioners, see Adrian Wilson, *The Making of Man-Midwifery: Childbirth in England, 1660–1770* (London: UCL Press, 1995); Jean Donnison, *Midwives and Medical Men: A History of Inter-Professional Rivalries and Women's Rights* (London: Heinemann, 1977; second edition New Barnet: Historical Publications, 1988); Hilary Marland (ed.), *The Art of Midwifery: Early Modern Midwives in Europe* (London and New York: Routledge, 1993, 1994); Hilary Marland and Anne-Marie Rafferty (eds), *Midwives, Society and Childbirth: Debates and Controversies in the Modern Period* (London and New York: Routledge, 1997); and, for a comparison with the US, Judith Walzer Leavitt, *Brought to Bed: Childbearing in America, 1750–1950* (New York and Oxford: Oxford University Press, 1986). Donnison's book also gives a detailed account of the campaign for midwife training and legislation in the late nineteenth century.

48 Irvine Loudon, 'Childbirth', in W.F. Bynum and Roy Porter (eds), *Companion Encyclopaedia of the History of Medicine*, vol. 2 (London and New York: Routledge, 1993), pp. 1050–71.

49 For changing terminology with respect to male midwifery practice, see Wilson, *The Making of Man-Midwifery*, pp. 164–5, 175–6. Ornella Moscucci gives a full and excellent description of the heightened emphasis on women's diseases and the evolution of gynaecology in *The Science of Woman: Gynaeocology and Gender in England 1800–1929* (Cambridge: Cambridge University Press, 1990). See also Ann Dally, *Women under the Knife: A History of Surgery* (London: Hutchinson Radius, 1991), and, for a comparison with North America, Ann Douglas Wood, '"The Fashionable Diseases": Women's Complaints and their Treatment in Nineteenth-Century America', and the response of Regina Markell Morantz, 'The Perils of Feminist History', in Judith Walzer Leavitt (ed.), *Women and Health in America* (Madison: University of Wisconsin Press, 1984), pp. 222–38, 239–45; G.J. Barker-Benfield, *The Horrors of the Half-Known Life* (New York: Harper and Row, 1976); Deborah Kuhn McGregor, *From Midwives to Medicine: The Birth of American Gynecology* (New Brunswick, NJ and London: Rutgers University Press, 1998), and Wendy Mitchinson, *The Nature of their Bodies: Women and their Doctors in Victorian Canada* (Toronto: University of Toronto Press, 1991).

50 Loudon, 'Childbirth', p. 1051.

51 Wilson, *The Making of Man-Midwifery*, chapter 11, pp. 145–58; Bronwyn Croxson, 'The Foundation and Evolution of the Middlesex Hospital's Lying-In Service, 1745–86', *Social History of Medicine*, 14 (2001), 27–57; Margaret Connor Versluysan, 'Midwives, Medical Men and "Poor Women Labouring of Child": Lying-in Hospitals in Eighteenth-Century London', in Helen Roberts (ed.), *Women, Health and Reproduction* (London: Routledge & Kegan Paul, 1981), pp. 18–49.

52 Moscucci, *The Science of Woman*, p. 81.

53 Irvine Loudon has calculated that there were 15,000 general practitioners in Great Britain by 1841, the ratio of all medical practitioners to the population 1:1,000, leading to claims that the profession was 'overstocked'. Irvine Loudon, *Medical Care and the General Practitioner 1750–1850* (Oxford: Clarendon Press, 1986), pp. 215, 216.

54 Anne Digby, *Making a Medical Living: Doctors and Patients in the English Market for Medicine, 1720–1911* (Cambridge: Cambridge University Press, 1994), chapter 9.

55 Patricia Branca, *Silent Sisterhood: Middle-Class Women in the Victorian Home* (London: Croom Helm, 1975), p. 65.

56 Digby, *Making a Medical Living*, p. 254.

57 Andrew Scull, *The Most Solitary of Afflictions: Madness and Society in Britain, 1700–1900* (New Haven and London: Yale University Press, 1993), especially chapter 6; Kathleen Jones, *Asylums and After: A Revised History of the Mental Health Services: From the Early 18th Century to the 1990s* (London: Athlone, 1993); and for Scotland, Jonathan Andrews, *'They're in the Trade … of Lunacy, They "cannot interfere" – they say': The Scottish Lunacy Commissioners and Lunacy Reform in Nineteenth-Century Scotland* (London: Wellcome Institute for the History of Medicine, Occasional Publications, No. 8, 1998).

58 Scull, *The Most Solitary of Afflictions*, p. 281.

59 William L.l. Parry-Jones, *The Trade in Lunacy: A Study of Private Madhouses in England in the Eighteenth and Nineteenth Centuries* (London: Routledge & Kegan Paul, 1972).

60 Showalter cites figures from the 1871 census when there were 1,182 female lunatics for every 1,000 male lunatics, and 1,242 female pauper lunatics for every 1,000 male pauper lunatics. Elaine Showalter, *The Female Malady: Women, Madness and English Culture, 1830–1980* (London: Virago, 1987, first published New York: Pantheon, 1985), p. 52.

61 See Joan Busfield, *Men, Women and Madness: Understanding Gender and Mental Disorder* (London: Macmillan, 1996), especially. ch. 7. David Wright, 'Delusions of Gender? Lay Identification and Clinical Diagnosis of Insanity in Victorian England', in Jonathan Andrews and Anne Digby (eds), *Sex and Seclusion, Class and Custody: Perspectives on Gender and Class in the History of British and Irish Psychiatry* (Amsterdam and New York: Rodopi, 2004), pp. 149–76. In the York Retreat women usually outnumbered men during the nineteenth century, but here, as elsewhere, this could be linked to women's lower mortality rates and longer period of stay, and to the fact that they outnumbered men in the general population. Women also outnumbered men in the Society of Friends, while the Friends subsidised treatment, an inducement for women to be sent to the Retreat. See Anne Digby, *Madness, Morality and Medicine: A Study of the York Retreat, 1796–1914* (Cambridge: Cambridge University Press, 1985), pp. 174–5.

62 Edinburgh University Library: Lothian Health Board Archive, Royal Edinburgh Hospital, LHB7/7/6, Annual Reports of the Royal Edinburgh Asylum, 1812–55: Physician's Annual Report for the Year 1855, p. 25. 'Climacteric change' refers to the menopause and 'secret vice' was a veiled reference to masturbation.

63 Charlotte MacKenzie, 'A Family Asylum: A History of the Private Madhouse at Ticehurst in Sussex, 1792–1917', unpublished PhD thesis, University of London, 1987, pp. 273, 504–5.

64 See e.g. Moscucci, *The Science of Woman*, ch. 1; Bruce Haley, *The Healthy Body and Victorian Culture* (Cambridge, Mass and London: Harvard University Press, 1978).

65 For an excellent summary of this process, see Digby, 'Women's Biological Straitjacket'. See also for the dominance of women by their reproductive systems, Jeffrey Weeks, *Sex, Politics and Society: The Regulation of Sexuality since 1800* (London: Longman, 1989).

66 See Moscucci, *The Science of Woman*, pp. 102–33; and Pat Jalland and John Hooper, *Women from Birth to Death: The Female Life Cycle in Britain 1830–1914* (Brighton: Harvester Press, 1986), for a cradle-to-grave assessment of the risk of being a woman.

67 John Burns, *The Principles of Midwifery; Including the Diseases of Women and Children*, seventh edition (London: Longman, Rees, Orme, Brown and Green, 1828). (The work was first published in 1809 and by 1837 had gone through nine editions.) Lochia is the postpartum discharge from the uterus.

68 See chapter 3 for the link between domestic ideology and puerperal insanity.

69 See Moscucci, *The Science of Woman*, pp. 112–27 for the fierce debate surrounding the 'speculum question' and pp. 135–40 for the controversy over

the surgical removal of the ovaries (quote on p. 138). Towards the end of the century, a number of gynaecologists were opposing surgical procedures and other radical forms of intervention. W.S. Playfair, for example, discouraged excessive 'local uterine treatment' and was reluctant to intervene in cases where he simply did not know what was wrong with his patients. See Hilary Marland, '"Uterine Mischief": W.S. Playfair and his Neurasthenic Patients', in Marijke Gijswijt-Hofstra and Roy Porter (eds), *Cultures of Neurasthenia From Beard to the First World War* (Amsterdam and New York: Rodopi, 2001), pp. 117–39. See also Barker-Benfield, *The Horrors of the Half-Known Life* and for a balanced discussion of the role of surgical gynaecology, Judith M. Roy, 'Surgical Gynaecology', in Rima Apple (ed.), *Women, Health, and Medicine in America* (New Brunswick, NJ: Rutgers University Press, 1990), pp. 173–95.

70 Moscucci, *The Science of Woman*, p. 128. Elaine Thomson, 'Women in Medicine in Late Nineteenth- and Early Twentieth-Century Edinburgh: A Case Study', unpublished PhD thesis, University of Edinburgh, 1998; and Judith Lockhart, '"Truly a Hospital for Women": The Birmingham and Midland Hospital for Women, 1871–1901', unpublished MA dissertation, University of Warwick, 2002 have also explored the relationship between poor health, lack of medical attention and gynaecological complaints.

71 Judith Walzer Leavitt's study of the shift from social childbirth to physician-attended deliveries in America has described the search for safer and less painful childbirth by women with debilitating and dangerous obstetrical histories, describing how 'the shadow of maternity' dominated the lives of women with large families and the impact of difficult labours, which could lead to devastating and painful conditions and deformities. Leavitt, *Brought to Bed*, chapter 1. See chapters 4 and 5 for the poor health of women admitted to asylums with puerperal insanity.

72 Alexander Hamilton, *A Treatise on the Management of Female Complaints*, seventh edition (Edinburgh: P. Hill and London: Underwood and Blacks, 1813; first published 1780), pp. 46–7. Cited Digby, 'Women's Biological Straitjacket', p. 197.

73 Burns, *The Principles of Midwifery*, third edition (London: Longman, Hurst, Rees, Orme, and Brown, 1814), p. 107. For details of the relationship of the female life cycle to insanity, see Showalter, *The Female Malady*.

74 Notably E.J. Tilt, *On the Preservation of the Health of Women at the Critical Periods of Life* (London: John Churchill, 1851), pp. 25–42, quote on p. 31.

75 There is a large secondary literature on hysteria, including Sander Gilman, Helen King, Roy Porter, George S. Rousseau and Elaine Showalter, *Hysteria Beyond Freud* (Berkeley, CA: University of California Press, 1993); Carroll Smith-Rosenberg, 'The Hysterical Woman: Sex Roles and Role Conflict in 19th-Century America', *Social Research*, 39 (1979), 652–78; Showalter, *The Female Malady*, chapter 6, and idem, *Hystories: Hysterical Epidemics and Modern Culture* (New York: Columbia University Press, 1997).

76 Thomas Laycock, *A Treatise on the Nervous Diseases of Women; Comprising an Inquiry into the Nature, Causes, and Treatment of Spinal and Hysterical Disorders* (London: Longman, Orme, Brown, Green, and Longmans, 1840), pp. 8–9.

77 For example, one case of 'hysteria' was admitted to Hanwell Asylum in 1840 compared with four cases connected to childbirth, three to 'milk

fever', and three related to pregnancy, out of 88 female admissions. Admissions with hysteria, especially to large pauper asylums like Hanwell, were unusual. Wellcome Trust Library: T.216.21, John Conolly, *The Report of the Resident Physician of the Hanwell Lunatic Asylum, Presented to the Court of Quarter Sessions for Middlesex, at the Michaelsmas Sessions, 1840*, p. 11.

78 See Helen King, *Hippocrates' Woman: Reading the Female Body in Ancient Greece* (London: Routledge, 1998), chapter 11 for changing ways of describing hysteria, quote on p. 205.

79 Ibid., p. 246.

80 See chapter 3 for Sara Coleridge's knowledge of moral insanity and female nervous disorders.

81 For networking and cultural links between mothers and midwives, see Doreen Evenden, *The Midwives of Seventeenth-Century London* (Cambridge: Cambridge University Press, 2000), and for the organisation of labour and the lying-in by midwives, Adrian Wilson, 'The Ceremony of Childbirth and Its Interpretation', in Valerie Fildes (ed.), *Women as Mothers in Pre-Industrial England* (London and New York: Routledge, 1990), pp. 68–107.

82 David Harley, 'Provincial Midwives in England: Lancashire and Cheshire, 1660–1760', in Marland (ed.), *The Art of Midwifery*, pp. 27–48 argues that women's changing tastes influenced the shift to male practitioners. See also, for the move to physician-attended births, Lewis, *In the Family Way*; and Amanda Vickery, *The Gentleman's Daughter: Women's Lives in Georgian England* (New Haven and London: Yale University Press, 1998), chapter 3.

83 François Mauriceau, *The Diseases of Women with Child, And in Child-bed*, trans. Hugh Chamberlen (London: John Darby, 1683), p. 299.

84 John Burns, *Popular Directions for the Treatment of the Diseases of Women and Children* (London: Longman, Hurst, Rees, Orme and Brown, 1811), Preface, p. iv.

85 Thomas Bull's *Hints to Mothers* quickly went through fourteen editions. See, for further details of health manuals, popular books and periodicals on motherhood, Branca, *Silent Sisterhood*, pp. 74–7, 82–4.

86 Thomas Bull, *Hints to Mothers, for the Management of Health during the Period of Pregnancy and Lying-In Room; with an Exposure of Popular Errors in Connection with Those Subjects*, sixteenth edition (London: Longmans, Green, and Co., 1865; first published 1837), pp. 3–4, 4–5.

87 Ibid., p. 45.

88 Leavitt and Walton have powerfully demonstrated these fears, revealed largely through women's own writing, in Judith Walzer Leavitt and Whitney Walton, '"Down to Death's Door": Women's Perceptions of Childbirth in America', in Leavitt (ed.), *Women and Health in America*, pp. 155–74. For the expression of fear by early modern women, see Patricia Crawford, 'The Construction and Experience of Maternity in Seventeenth-Century England' and Linda Pollock, 'Embarking on a Rough Passage: The Experience of Pregnancy in Early-Modern Society', in Fildes (ed.), *Women as Mothers*, pp. 3–38, 39–67.

89 Mears, *The Midwife's Candid Advice to the Fair Sex*, pp. 2, 4.

90 Robert Lee, *Lectures on the Theory and Practice of Midwifery, Delivered in the Theatre of St. George's Hospital* (London: Longman, Brown, Green, and

Longmans, 1844) (also reported in the *London Medical Gazette*, 1842–43), p. 1.

91 Michael Ryan, *A Manual of Midwifery and Diseases of Women and Children*, fourth edition (London: published by the author, 1841), p. 167.

92 Ibid., p. 334.

93 Nakamura has concluded that interest in the mental state of pregnant women had strong roots in the eighteenth century and that the nineteenth century provided no significant turning point in tone or context, but, though citing some interesting individual cases up to the eighteenth century, little in the way of evidence is provided to match the outpouring of material and efforts to develop an aetiology of puerperal insanity in the nineteenth century: Lisa Ellen Nakamura, 'Puerperal Insanity: Women, Psychiatry, and the Asylum in Victorian England, 1820–1895', unpublished PhD thesis, University of Washington, 1999, pp. 112–13, 133.

94 Burns, *The Principles of Midwifery*, third edition, p. 185.

2 Boundaries of Expertise and the Location of Puerperal Insanity

1 George Man Burrows, *Commentaries on the Causes, Forms, Symptoms, and Treatment, Moral and Medical, of Insanity* (London: Thomas and George Underwood, 1828), pp. 362–3.

2 The interpretation of the terms 'puerperal insanity', 'puerperal mania' and 'puerperal melancholia' in the nineteenth century can be confusing. All three were used and puerperal insanity was often regarded as synonymous with 'puerperal mania' and used interchangeably: Irvine Loudon, *Death in Childbirth: An International Study of Maternal Care and Maternal Mortality 1800–1950* (Oxford: Clarendon, 1992), p. 144.

3 As they also did in North America. See Nancy Theriot, 'Diagnosing Unnatural Motherhood: Nineteenth-Century Physicians and "Puerperal Insanity"', *American Studies*, 26 (1990), 69–88.

4 See Roy MacLeod, *Government and Expertise: Specialists, Administrators and Professionals, 1860–1919* (Cambridge and New York: Cambridge University Press, 1988), especially his introduction, quote on p. 10.

5 The substantial literature on moral management includes Anne Digby, *Madness, Morality and Medicine: A Study of the York Retreat, 1796–1914* (Cambridge: Cambridge University Press, 1985); idem, 'Moral Treatment at the Retreat, 1796–1846', in W.F. Bynum, R. Porter and M. Shepherd (eds), *Anatomy of Madness: Essays in the History of Psychiatry*, vol. II, *Institutions and Society* (London and New York: Tavistock, 1985), pp. 52–72; Andrew Scull, 'Moral Treatment Reconsidered: Some Sociological Comments on an Episode in the History of British Psychiatry', in idem (ed.), *Madhouses, Mad-Doctors and Madmen: The Social History of Psychiatry in the Victorian Era* (London: Athlone, 1981), pp. 105–20; idem, *The Most Solitary of Afflictions: Madness and Society in Britain, 1700–1900* (New Haven and London: Yale University Press, 1993), chapters 2–4; and Vieda Skultans, *English Madness: Ideas on Insanity 1580–1890* (London: Routledge & Kegan Paul, 1979), chapter 4. For

changes in theories of insanity and therapeutic responses over the nineteenth century, see Laurence J. Ray, 'Models of Madness in Victorian Asylum Practice', *Archives of European Sociology*, 22 (1981), 229–64.

6   This chapter also contributes to the growing literature that focuses on the treatment of mental disorder outside the walls of the asylum. See e.g. the essays in Peter Bartlett and David Wright (eds), *Outside the Walls of the Asylum: The History of Care in the Community 1750–2000* (London: Athlone, 1999).

7   Robert Gooch, *Observations on Puerperal Insanity* (London, 1820) (extracted from *Medical Transactions*, sixth volume of Royal College of Physicians, read at the College, 16 December 1819).

8   See W. Macmichael, *Lives of British Physicians* (London: John Murray, 1830), pp. 305–41 for Dr Henry Southey's account of Gooch's life and work; and *Dictionary of National Biography*.

9   Robert Gooch, *On Some of the Most Important Diseases Peculiar to Women; with Other Papers*, with a prefatory essay by Robert Ferguson, M.D. (London: New Sydenham Society, 1831); idem, *A Practical Compendium of Midwifery; Being the Course of Lectures on Midwifery, and on Diseases of Women and Infants, Delivered at St. Bartholomew's Hospital, by the Late Robert Gooch, M.D.* (London: Longman, Rees, Orme, Brown, and Green, 1831). For Gooch on puerperal fever, see Irvine Loudon, *The Tragedy of Childbed Fever* (Oxford: Oxford University Press, 2000), pp. 64–6; and idem (ed.), *Childbed Fever: A Documentary History* (New York and London: Garland, 1995), chapter 6.

10   'Robert Gooch (1784–1830)', in Richard Hunter and Ida Macalpine, *Three Hundred Years of Psychiatry 1535–1860* (London: Oxford University Press, 1963), pp. 796–800, quotes on p. 797.

11   See chapter 3 for detailed examples of Gooch's case histories and the disruption of the bourgeois household.

12   Macmichael, *Lives of British Physicians*, p. 341. The psychological doctor Dr Forbes Winslow also roundly praised the sound practical information included in Gooch's work on the diseases of females: *Lancet*, I (1830–31), p. 36. See also *Dictionary of National Biography*.

13   Gooch, *On Some of the Most Important Diseases Peculiar to Women*, p. 20.

14   Ibid., Preface, pp. il–l.

15   Gooch, *Observations*, pp. 3–4.

16   Ibid., pp. 4–6.

17   See Nancy M. Theriot, 'Negotiating Illness: Doctors, Patients and Families in the Nineteenth Century', *Journal of the History of Behavioral Sciences*, 37 (2001), 349–68, for the use of case notes by physicians to contribute to the formation of medical knowledge. See chapters 4 and 5 for an intensive discussion of case histories in asylums and private practice.

18   Henry Maunsell, *The Dublin Practice of Midwifery*, new revised edition (London: Longman, Brown, Green, Longmans, & Roberts, 1856), p. 252.

19   John Tricker Conquest, e.g., published his extensive study *Outlines of Midwifery* in 1820, shortly after Gooch's treatise appeared, followed by a second edition in 1821: J.T. Conquest, *Outlines of Midwifery, Developing its Principles and Practice; Intended as a Text Book for Students, and a Book of Reference for Junior Practitioners* (London: John Anderson, 1820). This

influential book included a summary of the symptoms and management of puerperal insanity. Like Gooch, Conquest had a flourishing obstetric practice and held an appointment as Physician-Accoucheur to the City of London Lying-in Hospital. Unlike Gooch, who was held in high regard as a teacher, Conquest was reputedly an unpopular lecturer and was persuaded in 1834 to resign his post as Lecturer in Midwifery to St Bartholomew's Hospital that he had inherited from Gooch in 1825. Fleetwood Churchill moved to Dublin after graduating as MD in Edinburgh in 1831. He helped establish the Western Lying-in Hospital in the city, where he also instructed students in midwifery, was Professor of Midwifery at the College of Physicians, and built up a successful career as author and obstetric practitioner. The chapters on puerperal mania in Churchill's textbooks *On the Theory and Practice of Midwifery* and *On the Diseases of Women*, both of which went through many editions, were widely cited: Fleetwood Churchill, *On the Theory and Practice of Midwifery* (London: Henry Renshaw, 1842); idem, *On the Diseases of Women; Including those of Pregnancy and Childbed* (Dublin: Fannin, 1849).

20 See Ornella Moscucci, *The Science of Woman: Gynaecology and Gender in England 1800–1929* (Cambridge: Cambridge University Press, 1990), especially chapter 2 for evolving professional organisations.

21 James Reid, 'On the Causes, Symptoms, and Treatment of Puerperal Insanity', *Journal of Psychological Medicine and Mental Pathology*, 1 (1848), 128–51, 284–94.

22 I. Loudon, 'Puerperal Insanity in the 19th Century', *Journal of the Royal Society of Medicine*, 81 (February 1988), 76–9, p. 76.

23 Burrows, *Commentaries*, pp. 362–408.

24 John Charles Bucknill and Daniel Hack Tuke, *A Manual of Psychological Medicine*, 1858 edition (New York and London: Hafner, 1968), pp. 235–9.

25 For psychiatrists' early attempts at organisation, see Trevor Turner, '"Not Worth Powder and Shot": The Public Profile of the Medico-Psychological Association, c.1851–1914', in G.E. Berrios and H. Freeman (eds), *150 Years of British Psychiatry 1841–1991* (London: Athlone, 1991), pp. 3–16. The courtroom became another arena where alienists could publicly demonstrate their competence to judge the cause and outcome of mental disorder, as they were called as expert witnesses in trials involving the insanity plea, including, as discussed in chapter 6, numerous cases of infanticide or concealment of pregnancy.

26 J.-E.-D. Esquirol, 'De l'aliénation mentale des nouvelles accouchés et des nourrices', *Annuaire Médico-Chirurgical de Hôpitaux de Paris*, 1 (1819), 600–32. His influential *Des Maladies mentales* was published in 1838 and translated in 1845 as *Mental Maladies: A Treatise on Insanity*, trans. E.K. Hunt (Philadelphia: Lea and Blanchard, 1845), including his chapter 'Mental Alienation of those Recently Confined, and of Nursing Women'.

27 L.-V. Marcé, *Traité de la folie des femmes enceintes, des nouvelles accouchés et des nourrices* (Paris: J.B. Baillière, 1858). Marcé seems, however, to be cited more in retrospect than by his nineteenth-century contemporaries.

28 John Conolly, 'Description and Treatment of Puerperal Insanity', Lecture XIII: 'Clinical Lectures on the Principal Forms of Insanity, Delivered in the Middlesex Lunatic-Asylum at Hanwell', *Lancet*, I (28 March 1846), 349–54,

quote on p. 349. For Conolly's career at Hanwell and advocacy of non-restraint, see Akihito Suzuki, 'The Politics and Ideology of Non-Restraint: The Case of the Hanwell Asylum', *Medical History*, 39 (1995), 1–17; and Andrew Scull, Charlotte MacKenzie and Nicholas Hervey, *Masters of Bedlam: The Transformation of the Mad-Doctoring Trade* (Princeton, NJ: Princeton University Press, 1996), chapter 3.

29 Conolly, 'Description and Treatment of Puerperal Insanity', p. 349.

30 Robert Boyd, 'Observations on Puerperal Insanity', *Journal of Mental Science*, 16 (1870), 153–65, on p. 153.

31 Alexander Morison, 'Medical News', *Medical Times & Gazette*, 7 (3 December 1853), 591. See Scull, MacKenzie and Hervey, *Masters of Bedlam*, chapter 5, for Morison's career outside the walls of the asylum.

32 Conolly, 'Description and Treatment of Puerperal Insanity', p. 350.

33 W. Tyler Smith, 'Lectures on Parturition, and the Principles & Practice of Obstetricy', *Lancet*, II (29 July 1848), 117–20, quote on p. 119.

34 Idem, 'Puerperal Mania', Lecture XXXIX: 'Lectures on the Theory and Practice of Obstetrics', *Lancet*, II (18 October 1856), 423–5, quote on p. 424. See also Hilary Marland, '"Destined to a Perfect Recovery": The Confinement of Puerperal Insanity in the Nineteenth Century', in Joseph Melling and Bill Forsythe (eds), *Insanity, Institutions and Society, 1800–1914* (London and New York: Routledge, 1999), pp. 137–56.

35 Loudon, 'Puerperal Insanity', p. 76. Modern psychiatric textbooks explain how puerperal and lactational psychoses were thought to be specific entities during the nineteenth century, but by the closing decades of the century, as will be explored in chapter 7, there was increased scepticism about the distinctiveness of the condition, and its specific link with childbirth, and psychiatrists were searching harder for links with predisposition, previous mental disorder and hereditary indications. Early twentieth-century psychiatrists, including Emil Kraepelin and Eugen Bleuler, regarded puerperal psychosis as indistinct from other psychoses, a view held by some practitioners today. Ian Brockington and his colleagues have outlined the way in which nosology denied puerperal psychosis its identity in the twentieth century: I.F. Brockington and R. Kumar (eds), *Motherhood and Mental Illness* (London and New York: Academic Press/Grune and Stratton, 1982).

36 Francis H. Ramsbotham, *The Principles and Practice of Obstetric Medicine and Surgery in Reference to the Process of Parturition*, third edition (London: John Churchill, 1851), p. 554.

37 Cited M.D. Macleod, 'An Address on Puerperal Insanity', *British Medical Journal*, II (7 August 1886), 239–42, on p. 241.

38 Bucknill and Tuke, *A Manual of Psychological Medicine*, pp. 235–6.

39 At the end of the century, one estimate suggested that as many as 1 in 4 deliveries were associated with mental disorder, while only 1 in 14 female asylum admissions were accounted for by this condition: C. Mercier, *A Textbook of Insanity* (London: Macmillan, 1902), p. 157. Cited Anne Digby, 'Women's Biological Straitjacket', in Susan Mendus and Jane Rendall (eds), *Sexuality and Subordination: Interdisciplinary Studies of Gender in the Nineteenth Century* (London and New York: Routledge, 1989), pp. 192–220, on p. 206.

40   David Wright, 'Delusions of Gender? Lay Identification and Clinical Diagnosis of Insanity in Victorian England', in Jonathan Andrews and Anne Digby (eds), *Sex and Seclusion, Class and Custody: Perspectives on Gender and Class in the History of British and Irish Psychiatry* (Amsterdam and New York: Rodopi, 2004), pp. 149–76, on p. 165.

41   Peter Bartlett, *The Poor Law of Lunacy: The Administration of Pauper Lunatics in Mid-Nineteenth-Century England* (London and Washington: Leicester University Press, 1999), pp. 171–2.

42   Loudon, 'Puerperal Insanity', p. 77; Richard Grundy, 'Observations upon Puerperal Insanity', *Journal of Psychological Medicine and Mental Pathology*, 13 (1860), 414–25, on p. 415.

43   Reid, 'On the Causes, Symptoms, and Treatment', pp. 140–1.

44   James Y. Simpson, *Clinical Lectures on the Diseases of Women* (Edinburgh: Adam and Charles Black, 1872), p. 556.

45   Reid, 'On the Causes, Symptoms, and Treatment', p. 142.

46   Loudon, 'Puerperal Insanity', p. 77.

47   West Yorkshire Archive Service, Wakefield, Annual Reports of the West Riding Pauper Lunatic Asylum, WRA, C85/108–10, 1833–67, 1868–79, 1880–86. For the admission of women to the County Asylums of Yorkshire, see Robert James Ellis, 'A Field of Practice or a Mere House of Detention? The Asylum and its Integration, with Special Reference to the County Asylums of Yorkshire, c.1844–1888', unpublished PhD thesis, University of Huddersfield, 2001.

48   Figures dropped away after 1867, with no admissions under the heading 'puerperal insanity' being listed between 1868 and 1878. Thereafter the figures increased again, with puerperal insanity accounting for around 5–10 per cent of female admissions during the 1880s.

49   Conolly, 'Description and Treatment of Puerperal Insanity', p. 349.

50   Churchill, *On the Diseases of Women*, fourth edition (Dublin: Fannin, 1857), p. 738.

51   Reid, 'On the Causes, Symptoms, and Treatment', pp. 134–5. Theriot's detailed survey of the American literature on puerperal insanity found similar accounts of patients' behaviour: Theriot, 'Diagnosing Unnatural Motherhood'.

52   John Burns, *The Principles of Midwifery; Including the Diseases of Women and Children*, seventh edition (London: Longman, Rees, Orme, Brown and Green, 1828), p. 533.

53   This is explored in detail in chapters 4 and 5.

54   Gooch, *On Some of the Most Important Diseases Peculiar to Women*, p. 54.

55   Burrows, *Commentaries*, pp. 363–4.

56   Reid, 'On the Causes, Symptoms, and Treatment', p. 143 (Reid's emphasis).

57   'J.C. Prichard and the Concept of "Moral Insanity"', with an introduction by G.E. Berrios, *History of Psychiatry*, 10 (1999), 111–26.

58   Reid, 'On the Causes, Symptoms, and Treatment', p. 143. In 1845 the 'apparent and assigned causes of disease' amongst women admitted to Bethlem Hospital included 'physical': puerperal, prolonged lactation, weaning and parturition, and 'moral': the birth of an illegitimate child and 'dread of approaching confinement'. Wellcome Trust Library, WLM28.BE5, B84, General Report of the Royal Hospitals of Bridewell and

Bethlem and of the House of Occupations, for the year ending 31 December 1845.

59 Esquirol, *Mental Maladies*.

60 Conolly, 'Description and Treatment of Puerperal Insanity', p. 350.

61 Burrows, *Commentaries*, pp. 389–90.

62 See chapters 5 and 7 for the relationship between heredity and puerperal insanity.

63 Prefacing reflex theories of nervous organisation in the Victorian period. See Moscucci, *The Science of Woman*, especially pp. 104–5; and Elaine Showalter, *The Female Malady: Women, Madness and English Culture, 1830–1980* (London: Virago, 1987, first published New York: Pantheon, 1985), chapter 5.

64 Gooch, *On Some of the Most Important Diseases Peculiar to Women*, p. 63.

65 Ibid.

66 Burns, *The Principles of Midwifery*, third edition (London: Longman, Hurst, Rees, Orme, and Brown, 1814), pp. 434–5.

67 Conquest, *Outlines of Midwifery*, p. 229.

68 Marshall Hall, *Commentaries on Some of the More Important of the Diseases of Females. In Three Parts* (London: Longman, Rees, Orme, Brown and Green, 1827), pp. 251–2. For Hall's work on the diseases of women, see Diana E. Manuel, *Marshall Hall 1790–1857* (Amsterdam and Atlanta, GA: Rodopi, 1996), pp. 64–89.

69 Marshall Hall, *Lectures on the Nervous System and its Diseases* (Philadelphia: E.L. Carey and A. Hart, 1836), p. 152. Cited Lisa Ellen Nakamura, 'Puerperal Insanity: Women, Psychiatry, and the Asylum in Victorian England, 1820–1895', unpublished PhD thesis, University of Washington, 1999, pp. 190–1.

70 Hall, *Commentaries*, pp. 248–9. For a detailed case study, see pp. 253–6.

71 Alexander Morison, *The Physiognomy of Mental Diseases*, second edition (London: Longman and Co. and S. Highley, 1843), p. 15.

72 Thomas Lightfoot, 'Puerperal Mania: Its Nature and Treatment', *Medical Times*, 21 (6 April 1850), 273–6, on p. 275.

73 Conolly, 'Description and Treatment of Puerperal Insanity', pp. 349–50.

74 Gooch, *On Some of the Most Important Diseases Peculiar to Women*, p. 61.

75 Macalpine and Hunter argued that puerperal insanity did not come to the notice of psychiatrists because cases were seen and treated by obstetricians and because no differentiation was made between the delirium of puerperal sepsis and mental illness proper: Hunter and Macalpine, *Three Hundred Years of Psychiatry*, pp. 796–7. The comments of seventeenth- and eighteenth-century physicians on mental disorders associated with childbirth, and particularly Haslam's figures for Bethlem Hospital between 1784 and 1794, cited in chapter 1, would seem to refute this claim.

76 Reid, 'On the Causes, Symptoms, and Treatment', p. 144 (Reid's emphasis).

77 Irvine Loudon has suggested that alleged cases of puerperal mania could have been suffering from puerperal fever, and this would help explain the not inconsiderable number of deaths from puerperal insanity, which he estimates as accounting for 2–3 per cent of maternal mortality in the nineteenth century. Loudon cites the report of the Registrar General for

the period 1872–76, which listed 573 deaths from puerperal insanity out of a total of 23,051 maternal deaths (2.5 per cent): Register General, *Annual Report for the Year 1876*, p. 244. Cited Loudon, *Death in Childbirth*, Appendix, Table 20, p. 570.

78 London Metropolitan Archives, General Lying-in Hospital London, H1/GL1/B19/1, Medical Officers' Case Book, 1827–28. A similar case is cited by Loudon of a woman recently delivered and with a fever of 103° F who attempted to leap out of the window of the Edinburgh Royal Maternity Hospital in 1873. She was restrained, but died later in delirium and her death was ascribed to puerperal mania. A post-mortem, however, revealed that she had pus in her peritoneal cavity, a clear indication of puerperal peritonitis, a severe inflammation of the peritoneum caused by infection in puerperal fever: Loudon, 'Puerperal Insanity', p. 77.

79 Edinburgh University Library: Lothian Health Board Archive, Royal Edinburgh Hospital, LHB7/51/6, Case Book, 1847–51, Agnes Hunter or Dunbar, admitted 9 June 1848, p. 180.

80 F.W. Mackenzie, 'On the Pathology and Treatment of Puerperal Insanity: Especially in Reference to its Relation with Anaemia', *London Journal of Medicine*, 3 (1851), 504–21, on p. 504. A number of obstetric physicians also associated puerperal mania with albuminuria (protein in the urine, a symptom of kidney disease), which was said to result in a particularly dangerous variant of the condition and in some cases convulsions. See Simpson, *Clinical Lectures*, pp. 561–6; Arthur Scott Donkin, 'On the Pathological Relation between Albuminuria and Puerperal Mania', *Edinburgh Medical Journal*, 8 (1862–63), 994–1004.

81 Mackenzie, 'On the Pathology and Treatment of Puerperal Insanity', p. 506.

82 Burns, *The Principles of Midwifery*, seventh edition, p. 533.

83 Conolly, 'Description and Treatment of Puerperal Insanity', p. 349.

84 Ramsbotham, *The Principles and Practice*, p. 555 (Ramsbotham's emphasis).

85 Though the notion that insanity was a disease of the body retained its edge well into the nineteenth century: William F. Bynum, 'Rationales for Therapy in British Psychiatry, 1780–1835', *Medical History*, 18 (1974), 317–34. See also Ray, 'Models of Madness in Victorian Asylum Practice', especially pp. 238–44 for management in the early Victorian period based on moral and physical aetiologies.

86 The rest cure was first set out by Silas Weir Mitchell in 1877 in *Fat and Blood: And How to Make Them* (Philadelphia: Lippincott, 1877). See, for a comparison of approaches between puerperal insanity and neurasthenia, Hilary Marland, '"Uterine Mischief": W.S. Playfair and his Neurasthenic Patients', in Marijke Gijswijt-Hofstra and Roy Porter (eds), *Cultures of Neurasthenia From Beard to the First World War* (Amsterdam and New York: Rodopi, 2001), pp. 117–39. *The Yellow Wallpaper* (1892) is the best-known account of the rest cure treatment for nervous collapse following childbirth: see Dale M. Bauer's edition of Charlotte Perkins Gilman, *The Yellow Wallpaper* (Boston and New York: Bedford, 1998), which also has extracts on the implementation of the rest cure and the relationship between nervous disorders and childbirth.

87 Thomas Denman, *Observations on the Rupture of the Uterus, on the Snuffles in Infants, and on Mania Lactea* (London: J. Johnson, 1810), p. 66.

88 Gooch, *Observations*, pp. 21–2.

89 Gooch, *On Some of the Most Important Diseases Peculiar to Women*, p. 80.

90 Ibid., p. 81.

91 *Lancet*, I (10 and 17 October 1829), 96, 122.

92 See Scull, *The Most Solitary of Afflictions*, p. 242 for opinions on the effectiveness of various medical treatments in cases of mania and melancholia as set out in the 1847 Report of the Commissioners in Lunacy. For asylum treatment in the nineteenth century, see also Digby, *Madness, Morality and Medicine*, chapter 6; and Charlotte MacKenzie, 'A Family Asylum: A History of the Private Madhouse at Ticehurst in Sussex, 1792–1917', unpublished PhD thesis, University of London, 1987, especially chapter 3: 3.

93 Ramsbotham, *The Principles and Practices*, p. 564.

94 See Virginia Berridge, *Opium and the People: Opiate Use and Drug Control Policy in Nineteenth and Early Twentieth Century England*, revised edition (London and New York: Free Association Books, 1999; first published New Haven and London: Yale University Press, 1987), chapter 6, for the use of opiates in medical practice. By the end of the century opiates were seen as a cause of mental illness and addiction to them as a form of insanity in itself.

95 Thomas J. Graham, *On the Diseases of Females; A Treatise Describing their Symptoms, Causes, Varieties, and Treatment. Including the Diseases and Management of Pregnancy and Confinement*, seventh edition (London: published for the author, 1861); first published as *The Diseases Peculiar to Females: A Treatise Illustrating their Symptoms, Causes, Varieties, and Treatment* (London: Simpkin and Marshall, 1834), pp. 263–4.

96 For techniques of counter-irritation, including issues, setons and blisters, see William Brockbank, *Ancient Therapeutic Arts* (London: William Heinemann, 1954). An issue involved placing a foreign body under the skin to cause a running sore, based on the idea that the body could thus be drained of noxious humours. Applying a seton (a skein of cotton passed below the skin and left with the ends protruding to promote drainage) or blisters would produce similar results.

97 Morison, *The Physiognomy of Mental Diseases*, second edition, pp. 16–17.

98 Boyd, 'Observations on Puerperal Insanity', p. 165.

99 Morison, *The Physiognomy of Mental Diseases*, second edition, p. 19.

100 J.B. Tuke, 'Cases Illustrative of the Insanity of Pregnancy, Puerperal Mania, and Insanity of Lactation', *Edinburgh Medical Journal*, 12 (1866–67), 1083–101, quote on p. 1093.

101 Anne Digby has pointed to the growing use of sedation at the York Retreat in the second quarter of the nineteenth century, which was also used as a means of reducing physical restraint: Digby, *Madness, Morality and Medicine*, pp. 128–9.

102 Tuke, 'Cases Illustrative', p. 1093.

103 Smith, 'Puerperal Mania', p. 423. For the debate on the use of chloroform in obstetrics, including Smith's stance and its association with overt sexuality, see A.J. Youngson, *The Scientific Revolution in Victorian Medicine*

(London: Croom Helm, 1979), chapter 3; and Mary Poovey, '"Scenes of an Indelicate Character": The Medical "Treatment" of Victorian Women', *Representations*, 14 (1986), 137–68. For a discussion of the link between chloroform and puerperal mania in Edinburgh, see chapter 4, pp. 115–16.

104 Ramsbotham, *The Principles and Practice*, p. 554.

105 'Insanity from Chloroform Employed in Parturition', *Journal of Psychological Medicine and Mental Pathology*, 3 (1850), 269–70.

106 Articles appearing in the *Association Medical Journal* (precursor of the *British Medical Journal*) in 1853 reveal the intensity of the debate following the publication of John Snow's article, 'On the Administration of Chloroform During Parturition' in June of that year (10 June 1853, 500–2). The discussion was taken up by the *Lancet* under its editor Wakley in connection with Queen Victoria's resort to chloroform in 1853.

107 Charles Kidd, 'On Chloroform and Some of its Clinical Uses', *London Medical Review or Monthly Journal of Medical and Surgical Science*, II (June 1862), 243–7, quote on p. 244 (emphasis added).

108 Graham, *On the Diseases of Females*, p. 264. See also A.T.H. Waters, 'On the Use of Chloroform in the Treatment of Puerperal Insanity', *Journal of Psychological Medicine and Mental Pathology*, 10 (1857), 123–35.

109 See Cath Quinn's essay for the use of photography in the late nineteenth century: 'Images and Impulses: Representations of Puerperal Insanity and Infanticide in Late Victorian England', in Mark Jackson (ed.), *Infanticide: Historical Perspectives on Child Murder and Concealment, 1550–2000* (Aldershot: Ashgate, 2002), pp. 193–215, and for a more general discussion of physiognomy, techniques of illustration and psychiatric photography, Sander L. Gilman, *Disease and Representation: Images of Illness from Madness to AIDS* (Ithaca, NY and London: Cornell University Press, 1988), chapter 2.

110 Alexander Morison, *The Physiognomy of Mental Diseases* (London, n.p., 1838) pamphlet edition.

111 Sander L. Gilman, *Seeing the Insane* (Lincoln and London: University of Nebraska Press, 1996; first published New York: John Wiley & Sons, 1982), p. 92.

112 Morison, *The Physiognomy of Mental Diseases*, second edition, p. 15.

113 John Conolly presented a series of four photographs of a patient suffering from puerperal mania in his lectures on physiognomy, as the first images to show the progress of a disorder through the various stages of mania, dementia, convalescence and recovery: John Conolly, 'The Physiognomy of Insanity', No. 8: 'Puerperal Mania', *Medical Times & Gazette*, 16 (19 June 1858), 623–5. See chapter 5 for a detailed discussion of this case.

114 Morison, *The Physiognomy of Mental Diseases*, pamphlet edition, plates VIII–X (no page numbers).

115 David Wright, 'Getting out of the Asylum: Understanding the Confinement of the Insane in the Nineteenth Century', *Social History of Medicine*, 10 (1997), 137–55, on p. 153.

116 Catherine Quinn, 'Include the Mother and Exclude the Lunatic. A Social History of Puerperal Insanity c.1860–1920', unpublished PhD thesis, University of Exeter, 2003, p. 243. Quinn gives several examples of transfers of women from local workhouses to asylums, but also points out that

workhouses with asylum wards could be acceptable refuges for cases of puerperal insanity (pp. 163, 244–6).

117 Richard Adair, Bill Forsythe and Joseph Melling, 'A Danger to the Public? Disposing of Pauper Lunatics in Late-Victorian and Edwardian England: Plympton St Mary Union and the Devon County Asylum, 1867–1914', *Medical History*, 42 (1998), 1–25, on pp. 18–19.

118 Morison, *The Physiognomy of Mental Diseases*, second edition, p. 15.

119 Howell Evans, 'Puerperal Mania Occurring Suddenly Three Days after Delivery Treated by Opiates and Purgatives', *Medical Times*, 15 (14 November 1846), 145.

120 See Joel Peter Eigen, *Witnessing Insanity: Madness and Mad-Doctors in the English Court* (New Haven and London: Yale University Press, 1995), pp. 142, 147–8; and chapter 6 for the role of surgeons and general practitioners in infanticide cases.

121 John Thomson, 'Statistical Report of Three Thousand Three Hundred Cases of Obstetricy', *Glasgow Medical Journal*, 3 (1855), 129–50.

122 Ibid., pp. 143–5.

123 Thomas Salter, 'Case of Puerperal Mania, Occurring at an Early Period of Utero-Gestation, and Relieved by Induced Abortion', *Provincial Medical and Surgical Journal* (30 June 1847), 346–8.

124 Ibid., p. 346.

125 Ibid., p. 347.

126 Ibid., p. 348.

127 Robert Lee, a physician specialising in midwifery and the diseases of women, wrote up an impressive number of his case histories, including several cases of insanity of pregnancy and puerperal mania. In cases of mania occurring during pregnancy Lee also considered the option of prematurely inducing his patients, particularly if they were almost full-term, but in some cases considered the patient too violent to carry out this procedure. See Robert Lee, *Three Hundred Consultations in Midwifery* (London: John Churchill and Sons, 1864).

128 Bucknill and Tuke, *A Manual of Psychological Medicine*, 1858 edition, p. 236.

129 London Metropolitan Archives; General Lying-in Hospital London, H1/GL1/B19/1, Medical Officers' Case Book, 1827–28.

130 General Lying-in Hospital London, H1/GL1/B19/8. Cited Nakamura, 'Puerperal Insanity', p. 243; and see Nakamura, pp. 241–5 for further examples of cases occurring in the late nineteenth century.

131 Mitchell Library, Glasgow: Greater Glasgow Health Board Archives, HB45/5/14, HB45/5/15, Glasgow Royal Maternity Hospital, Patient Registers, Indoor, 1855–66, 1866–81. The Edinburgh Royal Maternity Hospital also recorded similarly low numbers.

132 T.M. Madden, 'On Puerperal Mania', *British and Foreign Medico-Chirurgical Review*, XLVIII (1871), 477–95, on p. 477.

133 Ibid., pp. 489–90.

134 Ibid., p. 492. The emphasis on risk to the infant may have been a reflection of Irish lunacy law on dangerousness. The Dangerous Lunatics Act of 1838 established a close link between insanity and criminality. See Oonagh Walsh, 'Lunatic and Criminal Alliances in Nineteenth-Century Ireland', in Bartlett and Wright (eds), *Outside the Walls of the Asylum*, pp. 132–52.

135 E.B. Sinclair and G. Johnston, *Practical Midwifery: Comprising an Account of 13,748 Deliveries which Occurred in the Dublin Lying-in Hospital, during a Period of Seven Years, Commencing November 1847* (London: John Churchill, 1858), pp. 528–31.

136 Mackenzie, 'On the Pathology and Treatment of Puerperal Insanity', p. 517.

137 Smith, 'Puerperal Mania', p. 424.

138 Ibid.

139 Reid, 'On the Causes, Symptoms, and Treatment', p. 290.

140 Ramsbotham, *The Principles and Practice*, p. 568.

141 Gooch, *On Some of the Most Important Diseases Peculiar to Women*, p. 79.

142 Smith, 'Puerperal Mania', p. 425.

143 W.S. Playfair, *A Treatise on the Science and Practice of Midwifery*, second edition (London: Smith, Elder, 1878), p. 308.

144 Catherine Quinn has also pointed out that infants usually disappear from the medical record, even the small number born in asylums: 'Include the Mother and Exclude the Lunatic', p. 235.

145 Conolly, 'Description and Treatment of Puerperal Insanity', p. 349. See Suzuki, 'The Politics and Ideology of Non-Restraint'; and Scull, Mackenzie and Hervey, *Masters of Bedlam*, chapter 3 for Conolly's career and relationship with the asylum.

146 John Conolly, *An Inquiry Concerning the Indications of Insanity, with Suggestions for the Better Protection and Care of the Insane* (London: John Taylor, 1830), pp. 429, 427.

147 Ibid., p. 428.

148 See chapter 5 for the family as a barrier to cure and potential cause of puerperal insanity.

149 Admissions of congenital and epileptic insanity and general paralysis were also much higher in Clouston's establishment than comparable institutions. Wellcome Trust Library: T.216.21, Annual Reports of the Cumberland & Westmoreland Lunatic Asylum, 1863–70; *Carlisle Journal*, 18 April 1873, taken from Edinburgh University Library: Lothian Health Board Archive, Royal Edinburgh Hospital, LHB1/12/1, Press Cuttings Book, 1862–81.

150 Tuke, 'Cases Illustrative', p. 1092 (Tuke's emphasis). Many interested parties in Scotland were anxious to keep asylums as a last resort for mental disorder.

151 Ramsbotham, *The Principles and Practice*, p. 567.

152 Reid, 'On the Causes, Symptoms, and Treatment', p. 289 (Reid's emphasis).

153 Arthur H. M'Clintock and Samuel L. Hardy, *Practical Observations on Midwifery, and the Diseases Incident to the Puerperal State* (Dublin: Hodges and Smith, 1848), p. 68.

154 Smith, 'Puerperal Mania', p. 424. Smith also employed 'masterful' means of treating erotic and nervous symptoms of menopause, including injections of ice into the rectum and vagina and leeching the labia and cervix: W. Tyler Smith, 'The Climacteric Diseases in Women', *London Journal of Medicine*, 1 (1848), 601–9.

155 Twenty-four were removed uncured, five died and thirteen were still being treated: Bucknill and Tuke, *A Manual of Psychological Medicine*, p. 239.

156 Reid, 'On the Causes, Symptoms, and Treatment', pp. 148–9.
157 Madden, 'On Puerperal Mania', p. 485.
158 J.B. Tuke, 'On the Statistics of Puerperal Insanity as Observed in the Royal Edinburgh Asylum, Morningside', *Edinburgh Medical Journal*, 10 (1864–65), 1013–28, pp. 1026, 1019.
159 Ramsbotham, *The Principles and Practice*, p. 561.

3 Disordered Households: Puerperal Insanity and the Bourgeois Home

1 Robert Gooch, *A Practical Compendium of Midwifery; Being the Course of Lectures on Midwifery, and on the Diseases of Women and Infants, Delivered at St. Bartholomew's Hospital, by the Late Robert Gooch, M.D.* (London: Longman, Rees, Orme, Brown, and Greene, 1831), p. 290.
2 For constructions and contestations of notions of how gender shaped ideology, see Mary Poovey, *Uneven Developments: The Ideological Work of Gender in Mid-Victorian England* (London: Virago, 1989; first published Chicago: University of Chicago Press, 1988), especially chapter 1, pp. 1–23, quote on p. 10. For the framing of domestic ideology, see Lenore Davidoff and Catherine Hall, *Family Fortunes: Men and Women of the English Middle Class 1780–1850* (London: Hutchinson, 1987); and for links with Evangelicalism, Catherine Hall, *White, Male and Middle Class: Explorations in Feminism and History* (Cambridge: Polity, 1988), chapter 3, pp. 75–93.
3 With respect to domestic hygiene, the 'sanitary idea' and the bonding of social classes, see Mary Poovey, *Making a Social Body: British Cultural Formation, 1830–1864* (Chicago and London: University of Chicago Press, 1995), chapter 6, pp. 115–31; for Victorian women's work in sanitary reform aimed at urging working-class self-help, Perry Williams, 'The Laws of Health: Women, Medicine and Sanitary Reform, 1850–1890', in Marina Benjamin (ed.), *Science and Sensibility: Gender and Scientific Enquiry 1780–1945* (Oxford: Basil Blackwell, 1991), pp. 60–88; and for home visiting and charitable work among the poor, Anne Summers, 'A Home from Home – Women's Philanthropic Work in the Nineteenth Century', in Sandra Burman (ed.), *Fit Work for Women* (London: Croom Helm, 1979), pp. 33–63; and Frank Prochaska, *Women and Philanthropy in 19th Century England* (Oxford: Clarendon Press, 1980).
4 Andrew Scull, 'The Domestication of Madness', *Medical History*, 27 (1983), 233–48. See also Elaine Showalter, *The Female Malady: Women, Madness and English Culture, 1830–1980* (London: Virago, 1987; first published New York: Pantheon, 1985), chapter 1.
5 Sylvia D. Hoffert cites several such cases in the United States, including Elizabeth Dwight Cabot who in a letter to her sister written in 1859 described how 'Nothing can be as bad as that nervous depression – I had a little of it after the baby came & I had infinitely rather go through all of the physical torture of the proceeding, than have it over again': Elizabeth Dwight Cabot to Ellen Dwight Thistleton, 19 June 1859, sec. 2, box 2, folder 18, Elizabeth Dwight Cabot Letters, Hugh Cabot

Family Collection, Schlesinger Library. Cited Sylvia D. Hoffert, *Private Matters: American Attitudes toward Childbearing and Infant Nurture in the Urban North, 1800–1860* (Urbana and Chicago: University of Illinois Press, 1989), p. 112.

6   A. Walker, *The Holy Life of Mrs Elizabeth Walker* (London, 1690), pp. 63, 93; P. Stubbes, *A Christal Glasse for Christian Women* (London, 1591), pp. 4–5; E. Joceline, *The Mothers Legacie to her Unborne Child* (London, 1624), sig. a5. Cited in Patricia Crawford, 'The Construction and Experience of Maternity in Seventeenth-Century England', in Valerie Fildes (ed.), *Women as Mothers in Pre-Industrial England* (London and New York: Routledge, 1990), pp. 3–38, on p. 22.

7   Herts. Record Office, Panshanger MSS, Box 19, Shelf 226/51; Edward Hughes (ed.), *Letters of Spencer Cowper* (Surtees Society, 1956), pp. 6, 20–1, 66, 68. Cited in Randolph Trumbach, *The Rise of the Egalitarian Family: Aristocratic Kinship and Domestic Relations in Eighteenth-Century England* (New York: Academic Press, 1978), pp. 233–4.

8   'Countess of Kildare to Earl of Kildare', 15 May 1755, in *Correspondence of Emily, Duchess of Leinster (1731–1814)*, vol. I (ed. Brian Fitzgerald) (Dublin: Stationery Office, 1949), p. 17.

9   Stella Tillyard, *Aristocrats: Caroline, Emily, Louisa and Sarah Lennox 1740–1832* (London: Chatto & Windus, 1994), p. 232.

10 Ibid., pp. 230–1.

11 'Lady Caroline Fox to Countess of Kildare', 10 July 1760, in *Correspondence of Emily*, vol. I, p. 289.

12 'Lady Louisa Conolly to Duchess of Leinster', 4 September 1775, in *Correspondence of Emily, Duchess of Leinster (1731–1814)*, vol. III, ed. Brian Fitzgerald (Dublin: Stationery Office, 1957), p. 147.

13 Ibid., 22 May 1778, pp. 286–7. Tillyard (*Aristocrats*, p. 232) claims that childbearing brought Emily no psychological pain, but the correspondence of her sisters suggests otherwise, as they anticipated poor spirits and an abundance of tears as part of each lying-in.

14 Amanda Vickery, *The Gentleman's Daughter: Women's Lives in Georgian England* (New Haven and London: Yale University Press, 1998), p. 106. See chapter 3 of *The Gentleman's Daughter* for the many trials and tribulations of childbirth amongst Vickery's genteel women.

15 Ibid., pp. 88–9. William Smellie reported on a number of cases that he had attended where the women were stricken with anxiety after the death of their husbands. In 1747 he delivered a gentlewoman in labour with her first child, who 'a few days before had been so much affected with the sudden death of her husband, that she was seized with frequent faintings and great anxiety of mind'. Her 'pains' were weak and she continued three days with 'a kind of labour', 'yet, by encouraging and supporting her with cordials and nourishing things, and indulging her as much as possible with rest, she was safely delivered of a child, which seemed to have died soon after she heard the melancholy news of her husband's death': William Smellie, *A Treatise on the Theory and Practice of Midwifery*, vol. II, *Collection of Cases and Observations in Midwifery*, second edition (London: D. Wilson and T. Durham, 1757), No. II, Case I, 'From Anxiety and Grief', p. 300.

16   Judith Schneid Lewis, *In the Family Way: Childbearing in the British Aristocracy, 1760–1860* (New Brunswick, NJ: Rutgers, 1986), pp. 213–14. Amanda Vickery (*The Gentleman's Daughter*, p. 105) found no indication of postpartum disappointment when the baby was not a son and heir.

17   Edith, Marchioness of Londonderry (ed.), *Frances Anne: The Life and Times of Frances Anne, Marchioness of Londonderry, and Her Husband Charles, Third Marquess of Londonderry* (London: Macmillan, 1958), p. 128; Diary of the Earl of Verulam, 9 and 13 December 1822, Gorhambury MSS D/EV F46. Both cited in Lewis, *In the Family Way*, p. 214.

18   Diary of Francis Hugh Seymour, 3 November 1835, Seymour of Raglay MSS CR 114A/G 44. Cited ibid., pp. 214–15.

19   Lord Jermyn to Lady Elizabeth Drummond, 11 July 1835, Drummond of Cadland MSS B6/50/87. Cited ibid., p. 213.

20   'Queen Victoria to the King of the Belgians', 7 December 1841, 'Memorandum by Mr Anson', 26 December 1841, in John Raymond (ed.), *Queen Victoria's Early Letters*, revised edition (London: B.T. Batsford, 1963), pp. 73–4.

21   Cecil Woodham-Smith, *Queen Victoria: Her Life and Times*, vol. 1, *1819–1861* (London: Hamish Hamilton, 1972), p. 328.

22   Tyler Whittle, *Victoria and Albert at Home* (London and Henley: Routledge & Kegan Paul, 1980), p. 70.

23   'Queen Victoria to the Princess Royal', 9 March 1859, in Roger Fulford (ed.), *Dearest Child: Letters between Queen Victoria and the Princess Royal 1858–1861* (London: Evans, 1964), p. 165.

24   Queen Victoria used the word 'suffering' four times in a brief letter to her daughter following her delivery. Ibid., 29 January 1859, pp. 159–60.

25   For full details of the case, see Elizabeth Hamilton, *The Warwickshire Scandal* (Wilby, Norwich: Michael Russell, 1999).

26   J. Thompson Dickson, 'A Contribution to the Study of the So-Called Puerperal Insanity', *Journal of Mental Science*, 17 (1870), 379–90, p. 385. The Mordaunt case prompted Dickson to write this study, disputing the existence of puerperal insanity as a separate category.

27   'The Mordaunt Case in its Medical Aspects', *Medical Times & Gazette*, 1 (5 March 1870), 270–1, quote on p. 270.

28   Ibid., pp. 270–1.

29   Wellcome Trust Library (WTL), Manor House Asylum, Chiswick, 5725 Case Book, May 1870–October 1884, Lady Harriet Mordaunt, admitted 11 November 1871, pp. 74, 76; Hamilton, *The Warwickshire Scandal*, pp. 398–9.

30   Robert Gooch, *Observations on Puerperal Insanity* (London, 1820) (extracted from the sixth volume of *Medical Transactions*, Royal College of Physicians, read at the College, 16 December 1819), pp. 20–1.

31   Robert Gooch, *On Some of the Most Important Diseases Peculiar to Women; with Other Papers*, with a prefatory essay by Robert Ferguson, M.D. (London: The New Sydenham Society, 1831), p. 57.

32   Ibid., pp. 56–7.

33   Arthur H. M'Clintock and Samuel L. Hardy, *Practical Observations on Midwifery, and the Diseases Incident to the Puerperal State* (Dublin: Hodges and Smith, 1848), p. 67.

34   R.U. West, 'Fatal and Other Cases of Puerperal Mania', *Association Medical Journal*, 2 (11 August 1854), 716–18, quote on p. 718.
35   Ibid.
36   Ibid.
37   Ibid.
38   W.E. Image, 'Case of Melancholia Puerperalis Attonita', *London Medical Gazette*, 1 (1845), 281. Cited Lisa Ellen Nakamura, 'Puerperal Insanity: Women, Psychiatry, and the Asylum in Victorian England, 1820–1895', unpublished PhD thesis, University of Washington, 1999, p. 205.
39   James Y. Simpson, *Clinical Lectures on the Diseases of Women* (Edinburgh: Adam and Charles Black, 1872), pp. 566–7.
40   Fleetwood Churchill, *On the Diseases of Women; Including those of Pregnancy and Childbed*, fourth edition (Dublin: Fannin and Co., 1857), p. 737.
41   Francis H. Ramsbotham, *The Principles and Practice of Obstetric Medicine and Surgery in Reference to the Process of Parturition*, third edition (London: John Churchill, 1851), p. 561.
42   John Conolly, 'Description and Treatment of Puerperal Insanity', Lecture XIII: 'Clinical Lectures on the Principle Forms of Insanity, Delivered in the Middlesex Lunatic-Asylum at Hanwell', *Lancet*, I (28 March 1846), 349–54, on p. 351.
43   Gooch, *A Practical Compendium of Midwifery*, pp. 292–3, quotes on p. 292.
44   Ibid., p. 293.
45   Ibid.
46   Ibid.
47   Ibid., p. 294. See chapter 4, pp. 122–6 for the attitudes of the Edinburgh Asylum doctors to delusional behaviour.
48   Gooch, *Observations*, pp. 34–41.
49   Ibid., p. 36.
50   Ibid., pp. 38, 40.
51   Ibid., p. 41. This is an unusually positive depiction of a husband. See chapter 5 for more typical assessments.
52   W. Macmichael, *Lives of British Physicians* (London: John Murray, 1830), p. 341.
53   Gooch was also the trusted medical attendant of John Constable and his wife Maria; the painter suffered from extreme anxiety and bouts of deep gloom. Constable gave Gooch one of his paintings, a gift confined to only a few close friends. Personal communication Dr Conal Shields, 25 January 2003.
54   Gooch, *On Some of the Most Important Diseases Peculiar to Women*, p. 77; Gooch, *A Practical Compendium of Midwifery*, p. 290.
55   Thomas Trotter, *A View of the Nervous Temperament; being a Practical Inquiry into the Increasing Prevalence, Prevention, Treatment of those Diseases commonly called Nervous, Bilious, Stomach & Liver Complaints; Indigestion; Low Spirits, Gout, &c.* (Boston: Wright, Goodenow and Stockwell, 1808), p. 47.
56   Samuel Ashwell, *A Practical Treatise on the Diseases Peculiar to Women, Illustrated by Cases, Derived from Hospital and Private Practice* (London: Samuel Highley, 1845), pp. 731–2.

57 H. Hastings, 'Puerperal Mania Followed by Insanity', *Medical Times*, 19 (21 April 1849), 511.

58 Ibid. The Radcliffe Lunatic Asylum was built during the 1820s for higher-class lunatics, on Headington Hill, near Oxford. It was later renamed the Warneford Lunatic Asylum.

59 For removal to Ticehurst Asylum, see Charlotte MacKenzie, *Psychiatry for the Rich: A History of Ticehurst Private Asylum, 1792–1917* (London and New York: Routledge, 1992), chapter 4, 'Madness and the Victorian Family'.

60 Wellcome Trust Library (WTL), Manor House Asylum, Chiswick, 5725 Case Book, May 1870–October 1884, Ellen Johnston, admitted 22 July 1870, pp. 16–17.

61 WTL, Ticehurst House Hospital, 6366 6, Case Records, 1860–63, Harriet Chaplin, admitted 17 June 1861, pp. 141, 140.

62 Ibid., p. 156.

63 Ibid., 6365 5, Case Records, 1858–61, Eliza Gipps, admitted 10 April 1860, pp. 150–1. Trevor Turner in his retrospective analysis of the Ticehurst Asylum case books, labels Eliza Gipps as a definite case of schizophrenia, largely on the basis of her delusions: *A Diagnostic Analysis of the Casebooks of Ticehurst House Asylum, 1845–1890*, Psychological Medicine, Monograph Supplement 21 (Cambridge: Cambridge University Press, 1992), pp. 64–5.

64 WTL, Ticehurst House Hospital, 6366 6, Case Records, 1860–63, pp. 47–8.

65 Ibid., 6368 8, Case Records, 1862–74, p. 88.

66 Micael M. Clarke, *Thackeray and Women* (DeKalb, IL: North Illinois University Press, 1995), pp. 26–7.

67 Ann Monsarrat, *An Uneasy Victorian: Thackeray the Man 1811–1863* (London: Cassell, 1980), pp. 87–8; Clarke, *Thackeray and Women*, p. 26.

68 Gordon N. Ray (ed.), *The Letters and Private Papers of William Makepeace Thackery*, 4 vols (London: Oxford University Press, 1945), vol. 1, 'To Mrs. Shawe', 12 July 1838, pp. 366–7 (Thackeray's emphasis).

69 Ibid., 'To Mrs. Carmichael-Smyth', March 1839, p. 379.

70 Ibid., 'To Mrs. Carmichael-Smyth', 1 June 1840, p. 446.

71 Ibid., 'From Mrs. Thackeray to Mrs. Carmichael-Smyth', 4 August 1840, p. 462.

72 Ibid., 'To Mrs. Carmichael-Smyth', 20–21 August 1840, p. 463.

73 Ibid., pp. 464, 465.

74 Ibid., [?] August–1 September 1840, p. 469.

75 Ibid., 4–5 October 1840, p. 483, 17 September 1840, p. 474.

76 Ibid., 17 September 1840, p. 475.

77 Ibid., 19–20 September 1840, p. 477, 4–5 October 1840, p. 484.

78 Ibid., vol. 2, 'To Edward Fitzgerald', 10 January 1841, p. 3.

79 Ibid., 'To Mrs. Proctor', 5 April 1841, pp. 14–15.

80 Ibid., 28 May–5 June 1841, p. 23.

81 Ibid.

82 Ibid., 'To Mrs. Ritchie', 19 August 1841, p. 34.

83 Ibid., 'To Edward Fitzgerald', 13 September–October 1841, p. 37.

84 Ibid., 'To Mrs. Spencer', 10 February 1842, p. 41, 'To Edward Fitzgerald', 9 March 1842, p. 43.

85 Ibid.

86 Ibid., 'To Mrs. Carmichael-Smyth', 25–30 September 1842, p. 81.

87 Ibid., 28 November 1845, p. 217; Monsarrat, *An Uneasy Victorian*, p. 161. Mrs Bakewell may have been related to the Bakewell family who ran Spring Vale Asylum in Staffordshire; other family members were also involved in running private asylums.

88 Ray (ed.), *Letters* vol. 2, 6 August 1846, p. 243, 'To Jane Shawe', 19 September 1848, p. 431.

89 Gordon N. Ray, *Thackeray: The Uses of Adversity (1811–1846)* (London: Oxford University Press, 1955), p. 305.

90 For the powerful impact of Isabella's illness on Thackeray's writing as well as his take-over of the management of the household, see Clarke, *Thackeray and Women*, chapter 7.

91 Harry Ransom Humanities Research Center, The University of Texas at Austin, Coleridge, S. Misc.: Sara Coleridge, 'Diary of her Children's Early Years' (hereafter HRHRC, Diary), 27 August 1832. This may have been her uncle and sometime medical attendant, Dr Henry Southey, or her husband, Henry Nelson Coleridge.

92 Ibid., 12 September 1832.

93 Ibid., 20 September 1832.

94 Ibid., 2 and 11 December 1832.

95 *Minnow Among Tritons: Mrs. S.T. Coleridge's Letters to Thomas Poole*, ed. Stephen Potter (London: Nonesuch Press, 1934), no. 40, pp. 169–70.

96 Ibid., no. 38, p. 163.

97 HRHRC, Diary, 20 June 1833.

98 Ibid., 5 November 1833. Sara occasionally wrote her diary in the third person.

99 Ibid., 7 February 1834.

100 Katherine T. Meiners, 'Imagining Cancer: Sara Coleridge and the Environment of Illness', *Literature and Medicine*, 15 (1996), 48–63.

101 Samuel Taylor Coleridge lodged for many years with Mr and Mrs Gillman, who provided him with board and medical treatment, and were able to reduce his opium intake to manageable levels. See Kathleen Jones, *A Passionate Sisterhood: The Sisters, Wives and Daughters of the Lake Poets* (London: Virago, 1998), pp. 216, 230; Alexander W. Gillman, *The Gillmans of Highgate with Letters from Samuel Taylor Coleridge* (London: Elliot Stock, n.d. [1895]).

102 Anne Digby, *Making a Medical Living: Doctors and Patients in the English Market for Medicine, 1720–1911* (Cambridge: Cambridge University Press, 1994), pp. 174, 189–91.

103 Famously declaring this in the Report from the Select Committee on Medical Education, PP, 1834, Part 1, p. 17.

104 HRHRC, Diary, 27 May 1833.

105 James Cowles Prichard, *A Treatise on Insanity and Other Disorders Affecting the Mind* (London: Sherwood, Gilbert, and Piper, 1835), p. 315.

106 But in modest amounts compared with her father's intake. In 1801, according to Robert Southey, Coleridge was 'swilling' laudanum at the rate of 2 pints a week, which he combined with brandy and raw opium: Jones, *A Passionate Sisterhood*, p. 142. For opium usage in literary society in the early nineteenth century, see Virginia Berridge, *Opium and the People: Opiate Use and Drug Control Policy in Nineteenth and Early Twentieth*

Century England, revised edition (London and New York: Free Association Books, 1999; first published New Haven and London: Yale University Press, 1987), chapter 5.

107 HRHRC, Diary, 16 January 1833.

108 Ibid., 25 January 1833 (Coleridge's emphasis).

109 Sara Coleridge, 'Essay on Nervousness', 1834. The essay is reproduced in Bradford Keyes Mudge, *Sara Coleridge, A Victorian Daughter: Her Life and Essays* (New Haven and London: Yale University Press, 1980) and see Mudge, chapters 3 and 4 for Sara's illness. See chapter 2 for Gooch and Hall.

110 Letter, Sara Coleridge to Henry Nelson Coleridge, 3 November 1836. Cited Mudge, *Sara Coleridge*, p. 91.

111 Ibid., 13 November 1832. Cited in Mudge, *Sara Coleridge*, p. 87.

112 James Cowles Prichard, *A Treatise on Insanity* (London: Marchant, 1833), p. 14. Cited Vieda Skultans, *Madness and Morals: Ideas on Insanity in the Nineteenth Century* (London and Boston: Routledge & Kegan Paul, 1975), p. 182. See also G.E. Berrios, introduction to 'J.C. Prichard and the Concept of "Moral Insanity"', *History of Psychiatry*, 10 (1999), 111–26.

113 In 1855 Dr Fordyce Barker noted the 'curious fact' of thirteen cases of puerperal insanity occurring among the wives of physicians. This, he explained, was due to their 'education and quickness of intellect' and '[b]eginning, a new experience in life, and having access to their husband's books, they probably had read just enough on midwifery to fill their minds with apprehensions as to the horrors which might be in store for them, and thus developed the cerebral disturbances': Fordyce Barker, *The Puerperal Diseases: Clinical Lectures Delivered at Bellevue Hospital*, third edition (New York: Appleton, 1883). Cited in Dale M. Bauer's edition of Charlotte Perkins Gilman, *The Yellow Wallpaper* (Boston and New York: Bedford, 1998), pp. 184–5. *The Yellow Wallpaper* (1892) provides a vivid illustration of the tensions of women caught between motherhood and intellectual pursuits.

114 *The Letters of Robert Southey to John May, 1797 to 1838*, ed. Charles Ramos (Austin: Jenkins, 1976), p. 266. Cited Mudge, *Sara Coleridge*, p. 83.

115 There is a detailed account of Edith Southey's sad illness in the *Journal of Psychological Medicine and Mental Pathology*, 'The Wear and Tear of Literary Life; or, the Last Days of Robert Southey', 5 (1852), pp. 28–36 (based on *The Life and Correspondence of the late Robert Southey*, ed. Charles Cuthbert Southey, vol. VI (London: Longman, Brown, Green & Longmans, 1850)).

116 *The Letters of William and Dorothy Wordsworth*, rev. and ed. Alan G. Hill (Oxford: Clarendon Press, 1967), vol. 2, pp. 758–9. Cited Mudge, *Sara Coleridge*, p. 84.

117 Molly Lefebure, *Samuel Taylor Coleridge: A Bondage of Opium* (London: Victor Gollancz, 1974), p. 473.

118 *Dictionary of National Biography*: Robert Gooch, Henry Southey.

119 Macmichael, *The Lives of British Physicians*, pp. 305–41.

120 *New Letters of Robert Southey*, ed. Kenneth Curry, 2 vols (New York and London: Columbia University Press, 1965), vol. 2, p. 280.

121 HRHRC, Diary, 12 and 20 September 1832.

122   Showalter, *The Female Malady*, p. 5.

123   *Minnow Among Tritons*, no. 34, p. 147.

124   HRHRC, Diary, 25 February 1833.

125   Jones, *A Passionate Sisterhood*, p. 274.

126   Ibid., p. 281.

127   Letter, Sara Coleridge to Henry Nelson Coleridge, 16 October 1836. Cited Mudge, *Sara Coleridge*, p. 89.

128   Ibid., 19 October 1836. Cited Mudge, *Sara Coleridge*, p. 90. Though Oppenheim devotes a chapter to 'neurotic women' in her excellent study of 'shattered nerves', Sara Coleridge merits only one footnote reference (p. 344, n. 78): Janet Oppenheim, *'Shattered Nerves': Doctors, Patients, and Depression in Victorian England* (New York and Oxford: Oxford University Press, 1991).

129   William Pargeter, *Observations on Maniacal Disorders* (1792); reprint edited by Stanley W. Jackson (London and New York: Routledge, 1988), pp. 58–61, quote on p. 61.

130   Poovey, *Uneven Developments*, foreword, p. ix.

131   James Reid, 'On the Causes, Symptoms, and Treatment of Puerperal Insanity', *Journal of Psychological Medicine and Mental Pathology*, 1 (1848), 128–51, 284–94, quote on p. 293.

4   Incoherent, Violent and Thin: Patients and Puerperal Insanity in the Royal Edinburgh Asylum

1   Edinburgh University Library: Lothian Health Board Archive, Royal Edinburgh Hospital, LHB7/7/9, Annual Reports of the Royal Edinburgh Asylum, 1883–92 (hereafter Annual Reports): Physician-Superintendent's Annual Report for the Year 1883, pp. 13–14.

2   Other historians of psychiatry have worked extensively on the detailed and vivid case notes of the Royal Edinburgh Asylum as well as the institution's extraordinary collection of patients' letters during the period of Thomas Clouston's superintendence, 1873–1908: M. Barfoot and A. Beveridge, 'Madness at the Crossroads: John Home's Letters from the Royal Edinburgh Asylum, 1886–87', *Psychological Medicine*, 20 (1990), 263–84; idem, '"Our Most Notable Inmate": John Willis Mason at the Royal Edinburgh Asylum, 1864–1901', *History of Psychiatry*, 4 (1993), 159–208; Allan Beveridge, 'Madness in Victorian Edinburgh: A Study of Patients Admitted to the Royal Edinburgh Asylum under Thomas Clouston, 1873–1908', Parts I and II, *History of Psychiatry*, 6 (1995), 21–54, 133–56; idem, 'Life in the Asylum: Patients' Letters from Morningside, 1873–1908', *History of Psychiatry*, 9 (1998), 431–69; idem, 'Voices of the Mad: Patients' Letters from the Royal Edinburgh Asylum, 1873–1908', *Psychological Medicine*, 27 (1997), 899–908.

3   Roy Porter and Dorothy Porter, *In Sickness and in Health: The British Experience 1650–1850* (London: Forth Estate, 1988), p. 103.

4   However, see chapter 3 for Sara Coleridge's explanation of her nervous affliction following childbirth and other accounts by sufferers or their families, and for the best-known 'semi-autobiographical' account of a

woman suffering nervous collapse related to her recent delivery, Charlotte Perkins Gilman, *The Yellow Wallpaper* (1892), ed. Dale M. Bauer (Boston and New York: Bedford, 1998).

5   There is also a useful literature on the institution. See note 2 above and also Margaret Sorbie Thompson, 'The Mad, the Bad, and the Sad: Psychiatric Care in the Royal Edinburgh Asylum (Morningside) 1813–1894', unpublished PhD thesis, Boston University, 1984. For medical institutions in Edinburgh focusing on the health of women, though for a later period than examined here, see Elaine Thomson, 'Women in Medicine in Late Nineteenth- and Early Twentieth-Century Edinburgh: A Case Study', unpublished PhD thesis, University of Edinburgh, 1998; and Alison Nuttall, 'The Edinburgh Royal Maternity Hospital and The Medicalisation of Childbirth in Edinburgh 1844–1914: A Casebook-Centred Perspective', unpublished PhD thesis, University of Edinburgh, 2002.

6   J.B. Tuke, 'On the Statistics of Puerperal Insanity as Observed in the Royal Edinburgh Asylum, Morningside', *Edinburgh Medical Journal*, 10 (1864–65), 1013–28, quote on p. 1013.

7   Idem, 'Cases Illustrative of the Insanity of Pregnancy, Puerperal Mania, and Insanity of Lactation', *Edinburgh Medical Journal*, 12 (1866–67), 1083–101, quote on p. 1092.

8   Edinburgh University Library: Lothian Health Board Archive, Royal Edinburgh Hospital, LHB7/51, Case Books (hereafter CB): LHB7/51/9, 1851–55, Ann Denholm, admitted 7 September 1851, p. 5.

9   Ibid.

10 Ibid., entries 29 December 1851, 13 March, 14 April 1852, pp. 5–6.

11 Beveridge, 'Madness in Victorian Edinburgh', p. 22.

12 The East House building was designed by Robert Reid, and for an expenditure of £8,000 could accommodate only thirty well-to-do patients: Christine Stevenson, *Medicine and Magnificence: British Hospital and Asylum Architecture 1660–1815* (New Haven and London: Yale University Press, 2000), p. 206.

13 Also referred to as Medical Superintendent.

14 In 1844 the inmates of the City's Bedlam, part of the Charity Workhouse, were transferred to the Edinburgh Asylum, and, following the 1857 Scottish Lunacy Act, the Asylum took patients from the parishes of Edinburgh, Midlothian and Peeblesshire, while Orkney purchased presentation rights at the lowest rate of board to West House. See Thompson, 'The Mad, the Bad and the Sad', pp. 25–31 for the campaign to create asylum provision for paupers.

15 Guenter B. Risse, *Hospital Life in Enlightenment Scotland* (Cambridge: Cambridge University Press, 1986); Lisa Rosner, *Medical Education in the Age of Improvement* (Edinburgh: Edinburgh University Press, 1991).

16 D.K. Henderson, *The Evolution of Psychiatry in Scotland* (Edinburgh and London: E. & S. Livingstone, 1964), p. 54.

17 Referred to variously in annual reports as Assistant Physicians and Medical Assistants. They included John Batty Tuke who would go on to be Medical Superintendent at the Fife and Kinross Asylum, Thomas Clouston, Skae's successor in 1873, David Yellowlees, subsequently

Medical Superintendent at the Gartnavel Asylum in Glasgow, and John Sibbard, later a Lunacy Commissioner.

18 Although tensions would develop between the two and they competed to teach the same set of lectures. See Richard Hunter and Ida Macalpine, *Three Hundred Years of Psychiatry 1535–1860* (London: Oxford University Press, 1963), pp. 769–73, 1079–84; and *Dictionary of National Biography* for Morison, Laycock and Skae.

19 Henderson, *The Evolution of Psychiatry in Scotland*, p. 53; LHB7/7/6, Annual Report, 1840, pp. 11–12.

20 Beveridge, 'Life in the Asylum', p. 432; Henderson, *The Evolution of Psychiatry in Scotland*, p. 53.

21 Thompson, 'The Mad, the Bad, and the Sad', p. 86.

22 Jarvis Collection, American Antiquarian Society, Worcester, Mass: Edward Jarvis to Almira Jarvis written on the steam ship *Persia*, 10 August 1860. Cited Thompson, 'The Mad, the Bad, and the Sad', p. 60.

23 See Thomson, 'Women in Medicine', especially chapters 3 and 4.

24 Though Beveridge has shown how Clouston's 'Gospel of Fatness' also led to patients complaining about being stuffed with large quantities of coarse food: Beveridge, 'Life in the Asylum', p. 440.

25 See Roy Porter's ground-breaking article, 'The Patient's View: Doing Medical History from Below', *Theory and Society*, 14 (1985), 175–98; and e.g. Joan Lane, '"The Doctor Scolds Me": The Diaries and Correspondence of Patients in Eighteenth Century England', in Roy Porter (ed.), *Patients and Practitioners: Lay Perceptions of Medicine in Pre-Industrial Society* (Cambridge: Cambridge University Press, 1985), pp. 205–48; idem, *The Making of the English Patient* (Stroud: Sutton, 2000), especially chapter 3; L.M. Beier, *Sufferers and Healers: The Experience of Illness in Seventeenth-Century England* (London: Routledge & Kegan Paul, 1987).

26 David Wright, 'The Certification of Insanity in Nineteenth-Century England and Wales', *History of Psychiatry*, 9 (1998), 267–90; idem, 'The Discharge of Pauper Lunatics from County Asylums in Mid-Victorian England: The Case of Buckinghamshire, 1853–1872', in Joseph Melling and Bill Forsythe (eds), *Insanity, Institutions and Society, 1800–1914* (London and New York: Routledge, 1999), pp. 93–112.

27 Jonathan Andrews, 'Case Notes, Case Histories, and the Patient's Experience of Insanity at Gartnavel Royal Asylum, Glasgow, in the Nineteenth Century', *Social History of Medicine*, 11 (1998), 255–81.

28 Ibid., p. 255. Barfoot and Beveridge were the first to empirically work out the approach of Porter for two of Edinburgh Asylum's most famous patients, John Home and John Willis Mason: Barfoot and Beveridge, 'Madness at the Crossroads' and idem, '"Our Most Notable Inmate"'.

29 Akihito Suzuki, 'Framing Psychiatric Subjectivity: Doctor, Patient and Record-Keeping at Bethlem in the Nineteenth Century', in Melling and Forsythe (eds), *Insanity, Institutions and Society*, pp. 115–36.

30 Richard Adair, Bill Forsythe and Joseph Melling, 'A Danger to the Public? Disposing of Pauper Lunatics in Late-Victorian and Edwardian England: Plympton St Mary Union and the Devon County Asylum 1867–1914', *Medical History*, 42 (1998), 1–25; Joseph Melling, Bill Forsythe and Richard Adair, 'Families, Communities and the Legal Regulation of Lunacy in

Victorian England: Assessments of Crime, Violence and Welfare in Admissions to the Devon Asylum, 1845–1914', in Peter Bartlett and David Wright (eds), *Outside the Walls of the Asylum: A History of Care in the Community 1750–2000* (London and New Brunswick: Athlone, 1999), pp. 153–80. For families' decisions to commit relatives in North America, see Nancy Tomes, *A Generous Confidence: Thomas Story Kirkbride and the Art of Asylum-Keeping, 1840–1883* (Cambridge: Cambridge University Press, 1984); Constance M. McGovern, 'The Myths of Social Control and Custodial Oppression: Patterns of Psychiatric Medicine in Late Nineteenth-Century Institutions', *Journal of Social History*, 20 (1986), 3–23; and James E. Moran, 'Asylum in the Community: Managing the Insane in Antebellum America', *History of Psychiatry*, 9 (1998), 217–40.

31   Barfoot and Beveridge, 'Madness at the Crossroads', p. 263. See also the introduction to W.F. Bynum, R. Porter and M. Shepherd (eds), *The Anatomy of Madness: Essays in the History of Psychiatry*, vol. I, *People and Ideas* (London and New York: Tavistock, 1985), pp. 1–24.

32   Though Nakamura has made use of case notes, drawn particularly from Hanwell Asylum, in Lisa Ellen Nakamura, 'Puerperal Insanity: Women, Psychiatry, and the Asylum in Victorian England, 1820–1895', unpublished PhD thesis, University of Washington, 1999. Techniques of retrospective diagnosis have also been drawn on to analyse postpartum disorders in the Edinburgh Asylum: A. Rehman, 'Puerperal Insanity in the Nineteenth and Twentieth Centuries', unpublished MPhil thesis, University of Edinburgh, 1986 and A. Rehman, D. St. Clair and C. Platz, 'Puerperal Insanity in the Nineteenth and Twentieth Centuries', *British Journal of Psychiatry*, clvi (1990), 861–5 compared cases occurring amongst patients admitted between 1880–90 and 1971–80, concluding that there was an overall decline in the severity of puerperal psychosis by the 1970s, and also that manic episodes and severe symptoms, including delusions, declined, though general symptomatology was similar. Patient symptoms have been assessed and modern-day diagnoses applied, particularly that of schizophrenia, to disorders recorded in admissions and case records. See, for example, Trevor Turner, *A Diagnostic Analysis of the Casebooks of Ticehurst House Asylum, 1845–1890, Psychological Medicine*, Monograph Supplement 21 (Cambridge: Cambridge University Press, 1992); and Beveridge, 'Madness in Victorian Edinburgh', especially Part I, pp. 44–5.

33   Tuke's tables, based on his analysis of the case notes, have been reproduced at the end of the book (see Appendix): Tuke, 'On the Statistics'; and idem, 'Cases Illustrative'. During Thomas Clouston's period as Superintendent from 1873 to 1908 he also published on puerperal insanity: Thomas Clouston, *Clinical Lectures on Mental Diseases* (London: J. & A. Churchill, 1883), Lecture XV: 'Puerperal Insanity. Lactational Insanity. The Insanity of Pregnancy', pp. 501–30. See chapter 5 for Clouston on lactational insanity. The Edinburgh obstetrician James Simpson also wrote on cases of puerperal mania treated in the Edinburgh Royal Maternity Hospital: J.Y. Simpson, 'Clinical Lectures on the Diseases of Women', Lecture XXXII: 'On Puerperal Mania' and Lecture XXXIII: 'Puerperal Insanity: Its Connection with Albuminuria', *Medical Times & Gazette*, II (1 September and 10 November 1860), 201–2, 445–7;

idem, *Clinical Lectures on the Diseases of Women* (Edinburgh: Adam and Charles Black, 1872), pp. 555–84.

34 Porter, 'The Patient's View', p. 175. See also Roy Porter, *A Social History of Madness: Stories of the Insane* (London: Weidenfeld and Nicolson, 1987); idem (ed.), *The Faber Book of Madness* (London: Faber and Faber, 1991). Allan Ingram has also contributed extensively to the literature on the patient's view in the eighteenth century: Allan Ingram (ed.), *Patterns of Madness in the Eighteenth Century: A Reader* (Liverpool: Liverpool University Press, 1998); idem, *Voices of Madness: Four Pamphlets, 1683–1796* (Stroud: Sutton, 1997). See also Dale Peterson (ed.), *A Mad People's History of Madness* (Pittsburgh, PA: University of Pittsburgh Press, 1981).

35 Roy Porter, 'Hearing the Mad: Communication and Excommunication', in L. de Goei and J. Vijselaar (eds), *Proceedings of the First European Congress on the History of Psychiatry and Mental Health Care* (Rotterdam: Erasmus Publishing, 1993), pp. 338–52, quote on p. 338.

36 Harriet Nowell-Smith, 'Nineteenth-Century Narrative Case Histories: An Inquiry into Stylistics and History', *Canadian Bulletin of Medical History*, 12 (1995), 47–67, quote on p. 48. Garfinkel has also pointed to problems of working with clinical records, including absences in the information available which make them incomplete compilations: Harold Garfinkel, *Studies in Ethnomethodology* (Englewood Cliffs, NJ: Prentice-Hall, 1967), chapter 6.

37 Suzuki, 'Framing Psychiatric Subjectivity', p. 117.

38 Until the passing of the 1857 Lunacy (Scotland) Act, sheriffs had been charged with supervising provisions for the insane, including the issue of committal warrants and one verdict of insanity provided by a medical man. After the reforms of 1857 two medical certificates from doctors seeing the patient separately, which also recorded the grounds determining the verdict, were required for admission to Scottish asylums. A petition was also required, which for paupers was completed by the Inspector of the Poor, and, in the case of private patients, by a relative or friend of the patient. The two medical certificates and petition were sent to the sheriff who decided if the petition should be granted and the patient admitted. Once admitted, a copy of the sheriff's order and certificates had to be sent to the Commissioners in Lunacy within fourteen days. The sheriff's order lapsed after three years and then had to be renewed on an annual basis through an application by the asylum superintendent. The superintendent had the power to discharge any patients he considered to be sane. See Jonathan Andrews, 'They're in the Trade ... of Lunacy, They "cannot interfere" – they say': The Scottish Lunacy Commissioners and Lunacy Reform in Nineteenth-Century Scotland* (London: Wellcome Institute for the History of Medicine, Occasional Publications, No. 8, 1998), pp. 2–3, 33; Beveridge, 'Madness in Victorian Edinburgh', pp. 23–4. The English Lunatics Act of 1845, in contrast, continued the tradition of requiring the testimony of a single medical practitioner for pauper patients and two medical practitioners for private patients, though the magistrates were responsible for the running and visitation of county asylums: see Wright, 'The Certification of Insanity'; Kathleen Jones, *Asylums and After: A Revised History of the Mental Health Services: From the Early 18th Century to the 1990s* (London: Athlone, 1993).

39 Case Book Rules, posted into LHB7/51/37, CB, 1880–82.

40 Porter, 'Hearing the Mad', p. 345.

41 Richard Hunter and Ida Macalpine, *Psychiatry for the Poor, 1851. Colney Hatch Asylum, Friern Hospital 1973: A Medical and Social History* (London: Dawsons, 1974), p. 184.

42 See also chapter 2 for the visibility of puerperal insanity.

43 LHB7/7/6, Annual Report, 1839, p. 6. McKinnon, formerly of the Aberdeen Asylum, was selected from twenty medical gentlemen, the Managers deeming it expedient to select a medical practitioner: 'in the present advanced state of knowledge, with respect to the treatment of the insane, that this was requisite for the due conduct of the Institution ...'.

44 Volume 12, which appears to have covered East House private patients c.1854–67, is wanting.

45 LHB7/35/1–4, General Registers of Patients, 1851–57, 1858–63, 1863–69, 1870–76.

46 This compares poorly with Tuke's diligence in collecting 155 cases, though some, he admitted, were dubiously attributed cases of puerperal insanity: Tuke, 'On the Statistics', especially p. 1020.

47 LHB7/7/6 and 6A, Annual Reports, 1812–55, 1848–62. In the report for 1848 puerperal mania had attracted attention due to its relationship with two cases of infanticide admitted in that year: LHB7/7/6, Physician's Annual Report, 1848, pp.18–19. See chapter 6 for details of these cases.

48 LHB7/7/6, Annual Reports, 1812–55, LHB7/7/6A, Annual Reports, 1848–62, LHB7/7/7, Annual Reports, 1851–71.

49 Handwriting analysis has led to the conclusion that Skae added in the diagnosis. I would like to thank Michael Barfoot for pointing this out to me.

50 David Skae, 'A Rational and Practical Classification of Insanity', *Journal of Mental Science*, 9 (1863), 309–19; [the late] David Skae, 'The Morisonian Lectures on Insanity for 1873' (edited by T.S. Clouston), *Journal of Mental Science*, 19 (1873), 340–55, 20 (1874), 1–20.

51 Clouston, *Clinical Lectures*, p. 20 (Clouston's emphasis).

52 Bucknill and Tuke's text, with its details on classification and aetiology, appears to have been widely used in asylum practice, going into four editions between 1858 and 1879: John Charles Bucknill and Daniel Hack Tuke, *A Manual of Psychological Medicine* (Philadelphia: Blanchard & Lee, 1858).

53 See Laurence J. Ray, 'Models of Madness in Victorian Asylum Practice', *Archives of European Sociology*, 22 (1981), 229–64, especially pp. 238–44; and William F. Bynum, 'Rationales for Therapy in British Psychiatry, 1780–1835', *Medical History*, 18 (1974), 317–34.

54 For the sociologist Lindsay Prior, David Skae would be a past master at devising schemes to diagnostically slot his patients into, suggesting that the way psychiatric care is organised, described and treated 'serves both to reflect and constitute what such disorders are'. Texts produced by psychiatrists and classificatory admissions tables reveal not so much what mental illness was, Prior argues, but what theories psychiatrists held about mental illness: Lindsay Prior, *The Social Organisation of Mental Illness* (London: Sage, 1993), especially pp. 16, 1.

55   And was criticised by some psychiatrists for its lack of rigour. See e.g. J. Crichton-Browne, 'Skae's Classification of Mental Diseases', *Journal of Mental Science*, 21 (1875), 339–65.

56   LHB7/51/15, CB, 1862–65, Elizabeth Mawn or Whitewright, admitted 29 May 1864, p. 518.

57   Irvine Loudon has also made this point concerning the relative ease of labelling puerperal insanity, 'one of the few clearly recognized entities in 19th century psychiatry', though I would dispute his claim that such cases were frequently confused with puerperal fever: Irvine Loudon, 'Puerperal Insanity in the 19th Century', *Journal of the Royal Society of Medicine*, 81 (February 1988), 76–9, on p. 76; and see chapter 2, pp. 41–2.

58   Case Book Rules, posted into LHB7/51/37, CB, 1880–82. Cited Thompson, 'The Mad, the Bad, and the Sad', p. 141.

59   In 1869 Clouston appealed for a more systematic way of recording cases and in 1873 introduced uniform printed case books with recording schedules. This initiative was partly in response to the findings of a committee of the Medico-Psychological Association, set up in 1869 to devise a standardised method of recording clinical information. See T. Clouston, 'The Medical Treatment of Insanity', *Journal of Mental Science*, 16 (1870), 24–30; and also Beveridge, 'Madness in Victorian Edinburgh', p. 27.

60   Though sometimes with errors or inaccuracies, as pointed out by Barfoot and Beveridge, 'Madness at the Crossroads', p. 267.

61   Most of the case book entries recorded whether the woman was married or single, and the Scottish custom of continuing to use maiden names in addition to married names made married women distinctive (Elizabeth Mackie or Beveridge, etc.).

62   Beveridge, 'Madness in Victorian Edinburgh', p. 23.

63   Tuke, 'On the Statistics', pp. 1022–3, 1028. See chapter 5, pp. 151–3 for Clouston's comments on lactational insanity.

64   LHB7/51/15, CB, 1862–65, Elizabeth Mackie or Beveridge, admitted 15 April 1864, p. 490. Calcraft was public executioner in London from 1829 to 1874, a well-known and despised figure, not least because of his lack of skill on the scaffold. The Prince of Wales made frequent appearances in case notes in different guises, most often as a suitor.

65   LHB7/51/19, CB, 1867–76 (East House), Jane Duncanson or Melrose, admitted 26 December 1868, p. 175.

66   LHB7/51/13, CB, 1858–62, Margaret Harper, admitted 14 May 1861, p. 651.

67   LHB7/51/11, CB, 1855–58, Isabella Hay, admitted 24 April 1855, p. 73.

68   Ibid., entry 30 April 1855, pp. 73–4.

69   Ibid., entries 10 July, 25 August 1855, p. 74. See chapter 5 for more on the role of families as a potential cause of mental disorder.

70   LHB7/51/9, CB, 1851–55, Jean Main or Wright, admitted 9 December 1852, p. 291.

71   LHB7/51/11, CB, 1855–58, Eliza Paterson or Lumsden, admitted 21 July 1855, p. 115.

72   Ibid., entries 24 July, 10 August, 8 and 21 November 1855, pp. 115–16.

73   LHB7/51/15, CB, 1862–65, Agnes Hastie or Burns, admitted 13 June 1865, p. 730. No further details are given on the child's death.

74 LHB7/51/5, CB 1846–54, (East House), Jessie Eisdale, admitted 7 October 1854, entry 4 December 1854, pp. 1033–4.

75 The bodily state of patients was emphasised perhaps even more in other asylums, including the Warwick Country Lunatic Asylum discussed in the next chapter. There the vast majority of patients were paupers and most were admitted in a visibly poor state of health.

76 LHB7/51/9, CB, 1851–55, Janet Smith or Curle, admitted 1 April 1853, p. 411.

77 Ibid., entry 19 May 1853, p. 412.

78 Ibid., Jane Gardener, admitted 1 July 1853, p. 491. See chapter 5 for more on Jane Gardener's case.

79 LHB7/51/15, CB, 1862–65, Jane Stirling, admitted 17 January 1865, p. 664 (emphasis in case book). This is one of the few references, veiled or otherwise, to prostitution that I have come across.

80 See Sander L. Gilman, *Disease and Representation: Images of Illness from Madness to AIDS* (Ithaca, NY and London: Cornell University Press, 1988), pp. 36–9; and idem, *Seeing the Insane* (Lincoln and London: University of Nebraska Press, 1996; first published New York: John Wiley & Sons, 1982), especially pp. 72–101.

81 Photographs of patients were included in case books only from the late nineteenth century in Edinburgh (and Warwick), and then served a largely administrative, record-keeping function.

82 Alexander Morison, *The Physiognomy of Mental Diseases*, second edition (London: Longman and Co. and S. Highley, 1843), p. 15.

83 Tuke, 'Cases Illustrative', pp. 1090–1.

84 LHB7/51/15, CB, 1862–65, Jane Ferguson, admitted 24 September 1863, p. 367.

85 Ibid., entry for 10 October 1863, p. 367.

86 Tuke, 'Cases Illustrative', p. 1096.

87 Tuke, 'On the Statistics', p. 1019.

88 LHB7/51/9, CB, 1851–55, Janet Smith or Curle, admitted 1 April 1853, p. 411; LHB7/51/11, CB, 1855–58, Mary Sibbald, admitted 26 October 1855, p. 177.

89 And indeed few instances of 'hereditary predisposition' were noted more generally, e.g. only three out of 114 females admitted to the Asylum in 1855: LHB7/7/6, Annual Report, 1855, p. 25.

90 Tuke, 'On the Statistics', pp. 1025–8; idem, 'Cases Illustrative', p. 1089.

91 Tuke, 'On the Statistics', p. 1019.

92 LHB7/51/11, CB, 1855–58, Mary Sibbald, admitted 26 October 1855, p. 177.

93 LHB7/51/15, CB, 1862–65, Sarah Andrew, admitted 5 May 1863, p. 285.

94 LHB7/51/6, CB, 1847–51, Margaret Blackie, admitted 24 March 1851, p. 818.

95 Alison Nuttall has found low rates of forceps deliveries in the Edinburgh Maternity Hospital, even for prolonged labours, and only after 1890 was their use extended beyond obstructed deliveries, while other forms of intervention also remained low: Nuttall, 'The Edinburgh Royal Maternity Hospital', chapter 4.

96 Tuke, 'On the Statistics', p. 1026.

97 Ibid., pp. 1025, 1015.
98 LHB7/51/13, CB, 1858–62, Elizabeth Winks or Love, admitted 3 July 1859, readmitted 30 July 1859, 29 April 1861, 12 July 1861, entry 20 November 1862, pp. 177, 211, 642, 709. The horrendous case of Margaret Bell, who became insane following each of her pregnancies, is referred to below. See also chapter 2, pp. 56–7 for the case of a woman developing mania during several pregnancies.
99 Tuke, 'On the Statistics', p. 1020.
100 LHB7/51/15, CB, 1862–65, Rebecca Murray Inglis or Dobson, admitted 26 January 1864, p. 442; Tuke, 'Cases Illustrative', p. 1093.
101 LHB7/51/9, CB, 1851–55, Jean Main or Wright, admitted 9 December 1852, p. 291.
102 LHB7/51/15, CB, 1862–65, Isabella Evans or Martin, admitted 25 November 1863, p. 403. See chapter 3 for similar instances of shock triggering puerperal mania.
103 See chapter 2, pp. 48–9, for the debate on the dangers versus the benefits of the use of chloroform in labour.
104 Simpson, 'Clinical Lectures'; and idem, *Clinical Lectures*.
105 Ann Finlay or Petrie was admitted in December 1854, three weeks after a 'tedious' labour, when chloroform had been administered, though she was not labelled as suffering from puerperal mania. She was discharged recovered after six months: LHB7/51/9, CB, 1851–55, Ann Finlay or Petrie, admitted 13 December 1854, p. 849.
106 LHB7/7/6, Annual Report, 1855, p. 27. Alison Nuttall convincingly disputes the widespread use of chloroform in Edinburgh at this time, certainly in connection with deliveries in the Edinburgh Maternity Hospital. See Nuttall, 'The Edinburgh Royal Maternity Hospital', chapter 4.
107 Tuke, 'On the Statistics', p. 1019.
108 LHB7/51/9, CB, 1851–55, Mary Bird or Linton, admitted 7 August 1852, p. 231.
109 Tuke, 'On the Statistics', p. 1021. The fattening-up process, advocated by Skae, reached its peak under Clouston after 1873, together with emphasis on exercise and occupation. Clouston recommended in particular eggs and milk, and claimed the results in terms of improvement made up for the increased cost to the institution; 'a gain of two or three stone is quite common, and usually there is an immense advance along with this in nervous stability ... and in self-control, even if a complete recovery does not take place': LHB7/7/8, Annual Report, 1882, p. 22.
110 LHB7/51/11, CB, 1855–58, Mary Sibbald, admitted 26 October 1855, entries 27–29 October 1855, pp. 177–8.
111 LHB7/51/19, CB, 1867–76 (East House), Jane Duncanson or Melrose, admitted 26 December 1868, entries 4 January, 29 March 1868, pp. 175–7.
112 LHB7/51/13, CB, 1858–62, Margaret Harper, admitted 14 May 1861, entries 15 May–15 June 1861, pp. 651–3.
113 LHB7/7/7, Annual Report, 1869, p. 29.
114 LHB7/51/9, CB, 1851–55, Mary Bird or Linton, admitted 7 August 1852, entry 4 October 1852, pp. 231–2. The case notes occasionally criticised the way in which patients had been treated at home prior to admission; see also chapter 5, especially pp. 148–9, 158–9.

115 LHB7/51/6, CB, 1847–51, Elizabeth Robertson, admitted 17 April 1851, entry 25 May 1851, pp. 828–9. Robertson was not recorded in the case book as suffering from puerperal mania. Tarter emetic was used as a vomit; anodyne is a painkiller.

116 LHB7/51/15, CB, 1862–65, Agnes Watson, admitted 1 May 1865, entries 8 May–1 September 1865, pp. 704–7.

117 Tuke, 'Cases Illustrative'.

118 Ibid., pp. 1093–5, quote on p. 1094 (Rebecca Murray Inglis or Dobson, admitted 26 January 1864, LHB7/51/15, CB, 1862–65, p. 442).

119 LHB7/51/9, CB, 1851–55, Jean Main or Wright, admitted 9 December 1852, pp. 291–2.

120 Ibid., entry 28 February 1853, p. 292.

121 Tuke, 'Cases Illustrative', p. 1092. Cellulitis is inflammation of cellular tissues, usually due to infection, and can be life-threatening.

122 Tuke, 'On the Statistics', p. 1026. In cases of great severity mania can result in death from exhaustion, lack of sleep and food.

123 LHB7/51/9, CB, 1851–55, Ann McDonald or Grant, admitted 31 March 1852, entries 23 June 1852–2 August 1854, pp. 139–40, 574.

124 A beef extract, produced after 1865 and recommended for hospital use.

125 Tuke, 'Cases Illustrative', pp. 1097–8.

126 LHB7/51/11, CB, 1855–58, Mary Sibbald, admitted 26 October 1855, entries 15 and 25 March 1858, pp. 177–8, 190. Peritonitis is a severe inflammation of the peritoneum, the lining of the abdomen.

127 Tuke, 'On the Statistics', p. 1016.

128 LHB7/51/6, CB, 1847–51, Elizabeth Robertson, admitted 17 April 1851, entry 15 May 1851, p. 829.

129 Ibid., entries 14 and 20 August 1851, pp. 829, 833.

130 LHB7/51/15, CB, 1862–65, Agnes Bennett or Mathers, admitted 1 April 1862, entry 30 April 1862, p. 15.

131 LHB7/51/5, CB, 1846–54 (East House), Margaret Louisa Maitland or Moir, admitted 5 October 1851, p. 719. This case was not labelled insanity of pregnancy, but was, however, related to her current pregnancy, and also to a miscarriage and instance of severe depression during a previous pregnancy.

132 Ibid., entries 28 February, 15 March 1852, p. 722.

133 Ibid., entries 16 and 28 July 1852, pp. 722–3. See chapter 5 for the strains between husbands and their disturbed wives.

134 G.E. Berrios, 'Delusions as "Wrong Beliefs": A Conceptual History', *British Journal of Psychiatry*, 159, Supplement 14 (1991), 6–13.

135 See chapter 6 for women's struggles to resist these voices, urging them to harm or murder their infants, and also for delusions that they had killed their children.

136 LHB7/51/13, CB, 1858–62, Isabella James or Wood, admitted 12 February 1861, p. 585.

137 Ibid., entries 13 February, 11 April 1861, pp. 585–6.

138 LHB7/51/11, CB, 1855–58, Margaret Scott or Steele, admitted 26 July 1855, p. 119.

139 Ibid., entries 1 and 8 August 1855, pp. 119–20.

140 Ibid., 15 March 1856, 18 March, 29 December 1857, p. 120.

141   LHB7/51/13, CB, 1858–62, Catharine Clark or Grant, admitted 19 March 1862, p. 817.
142   Ibid., Jessie Jameson, admitted 7 January 1861, p. 563.
143   Ibid., entries 25 January, 15 March, 15 June 1861, p. 564.
144   Ibid., entries 16 September 1861, 15 March, 24 September 1862, pp. 564, 654.
145   Ibid., 7 and 25 January 1861, p. 563.
146   Ibid., Isabella James or Wood, admitted 12 February 1861, p. 585.
147   Tuke, 'On the Statistics', p. 1020.
148   Ibid., p. 1023.
149   Porter, 'Hearing the Mad', p. 347. Persaud has compared case notes with a study of a personal notebook kept by a Bethlem attendant in the 1820s, with the physicians often appearing to under-report bizarre and violent behaviour: Rajendra Persaud, 'A Comparison of Symptoms Recorded from the Same Patients by an Asylum Doctor and "a Constant Observer" in 1823: The Implications for Theories about Psychiatric Illness in History', *History of Psychiatry*, 3 (1992), 79–94.
150   Tuke, 'On the Statistics', pp. 1025–8.
151   Ibid., p. 1016.
152   LHB7/51/11, CB, 1855–58, Mary Cameron or Robertson, admitted 11 June 1855, p. 95.
153   Ibid., entry 15 September 1860, p. 96.
154   Ibid., LHB7/51/22, CB, 1870–73 (no date given), p. 67; LHB7/51/24, CB, [ ]–1873, entries 26 February 1881, 1 January 1888, pp. 390–1.
155   LHB7/51/15, CB, 1862–65, Margaret Wilson or Smith, admitted 3 December 1865, entry 27 February 1866, pp. 880–1.
156   Ibid., entries 19 March, 25 July 1866, p. 871.
157   Tuke, 'Cases Illustrative', pp. 1088–9.
158   LHB7/51/15, CB, 1862–65, Margaret Bell or Joss, admitted 5 June 1863, entry 16 July 1863, p. 303.
159   Ibid., entries 6 August, 10 and 30 September 1863, pp. 303–4.
160   Ibid., readmitted 19 September, entry 16 November 1864, p. 582.
161   Ibid., readmitted 22 May 1865, p. 714.
162   Ibid., entry 7 June 1865, p. 714.
163   Ibid., entries 12 September, 8 November 1865, pp. 714, 718.
164   Ibid., entry 5 September 1866, readmitted 5 March 1867, 23 January , entry 19 November 1869, p. 787; Tuke, 'Cases Illustrative', pp. 1088–9.
165   Letter of J.R. Turnbull, 21 May 1865. Pasted into LHB7/51/15, CB, 1862–65, p. 714.
166   Tuke, 'On the Statistics', p. 1015.
167   Tuke, 'Cases Illustrative', p. 1086.
168   Ibid., pp. 1086–7. See chapter 6, pp. 192–3 for details of Mary Oswald's long asylum career and continuing threat to her children.
169   Ibid., p. 1091.
170   LHB7/51/11, CB, 1855–58, Eliza Paterson or Lumsden, admitted 21 July 1855, entries 24 July, 15 September 1855, pp. 115–16.
171   LHB7/51/6, CB, 1847–51, Christina Nicol, admitted 5 August 1851, entries 13 and 23 August 1851, p. 910.
172   LHB7/51/15, CB, 1862–65, Agnes Watson, admitted 1 May 1865, p. 704.

173 LHB7/51/5, CB, 1846–54 (East House), Jessie Eisdale, admitted 7 October 1854, entries 15 and 16 January 1855, pp. 1033–4. Similar instances of self-harm involving needles and a crochet hook are referred to by Nakamura, 'Puerperal Insanity', p. 176.

174 See chapter 3 for further examples of subversion of household activities.

175 LHB7/51/9, CB, 1851–55, Jean Main or Wright, admitted 9 December 1852, pp. 291–2.

176 LHB7/51/6, CB, 1847–51, Isabella Hogg, admitted 20 August 1849, entries 19 May, 15 June 1853, p. 461.

177 Tuke, 'Cases Illustrative', p. 1093.

178 Tuke, 'On the Statistics', p. 1015.

179 Ibid., p. 1028.

180 More common were the efforts of family members to take the women home against the advice of the medical officers, which will be discussed in chapter 5.

181 Tuke, 'Cases Illustrative', pp. 1093–5.

182 LHB7/51/9, CB, 1851–55, Janet Smith or Curle, admitted 1 April 1853, entries 18 July, 26 October 1853, pp. 411–12, 448.

183 LHB7/51/6, CB, 1847–51, Elizabeth Robertson, admitted 17 April 1851, entries 25 May, 24 December 1851, pp. 828–9, 833.

5 Women, Doctors and Mental Disorder: Explaining Puerperal Insanity in the Nineteenth Century

1 John Conolly, 'The Physiognomy of Insanity', No. 8: 'Puerperal Mania', *Medical Times & Gazette*, 16 (19 June 1858), 623–5, quote on p. 624.

2 For Diamond's pioneering role in psychiatric photography, see Sander L. Gilman (ed.), *The Face of Madness: Hugh W. Diamond and the Origin of Psychiatric Photography* (Secaucus, NJ: Citadel, 1976); and Adrienne Burrows and Iwan Schumacher, *Portraits of the Insane: The Case of Dr Diamond* (London: Quartet Books, 1990). In a recent analysis of the illustration of Victorian patients in Bethlem Hospital, it has been argued that the pictures used in this and the other articles published by Conolly in the *Medical Times & Gazette* in 1858 and 1859 were taken by Henry Hering rather than Diamond and are of Bethlem patients: Colin Gale and Robert Howard, *Presumed Curable: An Illustrated Casebook of Victorian Psychiatric Patients in Bethlem Hospital* (Petersfield and Philadelphia: Wrightson Biomedical Publishing, 2003), pp. 12, 15. For the use of photography in depicting cases of puerperal insanity and infanticide in late Victorian England, see Cath Quinn, 'Images and Impulses: Representations of Puerperal Insanity and Infanticide in Late Victorian England', in Mark Jackson (ed.), *Infanticide: Historical Perspectives on Child Murder and Concealment, 1550–2000* (Aldershot: Ashgate, 2002), pp. 193–215.

3 Conolly, 'The Physiognomy of Insanity', p. 624.

4 Alexander Morison, *The Physiognomy of Mental Diseases*, second edition (London: Longman and Co. and S. Highley, 1843), p. 15. See chapter 2, especially pp. 50–3 for Morison's portrayals of puerperal mania.

5   Conolly, 'The Physiognomy of Insanity', p. 624.
6   Ibid., pp. 624–5.
7   Ibid., p. 624.
8   Elaine Showalter, *The Female Malady: Women, Madness and English Culture, 1830–1980* (London: Virago, 1987; first published New York: Pantheon, 1985), p. 87.
9   See ibid., pp. 86–90 for the photographic representation of mad women, quotes on pp. 87, 86.
10 Conolly, 'The Physiognomy of Insanity', p. 624.
11 Ibid., p. 625.
12 Sander L. Gilman, *Disease and Representation: Images of Illness from Madness to AIDS* (Ithaca, NY and London: Cornell University Press, 1988), pp. 39–40.
13 For Jackson's excellent study of photographing mental defectives, which draws extensively on the art historian John Tagg's work, see Mark Jackson, 'Images of Deviance: Visual Representations of Mental Defectives in Early Twentieth-Century Medical Texts', *British Journal of the History of Science*, 28 (1995), 319–37, quote on pp. 321–2.
14 See Sander L. Gilman, *Seeing the Insane* (Lincoln and London: University of Nebraska Press, 1996, first published New York: John Wiley & Sons, 1982), pp. 102–8, quote on p. 102.
15 J.-E.-D. Esquirol, *Mental Maladies: A Treatise on Insanity,* trans. E.K. Hunt (Philadelphia: Lea and Blanchard, 1845).
16 Conolly, 'The Physiognomy of Insanity', p. 624.
17 For the production and use of case notes on asylum populations, see chapter 4.
18 Robert Gooch was arguably the outstanding author of case notes on puerperal insanity. See Robert Gooch, *On Some of the Most Important Diseases Peculiar to Women; with Other Papers,* with a prefatory essay by Robert Ferguson, M.D. (London: The New Sydenham Society, 1831); and chapter 3 for examples of his accounts of cases.
19 Showalter, *The Female Malady,* p. 7. See also Jane Ussher, *Women's Madness: Misogyny or Mental Illness?* (New York, London, etc.: Harvester Wheatsheaf, 1991).
20 See chapter 2, especially pp. 38–43. See also Lisa Ellen Nakamura, 'Puerperal Insanity: Women, Psychiatry, and the Asylum in Victorian England, 1820–1895', unpublished PhD thesis, University of Washington, 1999, pp. 202–17 for social and moral causes of puerperal insanity in the late nineteenth century.
21 Showalter, *The Female Malady,* p. 55.
22 Mary Poovey, '"Scenes of an Indelicate Character": The Medical "Treatment" of Victorian Women', *Representations*, 14 (1986), 137–68, quote on p. 146.
23 Perhaps most actively in the US. See, for a summary of work in the field, Nancy Tomes, 'Historical Perspectives on Women and Mental Illness', in Rima D. Apple (ed.), *Women, Health, and Medicine in America* (New Brunswick, NJ: Rutgers University Press, 1990), pp. 143–71. Published in the same year as Showalter's book, and treating male and female patients together, Anne Digby provides a valuable profile of the experiences of

women in the York Retreat: *Madness, Morality and Medicine: A Study of the York Retreat, 1796–1914* (Cambridge: Cambridge University Press, 1985), especially chapter 8. See also Charlotte MacKenzie, *Psychiatry for the Rich: A History of Ticehurst Private Asylum, 1792–1917* (London and New York: Routledge, 1992), chapter 4 for the interactions of the Victorian family with a private asylum.

24 Ussher, *Women's Madness*, pp. 89–90.

25 Janet F. Saunders, 'Institutionalised Offenders: A Study of the Victorian Institution and its Inmates, with Special Reference to Late Nineteenth Century Warwickshire', unpublished PhD thesis, University of Warwick, 1983, p. 173. See also Constance M. McGovern, 'The Myths of Social Control and Custodial Oppression: Patterns of Psychiatric Medicine in Late Nineteenth-Century Institutions', *Journal of Social History*, 20 (1986), 3–23 for the poor state of female patients admitted to US asylums.

26 James Reid, 'On the Causes, Symptoms, and Treatment of Puerperal Insanity', *Journal of Psychological Medicine and Mental Pathology*, 1 (1848), 128–51, 284–94, p. 143.

27 Fleetwood Churchill, *On the Diseases of Women; Including those of Pregnancy and Childbed*, fourth edition (Dublin: Fannin and Co., 1857), p. 737.

28 See chapter 3 for the responses of Sara Coleridge and Isabella Thackeray to these responsibilities.

29 W.C. Ellis, *A Treatise on the Nature, Symptoms, Causes, and Treatment of Insanity, with Practical Observations on Lunatic Asylums, and a Description of the Pauper Lunatic Asylum for the County of Middlesex, at Hanwell* (London: Samuel Holdsworth, 1838), p. 241.

30 Samuel Ashwell, *A Practical Treatise on the Diseases Peculiar to Women, Illustrated by Cases, Derived from Hospital and Private Practice* (London: Samuel Highley, 1845), p. 723.

31 Gooch, *On Some of the Most Important Diseases Peculiar to Women*, p. 84.

32 See chapter 6, especially p. 176 for instances of this.

33 Gooch, *On Some of the Most Important Diseases Peculiar to Women*, p. 58. See chapter 4, p. 114 for the case of Margaret Blackie.

34 John Conolly, 'Description and Treatment of Puerperal Insanity', Lecture XIII: 'Clinical Lectures on the Principle Forms of Insanity, Delivered in the Middlesex Lunatic-Asylum at Hanwell', *Lancet*, I (28 March 1846), 349–54, quote on p. 351.

35 During the 1850s and 1860s some of the Warwick case books included printed sub-sections, but this was not systematic. Those that did separated notes into general physical appearance; skin, head and extremities; organs of digestion; alvine (abdominal) and urinary excretions; organs of circulation; organs of voice and respiration; functions of uterus; and sleep: e.g. Warwick County Record Office, Warwick County Lunatic Asylum (WCLA), CR 1664/617, Case Book (hereafter CB), 17 July 1852–31 December 1855.

36 WCLA, CR 2379/3, CB, 2 April 1864–13 May 1867, Emma Wall, admitted 9 August 1866 (1410) (page numbering is erratic in the Warwick case books and the admission number has been given instead of page references).

37 Ibid., Maria Alexander, admitted 21 December 1866 (1456).

38   Women were also admitted with more serious, sometimes fatal, ailments; see chapter 4, pp. 119–20 for cases in Edinburgh Asylum.
39   WCLA, CR 2379/3, CB, 2 April 1864–13 May 1867, Harriet Ashmore, admitted 26 May 1864 (1864).
40   Ibid., Emma Walker, admitted 8 November 1864 (1259).
41   Conolly, 'Description and Treatment of Puerperal Insanity', p. 353. This article gives examples of the effectiveness of non-restraint in individual cases of puerperal insanity. See chapter 2, pp. 60–1 for Conolly's changing views on the asylum as a place of cure.
42   Ellis, *A Treatise on the Nature, Symptoms, Causes, and Treatment of Insanity*, p. 60.
43   Edinburgh University Library: Lothian Health Board Archive, Royal Edinburgh Hospital, LHB7/51, Case Books (hereafter CB): LHB7/51/15, 1862–65, Helen McCorkle, admitted 26 September 1862, p. 157.
44   Ibid., entries 28 and 30 September, 1 October 1862, p. 157.
45   Though the Annual Reports had a modest circulation, they were reproduced or summed up in the local press. Edinburgh University Library: Lothian Health Board Archive, Royal Edinburgh Hospital, LHB7/7/7, Annual Reports of the Royal Edinburgh Asylum, 1851–71 (hereafter Annual Reports): Physician's Annual Report, 1866, pp. 17–18.
46   Ashwell, *A Practical Treatise on the Diseases Peculiar to Women*, pp. 731–2.
47   J.B. Tuke, 'On the Statistics of Puerperal Insanity as Observed in the Royal Edinburgh Asylum, Morningside', *Edinburgh Medical Journal*, 10 (1864–65), 1013–28, on p. 1022.
48   Ibid., pp.1022–3, 1028.
49   T.S. Clouston, *Clinical Lectures on Mental Diseases*, second edition (London: J. & A. Churchill, 1887), pp. 520–2.
50   Ibid., p. 524.
51   Ibid., p. 518.
52   Ibid., pp. 518–20.
53   Ibid., p. 523.
54   Thomas Clouston, 'Lecture on Mental Health', reprinted from the *Morningside Mirror*, February–March 1875, 1–4. Cited Elaine Thomson, 'Women in Medicine in Late Nineteenth- and Early Twentieth-Century Edinburgh: A Case Study', unpublished PhD thesis, University of Edinburgh, 1998, p. 132.
55   Thomson, 'Women in Medicine', pp. 172–3.
56   See chapter 4 for treatment at the Edinburgh Asylum in earlier decades. Elaine Thomson gives an excellent account of Clouston's methods of treatment and the rationale behind them at the Asylum and the Edinburgh Hospital for Women and Children. The weight of Thomson's evidence also suggests that Showalter may have been unduly harsh in focusing on Clouston's anxiety that women's overuse of brain energy would deplete their reproductive capacities. Thomson argues that Clouston is more complex in terms of understanding his patients' social circumstances and also in his support of female doctors: Thomson, 'Women in Medicine', pp. 130–6, 164–77, especially pp. 174–5; Showalter, *The Female Malady*, pp. 123, 126.
57   LHB7/7/7, Annual Reports, 1851–71.

58 LHB/7/51/15, CB, 1862–65, Janet Ewing or Watson, admitted 16 November 1865, pp. 864–5.

59 Ibid., Margaret Reid or Graham, admitted 20 November 1865, entry 2 December 1865, pp. 866–7.

60 Ibid., entry 24 February 1866, p. 867.

61 Anna Clark, *Women's Silence, Men's Violence: Sexual Assault in England, 1770–1845* (London: Pandora, 1987), p. 76.

62 WCLA, 1664/30 Annual Report of the Committee of Visitors of the County of Warwick Pauper Lunatic Asylum. For the Year 1869: Dr Parsey's Report, p. 11.

63 Conolly, 'Description and Treatment of Puerperal Insanity', p. 350.

64 T.M. Madden, 'On Puerperal Mania', *British and Foreign Medico-Chirurgical Review*, XLVII (1871), 477–95, p. 478.

65 Tuke, 'On the Statistics', p. 1019. This figure climbed to 25 per cent by the 1880s: Clouston, *Clinical Lectures*, p. 512.

66 WCLA, CR 2379/3, CB, 2 April 1864–13 May 1867, Mary Ann Margaret Turner, admitted 30 December 1865 (1353).

67 Ellis, *A Treatise on the Nature, Symptoms, Causes, and Treatment of Insanity*, p. 90.

68 See chapter 2, pp. 57–8 for admissions to maternity hospitals.

69 Reid, 'On the Causes, Symptoms, and Treatment', p. 146.

70 James Y. Simpson, *Clinical Lectures on the Diseases of Women* (Edinburgh: Adam and Charles Black, 1872), p. 567.

71 Ibid., pp. 555–6.

72 LHB7/51/15, CB, 1862–65, Isabella Pringle, admitted 23 July 1864, p. 548.

73 Clouston, *Clinical Lectures on Mental Diseases*, p. 505.

74 WCLA, CR 1664/619, CB, 4 January 1856–22 March 1861, Ann Standbridge, admitted 14 October 1856 (478).

75 LHB7/51/9, CB, 1851–55, Jane Gardener, admitted 1 July 1853, p. 491.

76 Ibid., entries 12 July, 28 December 1853, pp. 492, 584.

77 Akihito Suzuki, 'Framing Psychiatric Subjectivity: Doctor, Patient and Record-Keeping at Bethlem in the Nineteenth Century', in Joseph Melling and Bill Forsythe (eds), *Insanity, Institutions and Society, 1800–1914* (London and New York: Routledge, 1999), pp. 115–36.

78 Tuke, 'On the Statistics', p. 1021.

79 Ibid., p. 1023.

80 Ibid., p. 1021.

81 LHB7/7/9, Annual Report, 1885, p. 9.

82 Madden, 'On Puerperal Mania', p. 479.

83 Ibid., pp. 477, 495.

84 See chapter 2, pp. 58–63 for the debate on location.

85 LHB/7/51/15, CB, 1862–65, Margaret Bell, admitted 5 June 1863, entry 16 July 1863, p. 303. See chapter 4, pp. 127–9 for the details of Bell's asylum career.

86 Janet Saunders, 'Quarantining the Weak-Minded: Psychiatric Definitions of Degeneracy and the Late-Victorian Asylum', in W.F. Bynum, R. Porter and M. Shepherd (eds), *Anatomy of Madness: Essays in the History of Psychiatry*, vol. III, *The Asylum and Its Psychiatry* (London and New York: Tavistock, 1988), pp. 273–96; Andrew Scull, *The Most Solitary of Afflictions:*

Madness and Society in Britain, 1700–1900 (New Haven and London: Yale University Press, 1993), chapter 6.

87 Conolly, 'Description and Treatment of Puerperal Insanity', p. 350.

88 LHB7/51/9, CB, 1851–55, Annette Skirving, admitted 9 December 1852, pp. 289–90.

89 LHB7/51/15, CB, 1862–65, Margaret Grossert, admitted 14 September 1865, p. 808.

90 In the late nineteenth century, similar conclusions were reached for women suffering from neurasthenia: Hilary Marland, '"Uterine Mischief": W.S. Playfair and his Neurasthenic Patients', in Marijke Gijswijt-Hofstra and Roy Porter (eds), *Cultures of Neurasthenia From Beard to the First World War* (Amsterdam and New York: Rodopi, 2001), pp. 117–39; Barbara Sicherman, 'The Uses of a Diagnosis: Doctors, Patients, and Neurasthenia', *Journal of the History of Medicine*, 32 (1977), 33–54.

91 WCLA, CR 1664/619, CB, 4 January 1856–22 March 1861, Mary Ann Ball, admitted 28 November 1856 (494).

92 LHB7/51/11, CB, 1855–58, Isabella Hay, admitted 24 April 1855, entry 30 April 1855, pp. 73–4.

93 Robert Boyd, 'Observations on Puerperal Insanity', *Journal of Mental Science*, 16 (1870), 153–65, p. 156.

94 LHB7/51/15, CB, 1862–65, Helen McDonald, admitted 27 July 1864, p. 556.

95 Ibid., entry 31 December 1864, p. 567.

96 Ibid., Essie McKay, admitted 8 August 1865, p. 784.

97 James Young, 'Case of Puerperal Mania', *Edinburgh Medical Journal*, 13 (1867–68), 262–6, quote on p. 263.

98 Ellis, *A Treatise on the Nature, Symptoms, Causes, and Treatment of Insanity*, p. 92.

99 Madden, 'On Puerperal Mania', p. 489.

100 Ibid.

101 Ibid., p. 490.

102 Conolly, 'Description and Treatment of Puerperal Insanity', p. 350.

103 WCLA, CR 1664/617, CB, 17 July 1852–31 December 1855, Ann Taylor, admitted 14 April 1853 (182).

104 WCLA, CR 1664/621, CB, 13 May 1867–25 January 1870, Lois Eames, admitted 2 June 1868 (1600).

105 Though MacKenzie has claimed that family relationships were rarely named as a source of stress at Ticehurst Hospital: MacKenzie, *Psychiatry for the Rich*, p. 116.

106 Wellcome Trust Library, Ticehurst House Hospital, 6364 4 Case Records, 1856–60, Isabella Campbell Foster, admitted 17 April 1858, pp. 127–30, quote on p. 130.

107 Ibid., pp. 141–4, 6365 5, Case Records 1858–61, pp. 13–14, 23.

108 LHB7/51/15, CB, 1862–65, Agnes Bennett or Mathers, admitted 1 April 1862, entry 12 November 1862, readmitted 19 November 1862, pp. 15, 182.

109 Ibid., Agnes Jenkinson or Scott, admitted 14 July 1862, entries 15 December 1862, 24 February 1863, p. 103.

110 Ibid., Ellen Vernon or Redpath, admitted 7 April 1864, entry 1 July, 1 August 1864, pp. 484–5. The Royal Edinburgh Asylum case books record numerous similar removals.

111 A broader study of the relationship between female insanity and domestic environments has argued that, for some women, conditions in the asylum were preferable to those at home: Marjorie Levine-Clark, 'Dysfunctional Domesticity: Female Insanity and Family Relationships among the West Riding Poor in the Mid-Nineteenth-Century', *Journal of Family History*, 25 (2000), 341–61.

112 Several patients leaving the Manor House Asylum, Chiswick for example were removed to private residences, accompanied by companions: Wellcome Trust Library, Manor House Asylum, Chiswick, 5725 Case Book, May 1870–October 1884.

113 Peter Bartlett, *The Poor Law of Lunacy: The Administration of Pauper Lunatics in Mid-Nineteenth-Century England* (London and Washington: Leicester University Press, 1999), p. 171.

114 Showalter, *The Female Malady*, p. 5 and chapter 6. Carroll Smith-Rosenberg has argued that in the case of hysteria doctors could act as mediators as well as oppressors of their female patients: 'The Hysterical Woman: Sex Roles and Role Conflict in 19th Century America', *Social Research*, 39 (1979), 652–78.

115 Conolly, 'The Physiognomy of Insanity', p. 624.

6 Dangerous Mothers: Puerperal Insanity and Infanticide

1 *Warwick Advertiser*, 11 May 1867.

2 Ibid.

3 For literary representations of infanticide and the place of child murder in cultural debates and social practices, see Christine L. Krueger, 'Literary Defenses and Medical Prosecutions: Representing Infanticide in Nineteenth-Century Britain', *Victorian Studies*, 40 (1997), 271–94 and Josephine McDonagh, *Child Murder and British Culture 1720–1900* (Cambridge: Cambridge University Press, 2003); and for explanations of female crime in Victorian England, Lucia Zedner, *Women, Crime, and Custody in Victorian England* (Oxford: Clarendon Press, 1991), chapter 2.

4 *Lancet*, I (1858), p. 346. Cited George K. Behlmer, 'Deadly Motherhood: Infanticide and Medical Opinion in Mid-Victorian England', *Journal of the History of Medicine*, 34 (1979), 403–27, p. 403.

5 For populist coroners, see Ian A. Burney, *Bodies of Evidence: Medicine and the Politics of the English Inquest 1830–1926* (Baltimore: Johns Hopkins University Press, 2000) and Olive Anderson, *Suicide in Victorian and Edwardian England* (Oxford: Clarendon, 1987), pp. 15–40.

6 Editorial, *Daily Telegraph*, 5 August 1865, p. 4. Cited Tony Ward, 'Legislating for Human Nature: Legal Responses to Infanticide, 1860–1938', in Mark Jackson (ed.), *Infanticide: Historical Perspectives on Child Murder and Concealment, 1550–2000* (Aldershot: Ashgate, 2002), pp. 249–69, on p. 253.

7 'Child-Murder and Its Punishment', *Social Science Review*, New Series, 2 (1864), 452–9, quote on p. 452.

8 M.A. Baines, 'A Few Thoughts Concerning Infanticide', *Journal of Social Science*, 10 (1866), 535–40, quote on p. 535.

9   Meg Arnott, 'Infant Death, Child Care and the State: The Baby-Farming Scandal and the First Infant Life Protection Legislation of 1872', *Continuity and Change*, 9 (1994), 271–311; George K. Behlmer, *Child Abuse and Moral Reform in England, 1870–1908* (Stanford: Stanford University Press, 1982), especially chapter 2; Lionel Rose, *Massacre of the Innocents: Infanticide in Great Britain 1800–1939* (London: Routledge & Kegan Paul, 1986).

10 Ann R. Higginbotham, '"Sin of the Age": Infanticide and Illegitimacy in Victorian London', in Kristine Ottesen Garrigan (ed.), *Victorian Scandals: Representations of Gender and Class* (Athens: Ohio University Press, 1992), pp. 257–88.

11 'Child-Murder and Its Punishment', p. 452.

12 For more details on legislation, see Nigel Walker, *Crime and Insanity in England*, vol. 1: *The Historical Perspective* (Edinburgh: Edinburgh University Press, 1968), chapter 7; Rose, *Massacre of the Innocents*, especially chapter 8; Tony Ward, 'Legislating for Human Nature'.

13 'Child-Murder and Its Punishment', p. 453.

14 William Burke Ryan, *Infanticide: Its Law, Prevalence, Prevention and History* (London: J. Churchill, 1862).

15 Behlmer, 'Deadly Motherhood'; Rose, *Massacre of the Innocents*, especially chapters 5 and 6; Krueger, 'Literary Defenses and Medical Prosecutions'.

16 Ryan, *Infanticide*, pp. 45–6.

17 Mark Jackson, *New-Born Child Murder* (Manchester: Manchester University Press, 1996), pp. 119–20.

18 William Hunter, 'On the Uncertainty of the Signs of Murder, in the Case of Bastard Children', *Medical Observations and Inquiries*, 6 (1784), pp. 271, 278, 269, reprinted in William Cummin, *The Proof of Infanticide Considered: Including Dr. Hunter's Tract on Child Murder, with Illustrative Notes; and a Summary of the Present State of Medico-Legal Knowledge on that Subject* (London: Longman, Rees, Orme, Brown, Green, and Longman, 1836). See Dana Rabin, 'Bodies of Evidence, States of Mind: Infanticide, Emotion and Sensibility in Eighteenth-Century England', in Jackson (ed.), *Infanticide*, pp. 73–92 for medical responses in the eighteenth century.

19 Charles Servern, *First Lines of the Practice of Midwifery: To Which are Added Remarks on the Forensic Evidence Requisite in Cases of Foeticide and Infanticide* (London: S. Highley, 1831), p. 136.

20 Baines, 'A Few Thoughts Concerning Infanticide', p. 536.

21 Roger Smith, 'The Boundary between Insanity and Criminal Responsibility in Nineteenth-Century England', in Andrew Scull (ed.), *Madhouses, Mad-Doctors, and Madmen: The Social History of Psychiatry in the Victorian Era* (London: Athlone, 1981), pp. 363–84; Roger Smith, *Trial by Medicine: Insanity and Responsibility in Victorian Trials* (Edinburgh: Edinburgh University Press, 1981).

22 Joel Peter Eigen, '"I answer as a physician": Opinion as Fact in Pre-McNaughtan Insanity Trials', in Michael Clark and Catherine Crawford (eds), *Legal Medicine in History* (Cambridge: Cambridge University Press, 1994), pp. 167–99; Joel Peter Eigen, *Witnessing Insanity: Madness and Mad-Doctors in the English Court* (New Haven and London: Yale University Press, 1995).

23 William A. Guy and David Ferrier, *Principles of Forensic Medicine*, seventh edition, revised by William R. Smith (London: Henry Renshaw, 1895), p. 153.

24 For an exploration of recent developments in the field of infanticide, the law and postpartum psychiatric disorders, particularly in the US, see Margaret G. Spinelli, *Infanticide: Psychosocial and Legal Perspectives on Mothers Who Kill* (Washington, DC and London: American Psychiatric Publishing, 2003).

25 H. Walker, *Spirituall Experiences of Sundry Beleevers* (London, 1651), p. 37. Cited Anne Laurence, *Women in England 1500–1760: A Social History* (London: Phoenix, 1996), p. 80.

26 Jackson, *New-Born Child Murder*, pp. 120–3, quote on p. 121.

27 Rabin, 'Bodies of Evidence', p. 76.

28 Robert Gooch, *Observations on Puerperal Insanity* (London, 1820) (extracted from sixth volume of *Medical Transactions*, Royal College of Physicians, read at the College, 16 December 1819). See chapters 2 and 3 for Robert Gooch's work on puerperal insanity.

29 Eigen, *Witnessing Insanity*, p. 142.

30 Ibid. This case is taken from the Old Bailey Sessions Papers, 1822, case 811, 5th sess., 331.

31 See ibid.; Smith, *Trial by Medicine*, especially chapter 7; Walker, *Crime and Insanity in England*, vol. 1, chapter 7; Shelley Day, 'Puerperal Insanity: The Historical Sociology of a Disease', unpublished PhD thesis, University of Cambridge, 1985.

32 Fleetwood Churchill, *On the Diseases of Women; Including those of Pregnancy and Childbed*, fourth edition (Dublin: Fannin, 1857), p. 739.

33 W. Tyler Smith, 'Puerperal Mania', Lecture XXXIX: 'Lectures on the Theory and Practice of Obstetrics', *Lancet*, II (18 October 1856), 423–5, quote on p. 424.

34 John Conolly, 'Description and Treatment of Puerperal Insanity', Lecture XIII: 'Clinical Lectures on the Principal Forms of Insanity, Delivered in the Middlesex Lunatic-Asylum at Hanwell', *Lancet*, I (28 March 1846), 349–54, quote on p. 349.

35 Forbes Winslow, 'On Puerperal Insanity', *Journal of Psychological Medicine and Mental Pathology*, 12 (1859), 9–38, quote on p. 21.

36 T.S. Clouston, *Clinical Lectures on Mental Diseases*, second edition (London: J. & A. Churchill, 1887), p. 502.

37 Shelley Day argues that more attention was paid to symptoms of violence among women suffering from puerperal insanity in the course of the nineteenth century as it was highlighted increasingly in legal cases. Day, 'Puerperal Insanity', p. 174. Texts written in the early nineteenth century, however, including Gooch's publications, were already referring to the antipathy of mothers towards their families and offspring; as the volume of writing on the topic increased, so too do references to violence.

38 Eigen, however, cites a case of child murder tried at the Old Bailey in 1869, where the doctors testifying described the close link between puerperal melancholia and acts of violence, and argued that persons suffering melancholia did not have sufficient control to prevent them committing these acts: Joel Peter Eigen, *Unconscious Crime: Mental Absence and Criminal Responsibility in Victorian London* (Baltimore and London: Johns Hopkins University Press, 2003), pp. 76–8.

39   James Reid, 'On the Causes, Symptoms, and Treatment of Puerperal Insanity', *Journal of Psychological Medicine and Mental Pathology*, 1 (1848), 128–51, 284–340, quote on p. 135.
40   Ibid.
41   James Cowles Prichard, *On the Different Forms of Insanity in Relation to Jurisprudence, Designed for the Use of Persons concerned in Legal Questions Regarding Unsoundness of Mind* (London: Hippolyte Baillière, 1842), pp. 123–4. This case was also cited in Isaac Ray, *A Treatise on the Medical Jurisprudence of Insanity*, fourth edition (Boston: Little, Brown and Company, 1860), pp. 235–6.
42   See chapter 4, pp. 122–31 for instances of disturbed behaviour in the Royal Edinburgh Asylum.
43   Warwick County Record Office, Warwick County Lunatic Asylum (WCLA), CR 2379/3, Case Book (hereafter CB), 2 April 1864–13 May 1867, Mary Ann Margaret Turner, admitted 30 December 1865 (1353).
44   Ibid., Elizabeth Tandy, admitted 14 September 1866 (1420).
45   T.M. Madden, 'On Puerperal Mania', *British and Foreign Medico-Chirurgical Review*, XLVIII (1871), 477–95, on pp. 491–2.
46   Jonathan Andrews and Andrew Scull, *Customers and Patrons of the Mad-Trade: The Management of Lunacy in Eighteenth-Century London* (Berkeley, Los Angeles and London: University of California Press, 2003), pp. 70–1 (John Munro's 1766 Case Book, C–85–C–86).
47   Winslow, 'On Puerperal Insanity', p. 21.
48   Robert Gooch, *A Practical Compendium of Midwifery; Being the Course of Lectures on Midwifery, and on Diseases of Women and Infants, Delivered at St. Bartholomew's Hospital, by the Late Robert Gooch, M.D.* (London: Longman, Rees, Orme, Brown, and Green, 1831), p. 294.
49   Thomas Bull, *Hints to Mothers, for the Management of Health during the Period of Pregnancy and Lying-In Room; with an Exposure of Popular Errors in Connection with Those Subjects*, sixteenth edition (London: Longmans, Green, and Co, 1865; first published 1837), p. 4.
50   Ibid., p. 36.
51   This is the only case I have come across for the nineteenth century that makes such a specific association between maternal imaginings and puerperal insanity, and I am very grateful to Dr Frank Crompton for this reference, based on his work on the Worcester City and County Pauper Lunatic Asylum: Powick Lunatic Asylum Worcester, Patients' Books: Helen Thomas, admitted 2 June 1855 (case 388). See chapter 1 for more on the link between maternal imagination and mental disorder.
52   Robert Boyd, 'Observations on Puerperal Insanity', *Journal of Mental Science*, 16 (1870), 153–65, pp. 154, 156.
53   See, e.g., for the struggle to provide an organic explanation for the condition: F.W. Mackenzie, 'On the Pathology and Treatment of Puerperal Insanity: Especially in Reference to its Relation with Anaemia', *London Journal of Medicine*, 3 (1851), 504–21.
54   John Charles Bucknill, *Unsoundness of Mind in Relation to Criminal Acts* (London: Samuel Highley, 1854), pp. 91–2.
55   Joseph Wiglesworth made a similar link in the 1880s between unmarried motherhood and the risk of puerperal insanity, relating it firmly to brain

function: 'women who are prone to become pregnant illegitimately may also be, by their brain constitution, somewhat more liable than women in general to develop mania': Joseph Wiglesworth, 'Puerperal Insanity: An Analysis of Seventy-Three Cases of the Insanities of Pregnancy, Parturition, and Lactation', *Transactions of the Liverpool Medical Institution*, 6 (1885–86), 349–62, quote on p. 358.

56 Alfred Swaine Taylor, *The Principles and Practice of Medical Jurisprudence*, eighth edition (London: John Churchill, 1865), p. 1122.

57 Samuel Ashwell, *A Practical Treatise on the Diseases Peculiar to Women, Illustrated by Cases, Derived from Hospital and Private Practice* (London: Samuel Highley, 1845).

58 George Man Burrows, *Commentaries on the Causes, Forms, Symptoms, and Treatment, Moral and Medical, of Insanity* (London: Thomas and George Underwood, 1828), Commentary VI, 'Puerperal Insanity'.

59 Taylor, *The Principles and Practice of Medical Jurisprudence*, p. 1122 (emphasis added).

60 Ibid.

61 See, for the potential difficulties of separating out the delirium of puerperal fever from puerperal insanity, chapter 2, pp. 41–2.

62 'Chelmsford–Friday, March 10. Charge of Murder.–Acquittal on the Ground of Puerperal Insanity', *Journal of Psychological Medicine and Mental Pathology*, 1 (1848), 478–83; *Lancet*, I (1848), 318–19.

63 The breasts often bore the brunt of blame in infanticide trials as the precipitating cause of insanity, with medical witnesses claiming the mother to be in a state of temporary insanity because of an excessive flow of milk, or alternatively because the milk was suppressed, or simply because the mother had 'bad breasts'. In many cases the mother's continued efforts to breastfeed her child when she herself was debilitated and starving was seen as triggering mental breakdown.

64 *Lancet*, I (1848), p. 318.

65 Ibid.

66 Ibid., p. 319.

67 'Chelmsford–Friday, March 10. Charge of Murder', p. 481.

68 See, for the limited experience of some medical witnesses, Eigen, *Witnessing Insanity*, Appendix 2, 'Medical Witnesses who Testified at the Old Bailey about the Mental Condition of the Accused, 1760–1843', pp. 195–205.

69 See chapter 2, pp. 55–7 for general practitioners' treatment of puerperal insanity.

70 The *Lancet* reminded their readership that Judge Denman was the son of the esteemed obstetrician Thomas Denman, who, when writing on mental disorder related to childbirth, had given his opinion that the behaviour of the patient varies greatly – 'even in very bad cases there are lucid intervals, or a reasonableness on certain subjects, where the disorder would not be suspected': *Lancet*, I (1848), p. 319. See chapter 1, pp. 14–15 for Denman on mania lactea.

71 'Chelmsford–Friday, March 10. Charge of Murder', p. 480.

72 Bucknill, *Unsoundness of Mind*, pp. 86–7, 101.

73 *Warwick Advertiser*, 21 October 1865.

74   In 1856 William Tyler Smith explained how this would occur in first labours or if the head of the child was unusually large: at the point of birth women could 'lose their self-consciousness, or self-control, and commit, if allowed, extravagant acts, in these brief intervals of insanity': Smith, 'Puerperal Mania', p. 423. Others, as we have seen in chapter 2, disputed the existence of this form of insanity.

75   The phenomenon of birth into excrement also featured in forensic textbooks: see, for example, Johann Ludwig Casper, *Handbook of Forensic Medicine* (London: New Sydenham Society, 1864), which also supported the idea that the parturient could be completely ignorant of the act of delivery, and even married women could feel the urgent need to pass faeces, with the infant consequently being born into excrement and suffocated.

76   See chapter 5 for the conduct of families and husbands and the role they played in provoking insanity.

77   Warwick County Record Office, Quarter Sessions (QS), QS 26/1/61, County of Warwick, Calendar of Prisoners, Summer Assizes, 1856. The Assize records for Warwickshire are incomplete for the years 1859–69. The Assize proceedings have been surveyed in the local press in order to bridge this gap, but it is likely that many cases went unrecorded and in others scant information is given.

78   QS 26/1/3–63, Calendar of Prisoners, 1839–57.

79   Ibid., QS 26/1/39–58.

80   *Warwick Advertiser*, 2 August 1856.

81   Ibid.

82   Ibid., 26 May 1860.

83   Ibid., 28 April, 12 and 26 May 1860.

84   Ibid., 2 September 1865.

85   Ibid.

86   Ibid., 15 July 1865.

87   See also Cath Quinn, 'Images and Impulses: Representations of Puerperal Insanity and Infanticide in Late Victorian England', in Jackson (ed.), *Infanticide*, pp. 193–215 for the link between symptoms of infanticide and puerperal insanity in psychiatric photography.

88   [the late] David Skae, 'The Morisonian Lectures on Insanity for 1873' (edited by T.S. Clouston), *Journal of Mental Science*, 19 (1873), 340–55, 20 (1874), 1–20, quote on pp. 5–6.

89   Edinburgh University Library: Lothian Health Board Archive, Royal Edinburgh Hospital, LHB7/7/6A, Annual Reports of the Royal Edinburgh Asylum, 1848–62, Physician's Annual Report, 1848, p. 18.

90   Ibid., p. 19.

91   Ibid., p. 20.

92   *Warwick Advertiser*, 10 April 1858.

93   Ibid.

94   WCLA, CR 1664/619, CB, 4 January 1856–22 March 1861, Selina Cranmore, admitted 16 April 1858 (616).

95   *Warwick Advertiser*, 7 August 1858.

96   WCLA, CR 1664/619, Selina Cranmore; WCLA CR 1664/53, House Committee Book, 17 September 1858. I would like to thank David George for alerting me to this case.

97   Edinburgh University Library: Lothian Health Board Archive, Royal Edinburgh Hospital, LHB7/51, Case Books (hereafter CB): LHB7/51/6, 1847–51, Jane Anderson or Rutherford, admitted 6 October 1848, p. 244.

98   Ibid., pp. 244–5.

99   Ibid., entry 19 February 1849, p. 245. See Eigen, *Unconscious Crime*, pp. 78–9 for instances of failure of memory in infanticide trials linked to puerperal mania.

100 Edinburgh University Library: Lothian Health Board Archive, Royal Edinburgh Hospital, LHB7/14/3, Royal Edinburgh Asylum and Carlisle Asylum Papers, 1873–86, Medical Papers by Dr Clouston and his Assistants, 53: 'A Case of Child Murder', by James McLaren, Stirling District Lunacy Notes, 41–4, p. 41.

101 It was not clear whether she had used her fists or an instrument, if the child had been banged against the floor, or whether she had thrown her two-year-old daughter with great force against the infant's head.

102 LHB7/14/3, 'A Case of Child Murder', p. 42.

103 Ibid.

104 Ibid., p. 43.

105 Ibid., pp. 43–4.

106 WCLA, CR 2379/3, CB, 2 April 1864–13 May 1867, Hannah Harris, admitted 22 February 1866 (1374).

107 Jonathan Andrews has also referred to the ability to bear children as a factor in deciding on discharges from Broadmoor Asylum: 'The Boundaries of Her Majesty's Pleasure: Discharging Child-Murderers from Broadmoor and Perth Criminal Lunatic Department, c.1860–1920', in Jackson (ed.), *Infanticide*, pp. 216–48, especially pp. 231–4.

108 WCLA, CR 2379/3, Hannah Harris.

109 J.B. Tuke, 'Cases Illustrative of the Insanity of Pregnancy, Puerperal Mania, and Insanity of Lactation', *Edinburgh Medical Journal*, 12 (1866–67), 1083–101, p. 1086.

110 Ibid.

111 LHB7/51/15, CB, 1862–65, Mary Oswald or Carruthers, admitted 22 March 1864, p. 478.

112 Tuke, 'Cases Illustrative', p. 1086.

113 Ibid., p. 1087.

114 Ibid.

115 LHB7/51/22, CB, 1870–73, Mary Oswald or Carruthers, admitted 10 July 1871, p. 450.

116 Ibid., entry 22 August 1871, p. 451.

117 Ibid., Mary Oswald or Carruthers, readmitted 22 August 1872, p. 716, entry 15 October 1872, p. 717.

118 LHB7/51/24, CB, [ ]–1873, Mary Oswald or Carruthers, admitted 25 June 1873, pp. 5–6, entry 25 March 1874, p. 6.

119 Lucia Zedner has argued that prisons could be used instrumentally by women as a decent refuge where they could give birth or simply have shelter and freedom from care: *Women, Crime, and Custody*, pp. 170–2.

120 WCLA, CR 1664/266, Orders for Admission, 1860, No. 851: Order for Admission of Bridget Butler (including Secretary of State's Warrant 9 May 1860); WCLA, CR 1664/619, CB, 4 January 1856–22 March 1861, Bridget

Butler, admitted 11 May 1860 (847) (different admission numbers were given in the case book and admission order).

121 Ibid.
122 WCLA, CR 1664/280, Orders for Admission, 1867, No. 1504: Order for Admission of Elizabeth Barnwell.
123 *Warwick Advertiser*, 11 May 1867.
124 Ibid.
125 Ibid. The Coroner's Register records that George Barnwell was 'accidentally drowned', on 4 May 1867, at the Cape Lock on the Warwick Napton Canal: WCRO, CR 1769/45, Coroner's Register, 1837–1903. No inquisitions have survived for 1867.
126 WCLA, CR 1664/621, CB, 13 May 1867–25 January 1870, Elizabeth Barnwell, admitted 29 June 1867 (1504).
127 Ibid.
128 Zedner, *Women, Crime, and Custody*, p. 89.
129 A. James Hammerton, *Cruelty and Companionship: Conflict in Nineteenth-Century Married Life* (London and New York: Routledge, 1992), quote on p. 118.
130 Margot Finn, 'Working-Class Women and the Contest for Consumer Control in Victorian County Courts', *Past & Present*, 161 (1998), 116–54, quote on p. 119.
131 Martin J. Weiner, 'Alice Arden to Bill Sykes: Changing Nightmares of Intimate Violence in England, 1558–1869', *Journal of British Studies*, 40 (2001), 184–212.
132 Krueger, 'Literary Defenses and Medical Prosecutions'.
133 *Transactions of the National Association for the Promotion of Social Science 1866*, pp. 293–4; Zedner, *Women, Crime, and Custody*, p. 29.
134 Krueger, 'Literary Defenses and Medical Prosecutions', pp. 282, 284.
135 *Lancet*, I (1848), pp. 318–19.
136 See Colin Gale and Robert Howard, *Presumed Curable: An Illustrated Casebook of Victorian Psychiatric Patients in Bethlem Hospital* (Petersfield and Philadelphia: Wrightson Biomedical Publishing, 2003), p. 12. The practice of photographing Warwick and Edinburgh Asylum patients was only instituted much later in the nineteenth century.
137 Quinn, 'Images and Impulses' points out that only patients labelled as infanticides were consistently portrayed undertaking domestic tasks in this extensive photographic collection.
138 Zedner, *Women, Crime, and Custody*, p. 29.
139 See chapter 4 for a discussion of this combination of collapse, mania and enormous physical strength.

7   From Redemption to the Dark Age: The Demise of Puerperal Insanity

1 M.D. Macleod, 'An Address on Puerperal Insanity', *British Medical Journal*, II (7 August 1886), 239–42, quote on p. 241.
2 Michael Gelder, Richard Mayou and Philip Cowen (eds), *Shorter Oxford Textbook of Psychiatry*, fourth edition (Oxford: Oxford University Press,

2001), p. 500. For a comparison of cases occurring in Edinburgh Asylum in 1880–90 and 1971–80, see A. Rehman, 'Puerperal Insanity in the Nineteenth and Twentieth Centuries', MPhil thesis, University of Edinburgh, 1986.

3 Gelder, Mayou and Cowen (eds), *Shorter Oxford Textbook of Psychiatry*, pp. 499, 501.

4 I.F. Brockington, E.M. Schofield, P. Donnelly and C. Hyde, 'A Clinical Study of Post-Partum Psychosis', in M. Sandler (ed.), *Mental Illness in Pregnancy and the Puerperium* (Oxford: Oxford University Press, 1978), pp. 59–68, quote on p. 59. See also I.F. Brockington and R. Kumar (eds), *Motherhood and Mental Illness* (London and New York: Academic Press and Grune and Stratton, 1982).

5 Edward Shorter, *A History of Psychiatry: From the Era of the Asylum to the Age of Prozac* (New York, etc.: John Wiley, 1997), pp. 106–8.

6 Emil Kraepelin, *Lectures on Clinical Psychiatry* (London: Tindall and Cox, 1904), p. 129.

7 Gelder, Mayou and Cowen (eds), *Shorter Oxford Textbook of Psychiatry*, p. 500.

8 Ibid.

9 http://www.christianitymeme.org/yates.html: 'The Andrea Yates Case'.

10 Ibid.; http://www.time.com/nation/article/0,8599,195325,00.html: 'The Yates Odyssey'.

11 Maine à potû was an alternative term for pathological drunkenness.

12 W.H.O. Sankey, *Lectures on Mental Diseases* (London: John Churchill, 1866), p. 77.

13 J. Thompson Dickson, 'A Contribution to the Study of the So-Called Puerperal Insanity', *Journal of Mental Science*, 17 (1870), 379–90, quote on p. 380 (Dickson's emphasis).

14 Ibid., p. 389.

15 Quinn has also pointed out that the term 'puerperal insanity' was rarely used in the *Journal of Mental Science* after 1900: see Catherine Quinn, 'Include the Mother and Exclude the Lunatic. A Social History of Puerperal Insanity c.1860–1920', unpublished PhD thesis, University of Exeter, 2003, p.103. Nancy Theriot has argued that after the 1890s puerperal insanity was seen as a 'suspect' category in the United States 'and the emerging specialty of psychiatry emphasised the similarity between puerperal mania and any other mania': Nancy Theriot, 'Diagnosing Unnatural Motherhood: Nineteenth-Century Physicians and "Puerperal Insanity"', *American Studies*, 26 (1990), 69–88, quote on p. 79. Though he sportingly related several cases of 'so-called "puerperal insanity"', in 1901 the psychiatrist Edward Lane noted his embarrassment about presenting a paper on puerperal insanity and thus being 'asked to talk about something which I believed did not exist': Edward B. Lane, 'Puerperal Insanity', *Boston Medical and Surgical Journal*, 144 (20 June 1901), 606–9, quote on p. 606.

16 For Tuke's approach, see chapter 2, p. 47 and Chapter 4, p. 119.

17 Lisa Ellen Nakamura, 'Puerperal Insanity: Women, Psychiatry, and the Asylum in Victorian England, 1820–1895', University of Washington, 1999, especially pp. 202–16. Nakamura also, however, highlights the unsympathetic treatment meted out to maternity hospital patients by physicians insensitive to the anxieties and situation of unmarried mothers. In 1882,

for instance, Margaret Pearse delivered an illegitimate son in the General Lying-in Hospital in London. When she became 'rigid and staring' and refused to speak, the physician told the nurse to 'take no notice of her'. She became worse, was diagnosed with puerperal mania, tied to her bed and dosed massively with potassium bromide: p. 245.

18 Macleod, 'An Address on Puerperal Insanity', p. 241.

19 George H. Savage, *Insanity and Allied Neuroses: Practical and Clinical* (London and New York: Cassell, 1884), pp. 376–7.

20 Ibid., p. 377.

21 See e.g. Andrew Scull, *The Most Solitary of Afflictions: Madness and Society in Britain, 1700–1900* (New Haven and London: Yale University Press, 1993), chapters 6 and 7.

22 Richard Grundy, 'Observations upon Puerperal Insanity', *Journal of Psychological Medicine and Mental Pathology*, 13 (1860), 414–25, on pp. 414–15.

23 Warwick County Record Office, Warwick County Lunatic Asylum, CR 1664/654, Register of Admissions, Female Patients, 1886–90; Irvine Loudon, 'Puerperal Insanity in the 19th Century', *Journal of the Royal Society of Medicine*, 81 (February 1988), 76–9, p. 77.

24 W.F. Menzies, 'Puerperal Insanity: An Analysis of One Hundred and Forty Consecutive Cases', *American Journal of Insanity*, 50 (1893–94), 147–85, pp. 147, 149.

25 George H. Savage, 'Puerperal Insanity', in D. Hack Tuke, *A Dictionary of Psychological Medicine*, 2 vols (London: J. & A. Churchill, 1892), vol. 2, pp. 1034–42, quote on p. 1037. Quinn has also described how the late nine-teenth- and early twentieth-century emphasis on heredity undermined the disease label of puerperal insanity: 'Include the Mother and Exclude the Lunatic', pp. 89–105.

26 R. Percy Smith, 'Insanity of Pregnancy, the Puerperal Period, and Lactation', in *Quain's Dictionary of Medicine*, third edition (London: Longmans, Green, and Co., 1902), pp. 758–61, quote on p. 759. 'Neuropathic diathesis' refers to a constitutional predisposition to disease accompanied by a structural change in the brain.

27 Menzies, 'Puerperal Insanity', p. 161.

28 Joseph Wiglesworth, 'Puerperal Insanity: An Analysis of Seventy-Three Cases of the Insanities of Pregnancy, Parturition and Lactation', *Transactions of the Liverpool Medical Institution*, 6 (1885–86), 349–62, on pp. 358–9.

29 Savage, 'Puerperal Insanity', p. 1037.

30 However, these have been put to excellent use by historians of psychiatry interested in statistical evidence: see especially Allan Beveridge, 'Madness in Victorian Edinburgh: A Study of Patients Admitted to the Royal Edinburgh Asylum under Thomas Clouston, 1873–1908', Parts I and II, *History of Psychiatry*, 6 (1995), 21–54, 133–56.

31 Case Book Rules, posted into Edinburgh University Library: Lothian Health Board Archive, Royal Edinburgh Hospital, Case Book, LHB7/51/37, 1880–82. This shift towards physical diagnosis and 'objective' signs of disease, 'from words toward numbers and visual representations', was paral-leled in case notes on somatic disorders: Guenter B. Risse and John Harley

Warner, 'Reconstructing Clinical Activities: Patient Records in Medical History', *Social History of Medicine*, 5 (1992), 183–205, especially pp. 191–2.

32 A. Campbell Clark, 'Aetiology, Pathology, and Treatment of Puerperal Insanity', *Journal of Mental Science*, 33 (1887), 169–89, 372–9, 487–96, quotes on pp. 183, 182.

33 Ibid., p. 182.

34 Edinburgh University Library: Lothian Health Board Archive, Royal Edinburgh Hospital, LHB7/14/3, Royal Edinburgh Asylum and Carlisle Asylum Papers, 1873–86, Medical Papers by Dr Clouston and his Assistants, 62: 'Clinical Illustrations of Puerperal Insanity', by A. Campbell Clark, Reprinted (with additions) from the *Lancet*, vol. 2, 1883, Glasgow, 1884 (Printed David Maclure & Son), pp. 21, 26.

35 A. Campbell Clark, 'Clinical Illustrations of Puerperal Insanity', *Lancet*, II (21 July, 4 and 18 August 1883), 97–9, 180–1, 278–9, quote on p. 180.

Bibliography

Primary sources

Edinburgh University Library: Lothian Health Board Archive

Royal Edinburgh Hospital

Case Books
LHB7/51/1 (Male and Female), 1840–42
LHB7/51/2 (Male and Female), 1842–45
LHB7/51/4 (Female), 1843–47
LHB7/51/5 (Male and Female, East House), 1846–54
LHB7/51/6 (Female), 1847–51
LHB7/51/9 (Female), 1851–55
LHB7/51/11 (Female), 1855–58
LHB7/51/13 (Female), 1858–62
LHB7/51/15 (Female), 1862–65
LHB7/51/17 (Female), 1865–68
LHB7/51/19 (Male and Female, East House), 1867–76
LHB7/51/20 (Female), 1868–70
LHB7/51/22 (Female), 1870–73
LHB7/51/24 (Female), [ ]–1873
LHB7/51/26 (Female), 1874–75
LHB/7/51/37, 1880–82: Case Book Rules (pasted into book)
LHB7/35/1–4, General Registers of Patients, 1851–57, 1858–63, 1863–69, 1870–76
LHB7/7/6, Annual Reports, 1812–55
LHB7/7/6A, Annual Reports, 1848–62
LHB7/7/7, Annual Reports, 1851–71
LHB7/7/8, Annual Reports, 1872–82
LHB7/7/9, Annual Reports, 1883–92
LHB7/15/1, Regulations of the Royal Edinburgh Asylum for the Insane 1867
LHB7/12/1, Press Cuttings Book, 1862–81
LHB7/14/3, Royal Edinburgh Asylum and Carlisle Asylum Papers, 1873–86

Edinburgh Royal Maternity Hospital

LHB3/17/1, Special and Ordinary Case book, 1870–90
LHB3/18/3, Outdoor Case book, 1857–87
LHB3/18/4, Outdoor Case book, 1877–87

Mitchell Library, Glasgow: Greater Glasgow Health Board Archives

Glasgow Royal Maternity Hospital

HB45/5/14, Patient Registers, Indoor, 1855–66
HB45/5/15, Patient Registers, Indoor, 1866–81

Warwick County Record Office

Warwick County Lunatic Asylum

Case Books
CR 1664/617, 17 July 1852–31 December 1855
CR 1664/619, 4 January 1856–22 March 1861
CR 1664/620, 4 April 1861–24 March 1864
CR 1664/621, 13 May 1867–25 January 1870
CR 2379/2, 30 June 1852–1 April 1854
CR 2379/3, 2 April 1864–13 May 1867
CR 2379/1, Register of Admissions, 1852–75
CR 1664/654, Register of Admissions, Female Patients, 1886–90
CR 1664, 250–285, Orders for Admission, 1852–70
CR 1664/53, House Committee Book, 1858
CR 1664/30 Annual Reports of the Committee of Visitors, the Superintendent
  (W.H. Parsey) and from 1872 the Commissioners in Lunacy 1858–83

Court and Coroner's Records

QS 26/1, County of Warwick, Calendar of Prisoners
CR 1769/45, Coroner's Register, 1837–1903

London Metropolitan Archives

General Lying-in Hospital London

H1/GL1/B19/1, Medical Officers' Case Book, 1827–28
H1/GL1/B19/7, Medical Officers' Case Book, 1842–49
H1/GL1/B14/1, Delivery Book, 1828–56

Wellcome Trust Library

Ticehurst House Hospital

Case Records 6364 4, 1856–60; 6365 5, 1858–61; 6366 6, 1860–63; 6368 8, 1862–74

Manor House Asylum, Chiswick

5725 Case Book, May 1870–October 1884

T.216.21, Annual Reports of the Cumberland & Westmoreland Lunatic Asylum,
  1863–70
T.216.21, John Conolly, *The Report of the Resident Physician of the Hanwell Lunatic
  Asylum, Presented to the Court of Quarter Sessions for Middlesex, at the
  Michaelmas Sessions, 1840*
WLM28.BE5, B84, General Report of the Royal Hospitals of Bridewell and Bethlem
  and of the House of Occupations, for the year ending 31 December 1845

Harry Ransom Humanities Research Center, The University of Texas at Austin

Coleridge, S. Misc.: Sara Coleridge, 'Diary of her Children's Early Years'.
'Essay on Nervousness', 1834
(Consulted on microfilm Modern Records Centre, University of Warwick)

Medical journals

Association Medical Journal
British and Foreign Medico-Chirurgical Review
British Medical Journal
Edinburgh Medical Journal
Glasgow Medical Journal
Journal of Mental Science
Journal of Psychological Medicine and Mental Pathology
Lancet
London Journal of Medicine
London Medical Gazette
London Medical Review
Medical Times
Medical Times & Gazette
Provincial Medical and Surgical Journal
Psychological Medicine

Newspapers

The Times
Warwick Advertiser

Dictionaries

Dictionary of National Biography
A Dictionary of Psychological Medicine
Quain's Dictionary of Medicine

Contemporary books and articles

Ashwell, Samuel, *A Practical Treatise on the Diseases Peculiar to Women, Illustrated by Cases, Derived from Hospital and Private Practice* (London: Samuel Highley, 1845).

Baines, M.A., 'A Few Thoughts Concerning Infanticide', *Journal of Social Science*, 10 (1866), 535–40.

Barker, Fordyce, *The Puerperal Diseases: Clinical Lectures Delivered at the Bellevue Hospital*, third edition (New York: Appleton, 1883).

Boyd, Robert, 'Observations on Puerperal Insanity', *Journal of Mental Science*, 16 (1870), 153–65.

Bucknill, John Charles, *Unsoundness of Mind in Relation to Criminal Acts* (London: Samuel Highley, 1854).

Bucknill, John Charles and Daniel Hack Tuke, *A Manual of Psychological Medicine* 1858 edition (New York and London: Hafner, 1968).

Bull, Thomas, *Hints to Mothers, for the Management of Health during the Period of Pregnancy and Lying-In Room; with an Exposure of Popular Errors in Connection with Those Subjects*, sixteenth edition (London: Longmans, Green, and Co., 1865).

Burns, John, *Popular Directions for the Treatment of the Diseases of Women and Children* (London: Longman, Hurst, Rees, Orme and Brown, 1811).

Burns, John, *The Principles of Midwifery*, third edition (London: Longman, Hurst, Rees, Orme, and Brown, 1814).

Burns, John, *The Principles of Midwifery; Including the Diseases of Women and Children*, seventh edition (London: Longman, Rees, Orme, Brown and Green, 1828).

Burrows, George Man, *Commentaries on the Causes, Forms, Symptoms, and Treatment, Moral and Medical, of Insanity* (London: Thomas and George Underwood, 1828).

Casper, Johann Ludwig, *Handbook of Forensic Medicine* (London: New Sydenham Society, 1864).

'Chelmsford–Friday, March 10. Charge of Murder.–Acquittal on the Ground of Puerperal Insanity', *Journal of Psychological Medicine and Mental Pathology*, 1 (1848), 478–83.

'Child-Murder and Its Punishment', *Social Science Review*, New Series, 2 (1864), 452–9.

Churchill, Fleetwood, *On the Theory and Practice of Midwifery* (London: Henry Renshaw, 1842).

Churchill, Fleetwood, *On the Diseases of Women; Including those of Pregnancy and Childbed*, fourth edition (Dublin: Fannin, 1857).

Clark, A. Campbell, 'Clinical Illustrations of Puerperal Insanity', *Lancet*, II (21 July, 4 and 18 August 1883), 97–9, 180–1, 278–9.

Clark, A. Campbell, 'Aetiology, Pathology, and Treatment of Puerperal Insanity', *Journal of Mental Science*, 33 (1887), 169–89, 372–9, 487–96.

Clarke, John, *Practical Essays on the Management of Pregnancy and Labour; and on The Inflammatory and Febrile Diseases of Lying-in Women*, second edition (London: J. Johnson, 1806).

Clouston, T., 'The Medical Treatment of Insanity', *Journal of Mental Science*, 16 (1870), 24–30.

Clouston, T.S., *Clinical Lectures on Mental Diseases*, second edition (London: J.& A. Churchill, 1887).

Conolly, John, *An Inquiry Concerning the Indications of Insanity, with Suggestions for the Better Protection and Care of the Insane* (London: John Taylor, 1830).

Conolly, John, 'Description and Treatment of Puerperal Insanity', Lecture XIII: 'Clinical Lectures on the Principle Forms of Insanity, Delivered in the Middlesex Lunatic-Asylum at Hanwell', *Lancet*, I (28 March 1846), 349–54.

Conolly, John, 'The Physiognomy of Insanity', No. 8: 'Puerperal Mania', *Medical Times & Gazette*, 16 (19 June 1858), 623–5.

Conquest, J.T., *Outlines of Midwifery, Developing its Principles and Practice; Intended as a Text Book for Students, and a Book of Reference for Junior Practitioners* (London: John Anderson, 1820).

Crichton-Browne, J., 'Skae's Classification of Mental Diseases', *Journal of Mental Science*, 21 (1875), 339–65.

Cummin, William, *The Proof of Infanticide Considered: Including Dr. Hunter's Tract on Child Murder, with Illustrative Notes; and a Summary of the Present State of Medico-Legal Knowledge on that Subject* (London: Longman, Rees, Orme, Brown, Green, and Longman, 1836).

Denman, Thomas, *An Introduction to the Practice of Midwifery*, second edition (London: J. Johnson, 1801).

Denman, Thomas, *Observations on the Rupture of the Uterus, on the Snuffles in Infants, and on Mania Lactea* (London: J. Johnson, 1810).

Dickson, J. Thompson, 'A Contribution to the Study of the So-Called Puerperal Insanity', *Journal of Mental Science*, 17 (1870), 379–90.

Donkin, Arthur Scott, 'On the Pathological Relation between Albuminuria and Puerperal Mania', *Edinburgh Medical Journal*, 8 (1862–63), 994–1004.

Ellis, W.C., *A Treatise on the Nature, Symptoms, Causes, and Treatment of Insanity, with Practical Observations on Lunatic Asylums, and a Description of the Pauper Lunatic Asylum for the County of Middlesex, at Hanwell* (London: Samuel Holdsworth, 1838).

Esquirol, J.-E.-D., 'De l'aliénation mentale des nouvelles accouchés et des nourrices', *Annuaire Médico-Chirurgical de Hôpitaux de Paris*, 1 (1819), 600–32.

Esquirol, J.-E.-D., *Mental Maladies: A Treatise on Insanity*, trans. E.K. Hunt (Philadelphia: Lea and Blanchard, 1845).

Evans, Howell, 'Puerperal Mania Occurring Suddenly Three Days after Delivery Treated by Opiates and Purgatives', *Medical Times*, 15 (14 November 1846), 145.

Gillman, Alexander W., *The Gillmans of Highgate with Letters from Samuel Taylor Coleridge* (London: Elliot Stock, n.d. [1895]).

Gooch, Robert, *Observations on Puerperal Insanity* (London, 1820) (extracted from sixth volume of *Medical Transactions*, Royal College of Physicians, read at the College, 16 December 1819).

Gooch, Robert, *On Some of the Most Important Diseases Peculiar to Women; with Other Papers*, with a prefatory essay by Robert Ferguson, M.D. (London: The New Sydenham Society, 1831).

Gooch, Robert, *A Practical Compendium of Midwifery; Being the Course of Lectures on Midwifery, and on Diseases of Women and Infants, Delivered at St. Bartholomew's Hospital, by the Late Robert Gooch, M.D.* (London: Longman, Rees, Orme, Brown, and Green, 1831).

Graham, Thomas J., *On the Diseases of Females; A Treatise Describing their Symptoms, Causes, Varieties, and Treatment. Including the Diseases and Management of Pregnancy and Confinement*, seventh edition (London: published for the author, 1861).

Grundy, Richard, 'Observations upon Puerperal Insanity', *Journal of Psychological Medicine and Mental Pathology*, 13 (1860), 414–25.

Guy, William A. and David Ferrier, *Principles of Forensic Medicine*, seventh edition, revised by William R. Smith (London: Henry Renshaw, 1895).

Hall, Marshall, *Commentaries on Some of the More Important of the Diseases of Females. In Three Parts* (London: Longman, Rees, Orme, Brown and Green, 1827).

Hamilton, Alexander, *A Treatise on the Management of Female Complaints*, seventh edition (Edinburgh: P Hill and London: Underwood and Blacks, 1813).

Haslam, John, *Observations on Insanity: with Practical Remarks on the Disease, and An Account of the Morbid Appearances on Dissection* (London: F. and C. Rivington, 1798).

Hastings, H., 'Puerperal Mania Followed by Insanity', *Medical Times*, 19 (21 April 1849), 511.

Hunter, William, 'On the Uncertainty of the Signs of Murder, in the Case of Bastard Children', *Medical Observations and Inquiries*, 6 (1784), reprinted in

William Cummin, *The Proof of Infanticide Considered: Including Dr. Hunter's Tract on Child Murder, with Illustrative Notes; and a Summary of the Present State of Medico-Legal Knowledge on that Subject* (London: Longman, Rees, Orme, Brown, Green, and Longman, 1836).

'Insanity from Chloroform Employed in Parturition', *Journal of Psychological Medicine and Mental Pathology*, 3 (1850), 269–70.

Kidd, Charles, 'On Chloroform and Some of its Clinical Uses', *London Medical Review or Monthly Journal of Medical and Surgical Science*, II (June 1862), 243–7.

Kraepelin, Emil, *Lectures on Clinical Psychiatry* (London: Tindall and Cox, 1904).

Lane, Edward B., 'Puerperal Insanity', *Boston Medical and Surgical Journal*, 144 (20 June 1901), 606–9.

Laycock, Thomas, *A Treatise on the Nervous Diseases of Women; Comprising an Inquiry into the Nature, Causes, and Treatment of Spinal and Hysterical Disorders* (London: Longman, Orme, Brown, Green, and Longmans, 1840).

Leake, John, *A Lecture Introductory to the Theory and Practice of Midwifery* (London: R. Baldwin, 1773).

Lee, Robert, *Lectures on the Theory and Practice of Midwifery, Delivered in the Theatre of St. George's Hospital* (London: Longman, Brown, Green, and Longmans, 1844).

Lee, Robert, *Clinical Midwifery: Comprising the Histories of Five Hundred and Forty-Five Cases of Difficult, Preternatural, and Complicated Labour. With Commentaries*, second edition (London: John Churchill, 1848).

Lee, Robert, *Three Hundred Consultations in Midwifery* (London: John Churchill, 1864).

Lightfoot, Thomas, 'Puerperal Mania: It's Nature and Treatment', *Medical Times*, 21 (6 April 1850), 273–6.

Mackenzie, F.W., 'On the Pathology and Treatment of Puerperal Insanity: Especially in Reference to its Relation with Anaemia', *London Journal of Medicine*, 3 (1851), 504–21.

Macleod, M.D., 'An Address on Puerperal Insanity', *British Medical Journal*, II (7 August 1886), 239–42.

Macmichael, W., *Lives of British Physicians* (London: John Murray, 1830).

Madden, T. M., 'On Puerperal Mania', *British and Foreign Medico-Chirurgical Review*, XLVII (1871), 477–95.

Marcé, L.-V., *Traité de la folie des femmes enceintes, des nouvelles accouchés et des nourrices* (Paris: J.B. Baillière, 1858).

Maunsell, Henry, *The Dublin Practice of Midwifery* (London: Longman, Brown, Green, Longmans, & Roberts, 1856).

Mauriceau, Francois, *The Diseases of Women with Child, And in Child-bed*, trans. Hugh Chamberlen (London: John Darby, 1683).

M'Clintock, Arthur H. and Samuel L. Hardy, *Practical Observations on Midwifery, and the Diseases Incident to the Puerperal State* (Dublin: Hodges and Smith, 1848).

Mears, Martha, *The Midwife's Candid Advice to the Fair Sex; or the Pupil of Nature* (London: Crosby and Co. and R. Faudler, c.1805).

Menzies, W.F., 'Puerperal Insanity: An Analysis of One Hundred and Forty Consecutive Cases', *American Journal of Insanity*, 50 (1893–94), 147–85.

Mercier, C., *A Textbook of Insanity* (London: Macmillan, 1902).

Morison, Alexander, *The Physiognomy of Mental Diseases* (London: n.p., 1838), pamphlet edition.

Morison, Alexander, *The Physiognomy of Mental Diseases*, second edition (London: Longman and Co. and S. Highley, 1843).

Pargeter, William, *Observations on Maniacal Disorders* (1792); reprint edited by Stanley W. Jackson (London and New York: Routledge, 1988).

Playfair, W.S., *A Treatise on the Science and Practice of Midwifery* second edition (London: Smith, Elder, 1878).

Prichard, James Cowles, *A Treatise on Insanity and Other Disorders Affecting the Mind* (London: Sherwood, Gilbert, and Piper, 1835).

Prichard, James Cowles, *On the Different Forms of Insanity in Relation to Jurisprudence, Designed for the Use of Persons concerned in Legal Questions Regarding Unsoundness of Mind* (London: Hippolyte Baillière, 1842).

Ramsbotham, Francis H., *The Principles and Practice of Obstetric Medicine and Surgery in Reference to the Process of Parturition*, third edition (London: John Churchill, 1851).

Ray, Isaac, *A Treatise on the Medical Jurisprudence of Insanity*, fourth edition (Boston: Little, Brown and Company, 1860).

Reid, James, 'On the Causes, Symptoms, and Treatment of Puerperal Insanity', *Journal of Psychological Medicine and Mental Pathology*, 1 (1848), 128–51, 284–94.

Ryan, Michael, *A Manual of Midwifery and Diseases of Women and Children*, fourth edition (London, published by the author, 1841).

Ryan, William Burke, *Infanticide: Its Law, Prevalence, Prevention and History* (London: J. Churchill, 1862).

Salter, Thomas, 'Case of Puerperal Mania, Occurring at an Early Period of Utero-Gestation, and Relieved by Induced Abortion', *Provincial Medical and Surgical Journal* (30 June 1847), 346–8.

Sankey, W.H.O., *Lectures on Mental Diseases* (London: John Churchill, 1866).

Savage, George H., *Insanity and Allied Neuroses: Practical and Clinical* (London and New York: Cassell, 1884).

Savage, George H., 'Puerperal Insanity', in D. Hack Tuke, *A Dictionary of Psychological Medicine*, 2 vols (London: J. & A. Churchill, 1892), vol. 2, pp. 1034–42.

Severn, Charles, *First Lines of the Practice of Midwifery: To Which are Added Remarks on the Forensic Evidence Requisite in Cases of Foeticide and Infanticide* (London: S. Highley, 1831).

Sharp, Jane, *The Midwives Book. Or the Whole Art of Midwifry Discovered* (London: Simon Miller, 1671).

Sharp, Jane, *The Midwives Book. Or the Whole Art of Midwifry Discovered* (1671), ed. Elaine Hobby (New York and Oxford: Oxford University Press, 1999).

Simpson, J.Y., 'Clinical Lectures on the Diseases of Women': Lecture XXXII: 'On Puerperal Mania', and Lecture XXXIII: 'Puerperal Insanity: Its Connection with Albuminuria', *Medical Times & Gazette*, II (1 September and 10 November 1860), 201–2, 445–7.

Simpson, James Y., *Clinical Lectures on the Diseases of Women* (Edinburgh: Adam and Charles Black, 1872).

Sinclair, E.B. and G. Johnston, *Practical Midwifery: Comprising an Account of 13,748 Deliveries which Occurred in the Dublin Lying-in Hospital, during a Period of Seven Years, Commencing November 1847* (London: John Churchill, 1858).

Skae, David, 'A Rational and Practical Classification of Insanity', *Journal of Mental Science*, 9 (1863), 309–19.

[the late] Skae, David, 'The Morisonian Lectures on Insanity for 1873' (edited by T.S. Clouston), *Journal of Mental Science*, 19 (1873), 340–55, 20 (1874), 1–20.

Smellie, William, *A Treatise on the Theory and Practice of Midwifery*, 3 vols (London: D. Wilson and T. Durham, 1756, 1757, 1764).

Smith, R. Percy, 'Insanity of Pregnancy, the Puerperal Period, and Lactation', in *Quain's Dictionary of Medicine*, third edition (London: Longmans, Green, and Co., 1902), pp. 758–61.

Smith, W. Tyler, 'Lectures on Parturition, and the Principles & Practice of Obstetricy', *Lancet*, II (29 July 1848), 117–20.

Smith, W. Tyler, 'Puerperal Mania', Lecture XXXIX: 'Lectures on the Theory and Practice of Obstetrics', *Lancet*, II (18 October 1856), 423–5.

Taylor, Alfred Swaine, *The Principles and Practice of Medical Jurisprudence*, eighth edition (London: John Churchill, 1865).

Thomson, John, 'Statistical Report of Three Thousand Three Hundred Cases of Obstetricy', *Glasgow Medical Journal*, 3 (1855), 129–50.

Tilt, E.J., *On the Preservation of the Health of Women at the Critical Periods of Life* (London: John Churchill, 1851).

Trotter, Thomas, *A View of the Nervous Temperament; being a Practical Inquiry into the Increasing Prevalence, Prevention, Treatment of those Diseases commonly called Nervous, Bilious, Stomach & Liver Complaints; Indigestion; Low Spirits, Gout, &c.* (Boston: Wright, Goodenow and Stockwell, 1808).

Tuke, J.B., 'On the Statistics of Puerperal Insanity as Observed in the Royal Edinburgh Asylum, Morningside', *Edinburgh Medical Journal*, 10 (1864–65), 1013–28.

Tuke, J.B., 'Cases Illustrative of the Insanity of Pregnancy, Puerperal Mania, and Insanity of Lactation', *Edinburgh Medical Journal*, 12 (1866–67), 1083–101.

Waters, A.T.H., 'On the Use of Chloroform in the Treatment of Puerperal Insanity', *Journal of Psychological Medicine and Mental Pathology*, 10 (1857), 123–35.

West, R.U., 'Fatal and Other Cases of Puerperal Mania', *Association Medical Journal*, 2 (11 August 1854), 716–18.

Wiglesworth, Joseph, 'Puerperal Insanity: An Analysis of Seventy-Three Cases of the Insanities of Pregnancy, Parturition, and Lactation', *Transactions of the Liverpool Medical Institution*, 6 (1885–86), 349–62.

Winslow, Forbes, 'On Puerperal Insanity', *Journal of Psychological Medicine and Mental Pathology*, 12 (1859), 9–38.

Young, James, 'Case of Puerperal Mania', *Edinburgh Medical Journal*, 13 (1867–68), 262–6.

Secondary literature

Adair, Richard, Bill Forsythe and Joseph Melling, 'A Danger to the Public? Disposing of Pauper Lunatics in Late-Victorian and Edwardian England: Plympton St Mary Union and the Devon County Asylum, 1867–1914', *Medical History*, 42 (1998), 1–25.

Anderson, Olive, *Suicide in Victorian and Edwardian England* (Oxford: Clarendon, 1987).

Andrews, Jonathan, 'Case Notes, Case Histories, and the Patient's Experience of Insanity at Gartnavel Royal Asylum, Glasgow, in the Nineteenth Century', *Social History of Medicine*, 11 (1998), 255–81.

Andrews, Jonathan, *'They're in the Trade ... of Lunacy, They "cannot interfere" – they say': The Scottish Lunacy Commissioners and Lunacy Reform in Nineteenth-Century Scotland* (London: Wellcome Institute for the History of Medicine, Occasional Publications, No. 8, 1998).

Andrews, Jonathan, 'The Boundaries of Her Majesty's Pleasure: Discharging Child-Murderers from Broadmoor and Perth Criminal Lunatic Asylum, c. 1860–1920', in Mark Jackson (ed.), *Infanticide: Historical Perspectives on Child Murder and Concealment, 1550–2000* (Aldershot: Ashgate, 2002), pp. 216–48.

Andrews, Jonathan and Anne Digby (eds), *Sex and Seclusion, Class and Custody: Perspectives on Gender and Class in the History of British and Irish Psychiatry* (Amsterdam and New York: Rodopi, 2004).

Andrews, Jonathan and Andrew Scull, *Customers and Patrons of the Mad-Trade: The Management of Lunacy in Eighteenth-Century London* (Berkeley, Los Angeles and London: University of California Press, 2003).

Apple, Rima D. (ed.), *Women, Health, and Medicine in America* (New Brunswick, NJ: Rutgers University Press, 1990)

Arnott, Meg, 'Infant Death, Child Care and the State: The Baby-Farming Scandal and the First Infant Life Protection Legislation of 1872', *Continuity and Change*, 9 (1994), 271–311.

Bailin, Miriam, *The Sickroom and Victorian Fiction: The Art of Being Ill* (Cambridge: Cambridge University Press, 1994).

Barfoot, M. and A. Beveridge, 'Madness at the Crossroads: John Home's Letters from the Royal Edinburgh Asylum, 1886–87', *Psychological Medicine*, 20 (1990), 263–84.

Barfoot, M. and A. Beveridge, '"Our Most Notable Inmate": John Willis Mason at the Royal Edinburgh Asylum, 1864–1901', *History of Psychiatry*, 4 (1993), 159–208.

Barker-Benfield, G.J., *The Horrors of the Half-Known Life* (New York: Harper and Row, 1976).

Bartlett, Peter, *The Poor Law of Lunacy: The Administration of Pauper Lunatics in Mid-Nineteenth-Century England* (London and Washington: Leicester University Press, 1999).

Bartlett, Peter and David Wright (eds), *Outside the Walls of the Asylum: The History of Care in the Community 1750–2000* (London: Athlone, 1999).

Bauer, Dale M. (ed.), Charlotte Perkins Gilman, *The Yellow Wallpaper* (Boston and New York: Bedford, 1998).

Behlmer, George K., 'Deadly Motherhood: Infanticide and Medical Opinion in Mid-Victorian England', *Journal of the History of Medicine*, 34 (1979), 403–27.

Behlmer, George K., *Child Abuse and Moral Reform in England, 1870–1908* (Stanford: Stanford University Press, 1982).

Beier, L.M., *Sufferers and Healers: The Experience of Illness in Seventeenth-Century England* (London: Routledge & Kegan Paul, 1987).

Berridge, Virginia, *Opium and the People: Opiate Use and Drug Control Policy in Nineteenth and Early Twentieth Century England*, revised edition (London and New York: Free Association Books, 1999; first published New Haven and London: Yale University Press, 1987).

Berrios, G.E., 'Delusions as "Wrong Beliefs": A Conceptual History', *British Journal of Psychiatry*, 159, Supplement 14 (1991), 6–13.

Berrios, G.E., introduction to 'J.C. Prichard and the Concept of "Moral Insanity"', *History of Psychiatry*, 10 (1999), 111–26.

Beveridge, Allan, 'Madness in Victorian Edinburgh: A Study of Patients Admitted to the Royal Edinburgh Asylum under Thomas Clouston, 1873–1908', Parts I and II, *History of Psychiatry*, 6 (1995), 21–54, 133–56.

Beveridge, Allan, 'Voices of the Mad: Patients' Letters from the Royal Edinburgh Asylum, 1873–1908', *Psychological Medicine*, 27 (1997), 899–908.

Beveridge, Allan, 'Life in the Asylum: Patient's Letters from Morningside, 1873–1908', *History of Psychiatry*, 9 (1998), 431–69.

Bland, Lucy, *Banishing the Beast: English Feminism and Sexual Morality 1885–1914* (London: Penguin, 1995).

Branca, Patricia, *Silent Sisterhood: Middle-Class Women in the Victorian Home* (London: Croom Helm, 1975).

Brockbank, William, *Ancient Therapeutic Arts* (London: William Heinemann, 1954).

Brockington, I.F., E.M. Schofield, P. Donnelly and C. Hyde, 'A Clinical Study of Post-Partum Psychosis', in M. Sandler (ed.), *Mental Illness in Pregnancy and the Puerperium* (Oxford: Oxford University Press, 1978), pp. 59–68.

Brockington, I.F. and R. Kumar (eds), *Motherhood and Mental Illness* (London and New York: Academic Press/Grune and Stratton, 1982).

Burney, Ian, *Bodies of Evidence: Medicine and the Politics of the English Inquest 1830–1926* (Baltimore: Johns Hopkins University Press, 2000).

Burrows, Adrienne and Iwan Schumacher, *Portraits of the Insane: The Case of Dr Diamond* (London: Quartet Books, 1990).

Busfield, Joan, *Men, Women and Madness: Understanding Gender and Mental Disorder* (London: Macmillan, 1996).

Bynum, William F., 'Rationales for Therapy in British Psychiatry, 1780–1835', *Medical History*, 18 (1974), 317–34.

Bynum, W.F., R. Porter and M. Shepherd (eds), *The Anatomy of Madness: Essays in the History of Psychiatry*, vol. I, *People and Ideas*, vol. II, *Institutions and Society*, vol. III, *The Asylum and Its Psychiatry* (London and New York: Tavistock, 1985, 1988).

Clark, Anna, *Women's Silence, Men's Violence: Sexual Assault in England, 1770–1845* (London: Pandora, 1987).

Clarke, Micael M., *Thackeray and Women* (DeKalb: North Illinois Press, 1995).

Crawford, Patricia, 'The Construction and Experience of Maternity in Seventeenth-Century England', in Valerie Fildes (ed.), *Women as Mothers in Pre-Industrial England* (London and New York: Routledge, 1990), pp. 3–38.

Croxson, Brownyn, 'The Foundation and Evolution of the Middlesex Hospital's Lying-in Service, 1745–86', *Social History of Medicine*, 14 (2001), 27–57.

Curry, Kenneth (ed.), *New Letters of Robert Southey*, 2 vols (New York and London: Columbia University Press, 1965).

Dally, Ann, *Women under the Knife: A History of Surgery* (London: Hutchinson Radius, 1991).

Davidoff, Lenore and Catherine Hall, *Family Fortunes: Men and Women of the English Middle Class 1780–1850* (London: Hutchinson, 1987).

Digby, Anne, *Madness, Morality and Medicine: A Study of the York Retreat, 1796–1914* (Cambridge: Cambridge University Press, 1985).

Digby, Anne, 'Moral Treatment at the Retreat, 1796–1846', in W.F. Bynum, R. Porter and M. Shepherd (eds), *Anatomy of Madness: Essays in the History of Psychiatry*, vol. II, *Institutions and Society* (London and New York: Tavistock, 1985), pp. 52–72.

Digby, Anne, 'Women's Biological Straitjacket', in Susan Mendus and Jane Rendall (eds), *Sexuality and Subordination: Interdisciplinary Studies of Gender in the Nineteenth Century* (London and New York: Routledge, 1989), pp. 192–220.

Digby, Anne, *Making a Medical Living: Doctors and Patients in the English Market for Medicine, 1720–1911* (Cambridge: Cambridge University Press, 1994).

Donnison, Jean, *Midwives and Medical Men: A History of Inter-Professional Rivalries and Women's Rights* (London: Heinemann, 1977; second edition New Barnett: Historical Publications, 1988).

Eigen, Joel Peter, '"I answer as a physician": Opinion as Fact in Pre-McNaughton Insanity Trials', in Michael Clark and Catherine Crawford (eds), *Legal Medicine in History* (Cambridge: Cambridge University Press, 1994), pp. 167–99.

Eigen, Joel Peter, *Witnessing Insanity: Madness and Mad-Doctors in the English Court* (New Haven and London: Yale University Press, 1995).

Eigen, Joel Peter, *Unconscious Crime: Mental Absence and Criminal Responsibility in Victorian London* (Baltimore and London: Johns Hopkins University Press, 2003).

Evenden, Doreen, *The Midwives of Seventeenth-Century London* (Cambridge: Cambridge University Press, 2000).

Figlio, Karl, 'Chlorosis and Chronic Disease in Nineteenth-Century Britain: The Social Constitution of Somatic Illness in a Capitalist Society', *Social History*, 3 (1978), 167–97.

Fildes, Valerie (ed.), *Women as Mothers in Pre-Industrial England* (London and New York: Routledge, 1990).

Finn, Margot, 'Working-Class Women and the Contest for Consumer Control in Victorian County Courts', *Past & Present*, 161 (1998), 116–54.

Fitzgerald, Brian (ed.), *Correspondence of Emily, Duchess of Leinster (1731–1814)*, vols I–III (Dublin: Stationery Office, 1949, 1953, 1957).

Freeman, Phyllis R., Carley Rees Bogarad and Diane E. Sholomskas, 'Margery Kempe, a New Theory: The Inadequacy of Hysteria and Postpartum Psychosis as Diagnostic Categories', *History of Psychiatry*, 1 (1990), 169–90.

Fulford, Roger (ed.), *Dearest Child: Letters between Queen Victoria and the Princess Royal 1858–1861* (London: Evans, 1964).

Gale, Colin and Robert Howard, *Presumed Curable: An Illustrated Casebook of Victorian Psychiatric Patients in Bethlem Hospital* (Petersfield and Philadelphia: Wrightson Biomedical Publishing, 2003).

Garfinkel, Harold, *Studies in Ethnomethodology* (Englewood Cliffs, NJ: Prentice Hall, 1967).

Gelder, Michael, Richard Mayou and Philip Cowen (eds), *Shorter Oxford Textbook of Psychiatry*, fourth edition (Oxford: Oxford University Press, 2001).

Gilbert, Pamela K., *Disease, Desire, and the Body in Victorian Women's Popular Novels* (Cambridge: Cambridge University Press, 1997).

Gilman, Sander L. (ed.), *The Face of Madness: Hugh W. Diamond and the Origin of Psychiatric Photography* (Secaucus, NJ: Citadel, 1976).

Gilman, Sander L., *Seeing the Insane* (Lincoln and London: University of Nebraska Press, 1996; first published John Wiley & Sons: New York, 1982).

Gilman, Sander L., *Disease and Representation: Images of Illness from Madness to AIDS* (Ithaca, NY and London: Cornell University Press, 1988).

Gilman, Sander, Helen King, Roy Porter, George S. Rousseau and Elaine Showalter, *Hysteria Beyond Freud* (Berkeley, CA: University of California Press, 1993).

Haley, Bruce, *The Healthy Body and Victorian Culture* (Cambridge, Mass and London: Harvard University Press, 1978).

Hall, Catherine, *White, Male and Middle Class: Explorations in Feminism and History* (Cambridge: Polity, 1988).

Hamilton, Elizabeth, *The Warwickshire Scandal* (Wilby, Norwich: Michael Russell, 1999).

Hammerton, A. James, *Cruelty and Companionship: Conflict in Nineteenth-Century Married Life* (London and New York: Routledge, 1992).

Harley, David, 'Provincial Midwives in England: Lancashire and Cheshire, 1660–1760', in Hilary Marland (ed.), *The Art of Midwifery: Early Modern Midwives in Europe* (London and New York: Routledge, 1993, 1994), pp. 27–48.

Henderson, D.K., *The Evolution of Psychiatry in Scotland* (Edinburgh and London: E. & S. Livingstone, 1964).

Higginbotham, Ann R., '"Sin of the Age": Infanticide and Illegitimacy in Victorian London', in Kristine Ottesen Garrigan (ed.), *Victorian Scandals: Representations of Gender and Class* (Athens: Ohio University Press, 1992), pp. 257–88.

Hoffer, Peter C. and N.E.H. Hull, *Murdering Mothers: Infanticide in England and New England 1558–1803* (New York: New York University Press, 1981).

Hoffert, Sylvia D., *Private Matters: American Attitudes toward Childbearing and Infant Nurture in the Urban North, 1800–1860* (Urbana and Chicago: University of Illinois Press, 1998).

Holland, Eardley, 'The Princess Charlotte of Wales: A Triple Obstetric Tragedy', *Journal of Obstetrics & Gynaecology of the British Empire*, 58 (1951), 905–19.

Hunter, Richard and Ida Macalpine, *Three Hundred Years of Psychiatry 1535–1860* (London: Oxford University Press, 1963).

Hunter, Richard and Ida Macalpine, *Psychiatry for the Poor, 1851. Colney Hatch Asylum, Friern Hospital 1973: A Medical and Social History* (London: Dawsons, 1974).

Ingram, Allan (ed.), *Voices of Madness: Four Pamphlets, 1683–1796* (Stroud: Sutton, 1997).

Ingram, Allan (ed.), *Patterns of Madness in the Eighteenth Century: A Reader* (Liverpool: Liverpool University Press, 1998).

Jackson, Mark, 'Images of Deviance: Visual Representations of Mental Defectives in Early Twentieth-Century Medical Texts', *British Journal of the History of Science*, 28 (1995), 319–37.

Jackson, Mark, *New-Born Child Murder* (Manchester and New York: Manchester University Press, 1996).

Jackson, Mark (ed.), *Infanticide: Historical Perspectives on Child Murder and Concealment, 1550–2000* (Aldershot: Ashgate, 2002).

Jalland, Pat and John Hooper, *Women from Birth to Death: The Female Life Cycle in Britain 1830–1914* (Brighton: Harvester Press, 1986).

Jones, Kathleen, *Asylums and After: A Revised History of the Mental Health Services: From the Early 18th Century to the 1990s* (London: Athlone, 1993).

Jones, Kathleen, *A Passionate Sisterhood: The Sisters, Wives and Daughters of the Lake Poets* (London: Virago, 1998).

King, Helen, *Hippocrates' Woman: Reading the Female Body in Ancient Greece* (London: Routledge, 1998).

King, Helen, *The Disease of Virgins: Green Sickness, Chlorosis and the Problems of Puberty* (London and New York: Routledge, 2004).

Krueger, Christine L., 'Literary Defenses and Medical Prosecutions: Representing Infanticide in Nineteenth–Century Britain', *Victorian Studies*, 40 (1997), 271–94.

Lane, Joan, '"The Doctor Scolds Me": The Diaries and Correspondence of Patients in Eighteenth Century England', in Roy Porter (ed.), *Patients and Practitioners: Lay Perceptions of Medicine in Pre-Industrial Society* (Cambridge: Cambridge University Press, 1985), pp. 205–48.

Lane, Joan, *The Making of the English Patient* (Stroud: Sutton, 2000).

Laurence, Anne, 'Women's Psychological Disorders in Seventeenth-Century Britain', in Arina Angerman et al., *Current Issues in Women's History* (London and New York: Routledge, 1989), pp. 203–19.

Laurence, Anne, *Women in England 1500–1760: A Social History* (London: Phoenix, 1996).

Leavitt, Judith Walzer (ed.), *Women and Health in America* (Madison: University of Wisconsin Press, 1984).

Leavitt, Judith Walzer and Whitney Walton, '"Down to Death's Door": Women's Perceptions of Childbirth in America', in Judith Walzer Leavitt (ed.), *Women and Health in America* (Madison: University of Wisconsin Press, 1984), pp. 155–74.

Leavitt, Judith Walzer, *Brought to Bed: Childbearing in America, 1750–1950* (New York and Oxford: Oxford University Press, 1986).

Lefebure, Molly, *Samuel Taylor Coleridge: A Bondage of Opium* (London: Victor Gollancz, 1974).

Levine-Clark, Marjorie, 'Dysfunctional Domesticity: Female Insanity and Family Relationships among the West Riding Poor in the Mid-Nineteenth-Century', *Journal of Family History*, 25 (2000), 341–61.

Lewis, Jane, *Women in England 1870–1950: Sexual Divisions and Social Change* (Brighton: Wheatsheaf, 1984).

Lewis, Judith Schneid, *In the Family Way: Childbearing in the British Aristocracy, 1760–1860* (New Brunswick, NJ: Rutgers, 1986).

Loudon, I.S.L., 'Chlorosis, Anaemia, and Anorexia Nervosa', *British Medical Journal*, 281 (20–27 December 1980), 1669–75.

Loudon, Irvine, *Medical Care and the General Practitioner 1750–1850* (Oxford: Clarendon Press, 1986).

Loudon, Irvine, 'Puerperal Insanity in the 19th Century', *Journal of the Royal Society of Medicine*, 81 (February 1988), 76–9.

Loudon, Irvine, *Death in Childbirth: An International Study of Maternal Care and Maternal Mortality 1800–1950* (Oxford: Clarendon, 1992).

Loudon, Irvine, 'Childbirth', in W.F. Bynum and Roy Porter (eds), *Companion Encyclopedia of the History of Medicine*, vol. 2 (London and New York: Routledge, 1993), pp. 1050–71.

Loudon, Irvine (ed.), *Childbed Fever: A Documentary History* (New York and London: Garland, 1995).

London, Irvine, *The Tragedy of Childhood Fever* (Oxford: Oxford University Press, 2000).

Lunbeck, Elizabeth, *The Psychiatric Persuasion: Knowledge, Gender, and Power in Modern America* (Princeton, NJ: Princeton University Press, 1994).

McDonagh, Josephine, *Child Murder and British Culture 1720–1900* (Cambridge: Cambridge University Press, 2003).

MacDonald, Michael, *Mystical Bedlam: Madness, Anxiety, and Healing in Seventeenth-Century England* (Cambridge: Cambridge University Press, 1981).

MacKenzie, Charlotte, 'Women and Psychiatric Professionalization 1780–1914', in London Feminist History Collective (eds), *The Sexual Dynamics of History* (London: Pluto Press, 1983), pp. 107–19.

MacKenzie, Charlotte, *Psychiatry for the Rich: A History of Ticehurst Private Asylum, 1792–1917* (London and New York: Routledge, 1992).

MacLeod, Roy, *Government and Expertise: Specialists, Administrators and Professionals, 1860–1919* (Cambridge and New York: Cambridge University Press, 1988).

Manuel, Diana E., *Marshall Hall 1790–1857* (Amsterdam and Atlanta, GA: Rodopi, 1996).

Marland, Hilary (ed.), *The Art of Midwifery: Early Modern Midwives in Europe* (London and New York: Routledge, 1993, 1994).

Marland, Hilary and Anne-Marie Rafferty (eds), *Midwives, Society and Childbirth: Debates and Controversies in the Modern Period* (London and New York: Routledge, 1997).

Marland, Hilary, '"Destined to a Perfect Recovery": The Confinement of Puerperal Insanity in the Nineteenth Century', in Joseph Melling and Bill Forsythe (eds), *Insanity, Institutions and Society, 1800–1914* (London and New York: Routledge, 1999), pp. 137–56.

Marland, Hilary, 'At Home with Puerperal Mania: The Domestic Treatment of Insanity of Childbirth in the Nineteenth Century', in Peter Bartlett and David Wright (eds), *Outside the Walls of the Asylum: The History of Care in the Community 1750–2000* (London: Athlone Press, 1999), pp. 45–65.

Marland, Hilary '"Uterine Mischief": W.S. Playfair and his Neurasthenic Patients', in Marijke Gijswijt-Hofstra and Roy Porter (eds), *Cultures of Neurasthenia From Beard to the First World War* (Amsterdam and New York: Rodopi, 2001), pp. 117–39.

McGovern, Constance M., 'The Myths of Social Control and Custodial Oppression: Patterns of Psychiatric Medicine in Late Nineteenth-Century Institutions', *Journal of Social History*, 20 (1986), 3–23.

McGregor, Deborah Kuhn, *From Midwives to Medicine: The Birth of American Gynecology* (New Brunswick, NJ and London: Rutgers University Press, 1998).

Meiners, Katherine T., 'Imagining Cancer: Sara Coleridge and the Environment of Illness', *Literature and Medicine*, 15 (1996), 48–63.

Melling, Joseph and Bill Forsythe (eds), *Insanity, Institutions and Society, 1800–1914* (London and New York: Routledge, 1999).

Melling, Joseph, Bill Forsythe and Richard Adair, 'Families, Communities and the Legal Regulation of Lunacy in Victorian England: Assessments of Crime, Violence and Welfare in Admissions to the Devon Asylum, 1845–1914', in Peter Bartlett and David Wright (eds), *Outside the Walls of the Asylum: A History of Care in the Community 1750–2000* (London and New Brunswick, NJ: Athlone, 1999), pp. 153–80.

Mitchinson, Wendy, *The Nature of their Bodies: Women and their Doctors in Victorian Canada* (Toronto: University of Toronto Press, 1991).

Monsarrat, Ann, *An Uneasy Victorian: Thackeray the Man 1811–1863* (London: Cassell, 1980).

Moran, James E., 'Asylum in the Community: Managing the Insane in Antebellum America', *History of Psychiatry*, 9 (1998), 217–40.

Morantz, Regina Markell, 'The Perils of Feminist History', in Judith Walzer Leavitt (ed.), *Women and Health in America* (Madison: University of Wisconsin Press, 1984), pp. 239–45.

Moscucci, Ornella, *The Science of Woman: Gynaecology and Gender in England 1800–1929* (Cambridge: Cambridge University Press, 1990).

Mudge, Bradford Keyes, *Sara Coleridge, A Victorian Daughter: Her Life and Essays* (New Haven and London: Yale University Press, 1980).

Nowell-Smith, Harriet, 'Nineteenth-Century Narrative Case Histories: An Inquiry into Stylistics and History', *Canadian Bulletin of Medical History*, 12 (1995), 47–67.

O'Dowd, Michael J. and Elliot E. Philipp, *The History of Obstetrics and Gynaecology* (New York and London: Parthenon, 1994).

Oppenheim, Janet, *'Shattered Nerves': Doctors, Patients, and Depression in Victorian England* (New York and Oxford: Oxford University Press, 1991).

Parry-Jones, William Li., *The Trade in Lunacy: A Study of Private Madhouses in England in the Eighteenth and Nineteenth Centuries* (London: Routledge & Kegan Paul, 1972).

Persaud, Rajendra, 'A Comparison of Symptoms Recorded from the Same Patients by an Asylum Doctor and a "Constant Observer" in 1823: The Implications for Theories about Psychiatric Illness in History', *History of Psychiatry*, 3 (1992), 79–94.

Peterson, Dale (ed.), *A Mad People's History of Madness* (Pittsburgh, PA: University of Pittsburgh Press, 1981).

Pollock, Linda, 'Embarking on a Rough Passage: The Experience of Pregnancy in Early-Modern Society', in Valerie Fildes (ed.), *Women as Mothers in Pre-Industrial England* (London and New York: Routledge, 1990), pp. 39–67.

Poovey, Mary, '"Scenes of an Indelicate Character": The Medical "Treatment" of Victorian Women', *Representations*, 14 (1986), 137–68.

Poovey, Mary, *Uneven Developments: The Ideological Work of Gender in Mid-Victorian England* (London: Virago, 1989; first published Chicago: University of Chicago Press, 1988).

Poovey, Mary, *Making a Social Body: British Cultural Formation, 1830–1864* (Chicago and London: University of Chicago Press, 1995).

Porter, Roy, 'The Patient's View: Doing Medical History from Below', *Theory and Society*, 14 (1985), 175–98.

Porter, Roy, *A Social History of Madness: Stories of the Insane* (London: Weidenfeld and Nicolson, 1987).

Porter, Roy (ed.), *The Faber Book of Madness* (London: Faber and Faber, 1991).

Porter, Roy, 'Hearing the Mad: Communication and Excommunication', in L. de Goei and J. Vijselaar (eds), *Proceedings of the First European Congress on the History of Psychiatry and Mental Health Care* (Rotterdam: Erasmus Publishing, 1993), pp. 338–52.

Porter, Roy and Dorothy Porter, *In Sickness and in Health: The British Experience 1650–1850* (London: Forth Estate, 1988).

Potter, Stephen (ed.), *Minnow among Tritons: Mrs. S.T. Coleridge's Letters to Thomas Poole* (London: Nonesuch Press, 1934).

Prior, Lindsay, *The Social Organisation of Mental Illness* (London: Sage, 1993).

Prochaska, Frank, *Women and Philanthropy in 19th Century England* (Oxford: Clarendon Press, 1980).

Quinn, Cath, 'Images and Impulses: Representations of Puerperal Insanity and Infanticide in Late Victorian England', in Mark Jackson (ed.), *Infanticide: Historical Perspectives on Child Murder and Concealment, 1550–2000* (Aldershot: Ashgate, 2002), pp. 193–215.

Rabin, Dana, 'Bodies of Evidence, States of Mind: Infanticide, Emotion and Sensibility in Eighteenth-Century England', in Mark Jackson (ed.), *Infanticide: Historical Perspectives on Child Murder and Concealment, 1550–2000* (Aldershot: Ashgate, 2002), pp. 73–92.

Ray, Gordon N. (ed.), *The Letters and Private Papers of William Makepeace Thackery*, 4 vols (London: Oxford University Press, 1945), vols 1 and 2.

Ray, Gordon N., *Thackeray: The Uses of Adversity (1811–1846)* (London: Oxford University Press, 1955).

Ray, Laurence J., 'Models of Madness in Victorian Asylum Practice', *Archives of European Sociology*, 22 (1981), 229–64.

Raymond, John (ed.), *Queen Victoria's Early Letters*, revised. edition (London: B.T. Batsford, 1963).

Rehman, A., D. St. Clair and C. Platz, 'Puerperal Insanity in the Nineteenth and Twentieth Centuries', *British Journal of Psychiatry*, clvi (1990), 861–5.

Ripa,Yannick, *Women and Madness: The Incarceration of Women in Nineteenth-Century France* (Cambridge: Polity Press, 1990).

Risse, Guenter B., *Hospital Life in Enlightenment Scotland* (Cambridge: Cambridge University Press, 1986).

Risse, Guenter B. and John Harley Warner, 'Reconstructing Clinical Activities: Patient Records in Medical History', *Social History of Medicine*, 5 (1992), 183–205.

Romito, Patrizia, 'Postpartum Depression and the Experience of Motherhood', *Acta Obstetricia et Gynecologica Scandinavica*, 69, Supplement 154 (1990), 7–37.

Roodenburg, Herman W., 'The Maternal Imagination: The Fears of Pregnant Women in Seventeenth-Century Holland', *Journal of Social History*, 21 (1988), 701–16.

Rose, Lionel, *Massacre of the Innocents: Infanticide in Great Britain 1800–1939* (London: Routledge & Kegan Paul, 1986).

Rosner, Lisa, *Medical Education in the Age of Improvement* (Edinburgh: Edinburgh University Press, 1991).

Ross, Ellen, *Love and Toil: Motherhood in Outcast London, 1870–1918* (New York and Oxford: Oxford University Press, 1993).

Roy, Judith M., 'Surgical Gynaeocology', in Rima Apple (ed.), *Women, Health, and Medicine in America* (New Brunswick, NJ: Rutgers University Press, 1990), pp. 173–95.

Saunders, Janet, 'Quarantining the Weak-Minded: Psychiatric Definitions of Degeneracy and the Late-Victorian Asylum', in W.F. Bynum, R. Porter and M. Shepherd (eds), *Anatomy of Madness: Essays in the History of Psychiatry*, vol. III, *The Asylum and its Psychiatry* (London and New York: Tavistock, 1988), pp. 273–96.

Scull, Andrew, 'Moral Treatment Reconsidered: Some Sociological Comments on an Episode in the History of British Psychiatry', in idem (ed.), *Madhouses,*

Mad-Doctors and Madmen: The Social History of Psychiatry in the Victorian Era (London: Athlone, 1981), pp. 105–20.

Scull, Andrew, 'The Domestication of Madness', *Medical History*, 27 (1983), 233–48.

Scull, Andrew, *The Most Solitary of Afflictions: Madness and Society in Britain, 1700–1900* (New Haven and London: Yale University Press, 1993).

Scull, Andrew, Charlotte MacKenzie and Nicholas Hervey, *Masters of Bedlam: The Transformation of the Mad-Doctoring Trade* (Princeton, NJ: Princeton University Press, 1996).

Shorter, Edward, *A History of Psychiatry: From the Era of the Asylum to the Age of Prozac* (New York, etc.: John Wiley, 1997).

Showalter, Elaine, *The Female Malady: Women, Madness and English Culture, 1830–1980* (London: Virago, 1987, first published New York: Pantheon, 1985).

Showalter, Elaine, *Hystories: Hysterical Epidemics and Modern Culture* (New York: Columbia University Press, 1997).

Sicherman, Barbara, 'The Uses of a Diagnosis: Doctors, Patients, and Neurasthenia', *Journal of the History of Medicine*, 32 (1977), 33–54.

Skultans, Vieda, *Madness and Morals: Ideas on Insanity in the Nineteenth Century* (London and Boston: Routledge & Kegan Paul, 1975).

Skultans, Vieda, *English Madness: Ideas on Insanity 1580–1890* (London: Routledge & Kegan Paul, 1979).

Small, Helen, *Love's Madness: Medicine, the Novel, and Female Insanity, 1800–1865* (Oxford: Clarendon, 1996).

Smith, Roger, 'The Boundary between Insanity and Criminal Responsibility in Nineteenth-Century England', in Andrew Scull (ed.), *Madhouses, Mad-Doctors and Madmen: The Social History of Psychiatry in the Victorian Era* (London: Athlone, 1981), pp. 363–84.

Smith, Roger, *Trial by Medicine: Insanity and Responsibility in Victorian Trials* (Edinburgh: Edinburgh University Press, 1981).

Smith-Rosenberg, Carroll, 'The Hysterical Woman: Sex Roles and Role Conflict in 19th Century America', *Social Research*, 39 (1979), 652–78.

Spinelli, Margaret G., *Infanticide: Psychosocial and Legal Perspectives on Mothers Who Kill* (Washington, DC and London: American Psychiatric Publishing, 2003).

Stevenson, Christine, *Medicine and Magnificence: British Hospital and Asylum Architecture 1660–1815* (New Haven and London: Yale University Press, 2000).

Summers, Anne, 'A Home from Home – Women's Philanthropic Work in the Nineteenth Century', in Sandra Burman (ed.), *Fit Work for Women* (London: Croom Helm, 1979), pp. 33–63.

Suzuki, Akihito, 'The Politics and Ideology of Non-Restraint: The Case of the Hanwell Asylum', *Medical History*, 39 (1995), 1–17.

Suzuki, Akihito, 'Framing Psychiatric Subjectivity: Doctor, Patient and Record-Keeping at Bethlem in the Nineteenth Century', in Joseph Melling and Bill Forsythe (eds), *Insanity, Institutions and Society, 1800–1914* (London and New York: Routledge, 1999), pp. 115–36.

Theriot, Nancy, 'Diagnosing Unnatural Motherhood: Nineteenth-Century Physicians and "Puerperal Insanity"', *American Studies*, 26 (1990), 69–88.

Theriot, Nancy, 'Negotiating Illness: Doctors, Patients and Families in the Nineteenth Century', *Journal of the History of Behavioral Sciences*, 37 (2001), 349–68.

Tillyard, Stella, *Aristocrats: Caroline, Emily, Louisa and Sarah Lennox 1740–1832* (London: Chatto & Windus, 1994).

Todd, Dennis, *Imagining Monsters: Miscreations of the Self in Eighteenth-Century England* (Chicago and London: University of Chicago Press, 1995).

Tomes, Nancy, *A Generous Confidence: Thomas Story Kirkbride and the Art of Asylum-Keeping, 1840–1883* (Cambridge: Cambridge University Press, 1984).

Tomes, Nancy, 'Historical Perspectives on Women and Mental Illness', in Rima D. Apple (ed.), *Women, Health, and Medicine in America* (New Brunswick, NJ: Rutgers University Press, 1990), pp. 143–71.

Trumbach, Randolph, *The Rise of the Egalitarian Family: Aristocratic Kinship and Domestic Relations in Eighteenth-Century England* (New York: Academic Press, 1978).

Turner, Trevor, '"Not Worth Powder and Shot": The Public Profile of the Medico-Psychological Association, c.1851–1914', in G.E. Berrios and H. Freeman (eds), *150 Years of British Psychiatry 1841–1991* (London: Athlone, 1991), pp. 3–16.

Turner, Trevor, *A Diagnostic Analysis of the Casebooks of Ticehurst House Asylum, 1845–1890, Psychological Medicine*, Monograph Supplement 21 (Cambridge: Cambridge University Press, 1992).

Ulrich, Laurel Thatcher, *A Midwife's Tale: The Life of Martha Ballard, Based on Her Diary, 1785–1812* (New York: Alfred Knopf, 1990).

Ussher, Jane, *Women's Madness: Misogyny or Mental Illness?* (New York, London, etc.: Harvester Wheatsheaf, 1991).

Versluysan, Margaret Connor, 'Midwives, Medical Men and "Poor Women Labouring of Child": Lying-In Hospitals in Eighteenth-Century London', in Helen Roberts (ed.), *Women, Health and Reproduction* (London: Routledge & Kegan Paul, 1981), pp. 18–49.

Vickery, Amanda, *The Gentleman's Daughter: Women's Lives in Georgian England* (New Haven and London: Yale University Press, 1998).

Walker, Nigel, *Crime and Insanity in England*, vol. 1: *The Historical Perspective* (Edinburgh: Edinburgh University Press, 1968).

Walsh, Oonagh, 'Lunatic and Criminal Alliances in Nineteenth-Century Ireland', in Peter Bartlett and David Wright (eds), *Outside the Walls of the Asylum: The History of Care in the Community 1750–2000* (London: Athlone, 1999), pp. 132–52.

Ward, Tony, 'Legislating for Human Nature: Legal Responses to Infanticide, 1860–1938', in Mark Jackson (ed.), *Infanticide: Historical Perspectives on Child Murder and Concealment, 1550–2000* (Aldershot: Ashgate, 2002), pp. 249–69.

Weeks, Jeffrey, *Sex, Politics and Society: The Regulation of Sexuality since 1800* (London: Longman, 1989).

Weiner, Martin J., 'Alice Arden to Bill Sykes: Changing Nightmares of Intimate Violence in England, 1558–1869', *Journal of British Studies*, 40 (2001), 184–212.

Whittle, Tyler, *Victoria and Albert at Home* (London and Henley: Routledge & Kegan Paul, 1980).

Williams, Perry, 'The Laws of Health: Women, Medicine and Sanitary Reform, 1850–1890', in Marina Benjamin (ed.), *Science and Sensibility: Gender and Scientific Enquiry 1780–1945* (Oxford: Basil Blackwood, 1991), pp. 60–88.

Wilson, Adrian, 'The Ceremony of Childbirth and Its Interpretation', in Valerie Fildes (ed.), *Women as Mothers in Pre-Industrial England* (London and New York: Routledge, 1990), pp. 68–107.

Wilson, Adrian, *The Making of Man-Midwifery: Childbirth in England, 1660–1770* (London: UCL Press, 1995).

Windeatt, B.A., translation and introduction to *The Book of Margery Kempe* (London: Penguin, 1985).

Wood, Ann Douglas, '"The Fashionable Diseases": Women's Complaints and their Treatment in Nineteenth-Century America', in Judith Walzer Leavitt (ed.), *Women and Health in America* (Madison: University of Wisconsin Press, 1984), pp. 222–38.

Wood, Jane, *Passion and Pathology in Victorian Fiction* (Oxford: Oxford University Press, 2001).

Woodham-Smith, Cecil, *Queen Victoria: Her Life and Times*, vol. 1, *1819–1861* (London: Hamish Hamilton, 1972).

Wright, David, 'Getting out of the Asylum: Understanding the Confinement of the Insane in the Nineteenth Century', *Social History of Medicine*, 10 (1997), 137–55.

Wright, David, 'The Certification of Insanity in Nineteenth-Century England and Wales', *History of Psychiatry*, 9 (1998), 267–90.

Wright, David, 'The Discharge of Pauper Lunatics from County Asylums in Mid-Victorian England: The Case of Buckinghamshire, 1853–1872', in Joseph Melling and Bill Forsythe (eds), *Insanity, Institutions and Society, 1800–1914* (London and New York: Routledge, 1999), pp. 93–112.

Wright, David, 'Delusions of Gender? Lay Identification and Clinical Diagnosis of Insanity in Victorian England', in Jonathan Andrews and Anne Digby (eds), *Sex and Seclusion, Class and Custody: Perspectives on Gender and Class in the History of British and Irish Psychiatry* (Amsterdam and New York: Rodopi, 2004), pp. 149–76.

Youngson, A.J., *The Scientific Revolution in Victorian Medicine* (London: Croom Helm, 1979).

Zedner, Lucia, *Women, Crime, and Custody in Victorian England* (Oxford: Oxford University Press, 1991).

Unpublished theses and dissertations

Day, Shelley, 'Puerperal Insanity: The Historical Sociology of a Disease', PhD thesis, University of Cambridge, 1985.

Ellis, Robert James, 'A Field of Practice or a Mere House of Detention? The Asylum and its Integration, with Special Reference to the County Asylums of Yorkshire, c.1844–1888', PhD thesis, University of Huddersfield, 2001.

Lockhart, Judith, '"Truly a Hospital for Women": The Birmingham and Midland Hospital for Women, 1871–1901', MA dissertation, University of Warwick, 2002.

MacKenzie, Charlotte, 'A Family Asylum: A History of the Private Madhouse at Ticehurst in Sussex, 1792–1917', PhD thesis, University of London, 1987.

Nakamura, Lisa Ellen, 'Puerperal Insanity: Women, Psychiatry, and the Asylum in Victorian England, 1820–1895', PhD thesis, University of Washington, 1999.

Nuttall, Alison, 'The Edinburgh Royal Maternity Hospital and The Medicalisation of Childbirth in Edinburgh 1844–1914: A Casebook-Centred Perspective', PhD thesis, University of Edinburgh, 2002.

Quinn, Catherine, 'Representations of Puerperal Insanity in England and Scotland, 1850–1900', MA dissertation, University of Manchester, 1998.

Quinn, Catherine, 'Include the Mother and Exclude the Lunatic. A Social History of Puerperal Insanity c.1860–1920', University of Exeter, 2003.

Rehman, A., 'Puerperal Insanity in the Nineteenth and Twentieth Centuries', MPhil thesis, University of Edinburgh, 1986.

Ryan, Jennifer, 'Confinement after Confinement: The Status of "Puerperal Insanity" in the Nineteenth and Twentieth Centuries', Intercalated BSc dissertation, Wellcome Centre London, 1999.

Saunders, Janet F., 'Institutionalised Offenders: A Study of the Victorian Institution and its Inmates, with Special Reference to Late Nineteenth Century Warwickshire', PhD thesis, University of Warwick, 1983.

Thompson, Margaret Sorbie, 'The Mad, the Bad, and the Sad: Psychiatric Care in the Royal Edinburgh Asylum (Morningside) 1813–1894', PhD thesis, Boston University, 1984.

Thomson, Elaine, 'Women in Medicine in Late Nineteenth- and Early Twentieth-Century Edinburgh: A Case Study', PhD thesis, University of Edinburgh, 1998.

Web-sites

http://www.christianitymeme.org/yates.html: 'The Andrea Yates Case'.

http://www.time.com/nation/article/0,8599,195325,00.html: 'The Yates Odyssey'.

Index

treatment 44, 45
violence in women 261 n38
see also puerperal insanity
memoirs of patients with puerperal
insanity 95
menopause symptoms of treated
234 n154
menstruation
and hysteria 21
as sign of improvement in
puerperal insanity 47
mental diseases/disturbance/illness
associated with childbirth 12, 20,
39, 202–3, 224 n93, 227 n39,
263 n63
classification 33, 104, 205, 247
n54
committal to asylums 246 n38
delirium in 229 n75
physiognomy to diagnose 49–50
supervising provisions for insane
246 n38
and theories of degeneration 160,
207
see also hereditary factors in
puerperal insanity
in wealthy families 67
see also puerperal insanity
Metropolitan Dispensary, London
49
middle-class women
admission to asylums 54, 98, 206
bonding with physicians 66
in Royal Edinburgh Asylum 98
midwifery
books on 11–13, 14–15, 20, 23–6
development in nineteenth century
15–16, 17–18, 32
division in views from alienists
29, 39
domestic treatment of puerperal
insanity 34, 58–60
entrance of men into 17, 26
terminology 220 n49
midwives
challenged by men-midwives 23,
26
on fear surrounding childbirth
11, 25

views on causes of puerperal
insanity 39, 141, 143–4
views on treatment of puerperal
insanity 54, 58–9
see also men-midwives
miscarriage 145
Moir, Louisa *see* Maitland, Louisa
Moncrieffe, Harriett 69–70
monthly nurses 23, 62
moral insanity 40, 89–90, 91, 155
moral management 29, 61–2, 76,
121–2, 126, 133
Morison, Alexander 33–4, 41, 47,
50, 54, 55, 98, 111–12, 137
Morningside Asylum *see* Royal
Edinburgh Asylum
morphine treatment for puerperal
insanity 47, 88, 118–19, 196,
206
mortality of women with puerperal
insanity 229 n77
see also death of husband
motherhood
anxieties in 11, 24–5, 68, 124,
142–50, 176–7, 202
attitude to in 19th century 16–17,
24
challenge of 144, 166
collapse of ability 133
risks 201
Victorian ideal 8, 144, 202
see also childbirth
Munro, John 176
Murray's Royal Asylum, Perth 162
mustard poultice treatment for
puerperal insanity 46

Nakamura, Lisa Ellen 206, 245 n32,
267–8 n17
Napier, Richard 9–10
narcotic treatment for puerperal
insanity 45, 46–7, 88
National Association for the
Promotion of Social Science 169
neurasthenia, treatment of 44, 258
n90
Nevinson, Dr 89
Nicholls, Elizabeth 57
Nicol, Christina 130